Elements of Electrical Machines

Elements of Electrical Machines

Pradip Kumar Sadhu PhD
Professor
Department of Electrical Engineering
Indian Institute of Technology (ISM)
Dhanbad, Jharkhand

Soumya Das PhD
Assistant Professor
Department of Electrical Engineering
University Institute of Technology
Burdwan University, West Bengal

Shiv Prakash Bihari ME
Assistant Professor
Department of Electrical Engineering
Inderprastha Engineering College
Ghaziabad, Uttar Pradesh

CBS Publishers & Distributors Pvt Ltd

New Delhi • Bengaluru • Chennai • Kochi • Kolkata • Mumbai
Bhopal • Bhubaneswar • Hyderabad • Jharkhand • Nagpur • Patna • Pune • Uttarakhand
• Dhaka (Bangladesh) • Kathmandu (Nepal)

Elements of
Electrical Machines

ISBN: 978-93-89396-20-1

Copyright © Authors and Publisher

First Edition: **2020**

Published by Satish Kumar Jain and produced by Varun Jain for
CBS Publishers & Distributors Pvt Ltd
4819/XI Prahlad Street, 24 Ansari Road, Daryaganj, New Delhi 110 002, India
Ph: 23289259, 23266861, 23266867 Website: www.cbspd.com
Fax: 011-23243014   e-mail: delhi@cbspd.com; cbspubs@airtelmail.in

Corporate Office: 204 FIE, Industrial Area, Patparganj, Delhi 110 092
Ph: 4934 4934   Fax: 4934 4935   e-mail: publishing@cbspd.com; publicity@cbspd.com

Branches

- **Bengaluru:** Seema House 2975, 17th Cross, K.R. Road, Banasankari 2nd Stage, Bengaluru 560 070, Karnataka
  Ph: +91-80-26771678/79   Fax: +91-80-26771680   e-mail: bangalore@cbspd.com
- **Chennai:** 7, Subbaraya Street, Shenoy Nagar, Chennai 600 030, Tamil Nadu
  Ph: +91-44-26680620/26681266   Fax: +91-44-42032115   e-mail: chennai@cbspd.com
- **Kochi:** 68/1534, 35, 36 Power House Road, Opp KSEB Power House, Ernakulam 682 018, Kochi, Kerala
  Ph: +91-484-4059061-65   Fax: +91-484-4059065   e-mail: kochi@cbspd.com
- **Kolkata:** 6/B, Ground Floor, Rameswar Shaw Road, Kolkata 700 014, West Bengal
  Ph: +91-33-22891126, 22891127, 22891128   e-mail: kolkata@cbspd.com
- **Mumbai:** 83-C, Dr E Moses Road, Worli, Mumbai 400018, Maharashtra
  Ph: +91-22-24902340/41   Fax: +91-22-24902342   e-mail: mumbai@cbspd.com

Representatives

• **Bhopal**	0-8319310552	• **Bhubaneswar**	0-9911037372	• **Hyderabad**	0-9885175004]
• **Jharkhand**	0-9811541605	• **Nagpur**	0-9421945513	• **Patna**	0-9334159340
• **Pune**	0-9623451994	• **Uttarakhand**	0-9716462459		
• **Dhaka (Bangladesh)**	01912-003485	• **Kathmandu (Nepal)**	977-9818742655		

Printed at: Mudrak Printers, Noida, UP, India

Preface

*E*lements of Electrical Machines has been designed as a textbook for electrical engineering as well as nonelectrical electrical engineering students pursuing their studies in mechanical, mining, textile, chemical, industrial, environmental, aerospace, electronics and communication, computer science, information technology in BE/BTech, AMIE, Diploma, as well as for those preparing for other competitive examinations in India and overseas. It is equally helpful for aspiring engineers to understand the theoretical aspects. This book is easy to comprehend and self-explanatory in its direct approach.

The book is designed to cover an extensive range of topics under electrical machines. It consists of eight chapters: 1. DC machines, 2. Single phase transformer, 3. Three phase transformer, 4. Three phase induction motors, 5. Single phase induction motors, 6. Synchronous generator, 7. Synchronous motors and 8. General introduction of special machines.

The presentation of the subject matter is very systematic and the language of the text is lucid and direct. We lay no claim to the original research in preparing the book. Liberal use of the materials available in the works of renowned authors has been made. However, we have tried to approach a huge amount of material available from primary and secondary sources into coherent body of description and analysis.

We welcome constructive criticism and will be grateful for any appraisal by the readers.

Pradip Kumar Sadhu
Soumya Das
Shiv Prakash Bihari

Acknowledgements

We are fortunate to receive useful comments and suggestions from students, which have helped in improving the technical content and clarity. We are greatful to all of them.

We are indebted to readers in the academia and industry, worldwide, for their invaluable feedback and for taking the trouble to draw our attention to the improvements required.

We also thank the reviewers, who took time from their busy schedules to send us suggestions.

Most importantly, we would like to thank the staff of CBS Publishers & Distributors, for their support in making this project a reality. We are grateful to the authorities of Inderprashtha Engineering College, Ghaziabad, Indian Institute of Technology (Indian School of Mines), Dhanbad and University Institute of Technology, Burdwan University, for providing all the facilities in writing this book.

Finally, we are grateful to our families for their love, tolerance, patience and support throughout this time consuming project. Readers of the book are welcome to send their comments and feedback.

Pradip Kumar Sadhu
Soumya Das
Shiv Prakash Bihari

Contents

DC Machines

1.1 INTRODUCTION

Rotating electrical machines can be broadly classified into two categories: DC machines and AC machines. All those machines, which either generate or use DC supply falls under the category of DC machines. These machines are further divided into two categories, i.e. DC generators and DC motors. The advantages of DC machine includes higher starting torque, quick starting and stopping, reversing, variable speeds with voltage input and are easier and cheaper to control.

1.2 CONCEPT OF ELECTRO–MECHANICAL ENERGY CONVERSION

It refers to the conversion of electrical energy into mechanical energy or *vice versa*. This energy conversion is achieved through a magnetic medium as the intermediate stage between electrical and mechanical forms of energy.

The device used to convert mechanical energy to electrical energy is known as *generator*, while the device which converts electrical energy to mechanical energy is called as *motor*. In generator, mechanical energy is provided by a prime-mover, such as, a diesel engine, this mechanical energy rotates the armature of generator in magnetic field to get electrical output. While in motor, electrical energy is provided to armature conductors placed in a magnetic field to produce a mechanical torque. The block diagram representation of electro–mechanical energy conversion is shown in Fig. 1.1.

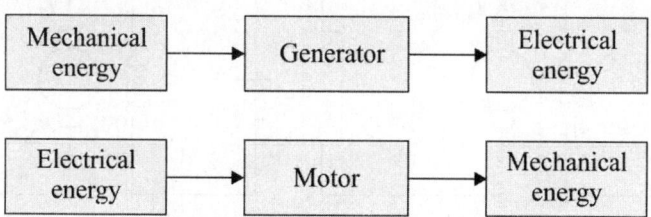

Fig. 1.1: Diagrammatic representation of concept of electro–mechanical energy conversion

1.3 TYPES OF DC MACHINES

Based on type of excitation used in field winding to set up magnetic field, DC machines can be classified as shown in Fig. 1.1(*a*).

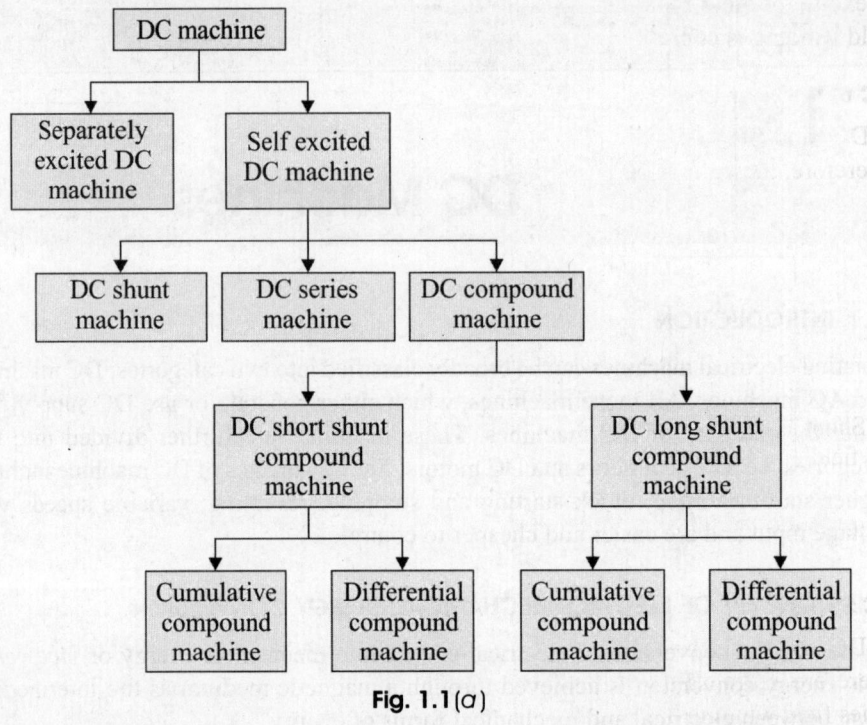

Fig. 1.1(a)

1.3.1 Separately Excited DC Machine

When the field coils of a DC machine are excited or energised from an external DC source, the machine is known as separately excited machine (Fig. 1.2).

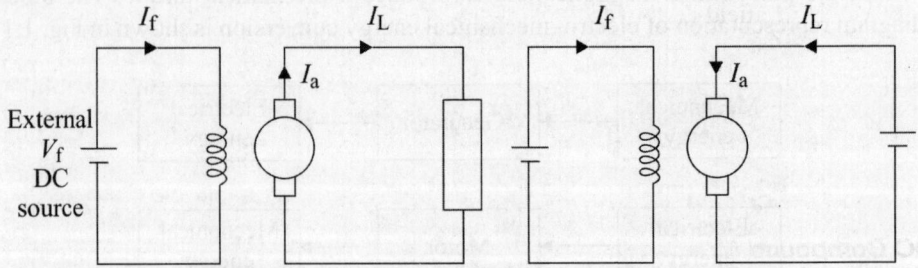

Fig. 1.2: Circuit diagram of separately excited (a) DC generator and (b) DC motor

1.3.2 Self Excited DC Machine

In self excited DC machine, the field coils are energised from the current supplied by the generator itself (in case of DC generator) or the current supplied by DC source connected with armature (in case of DC motor). Therefore, no external source is required to excite the field winding. These machines are further classified based on the way field winding is connected with the armature winding.

DC Shunt Machine

In DC shunt machine field winding is connected in parallel with the armature winding. Therefore, the voltage across armature and field windings remain same (Fig. 1.3).

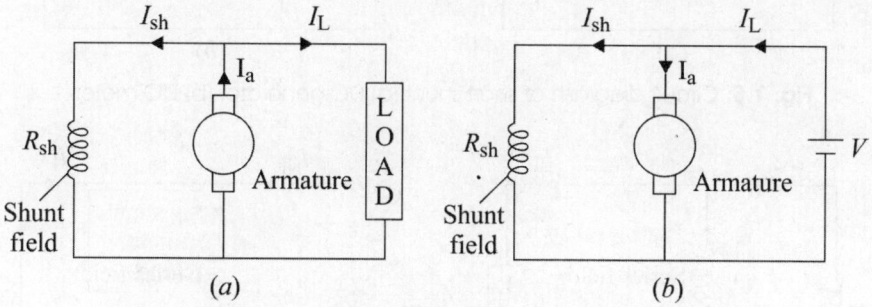

Fig. 1.3: Circuit diagram of DC shunt (a) Generator (b) Motor

DC Series Machine

In DC series machine, field winding is connected in series with armature circuit. Therefore, current flowing through armature and field windings remain same. As the series field winding has to carry large armature current, it is made of thick conductors with less number of turns (Fig. 1.4).

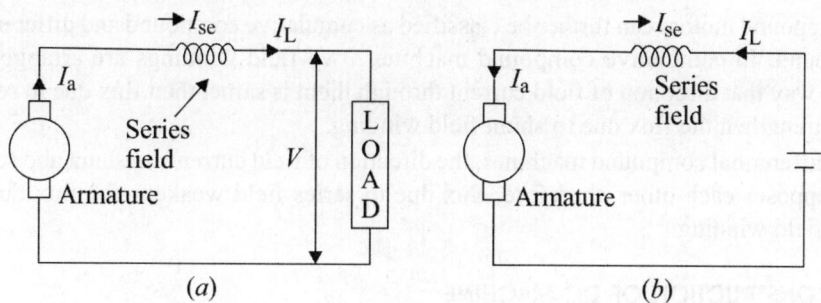

Fig. 1.4: Circuit diagram of DC series (a) Generator (b) Motor

DC Compound Machine

In DC compound machines, two field coils are used. One of these coils is connected in parallel with the armature, while other is connected in series with the armature. Based on the connection of shunt winding with armature, these machines are classified as short shunt or long shunt.

In short shunt machines, shunt winding is connected in parallel with armature winding, as shown in Fig. 1.5. In long shunt machines, shunt winding is connected in parallel with armature and series field winding as represented in Fig. 1.6.

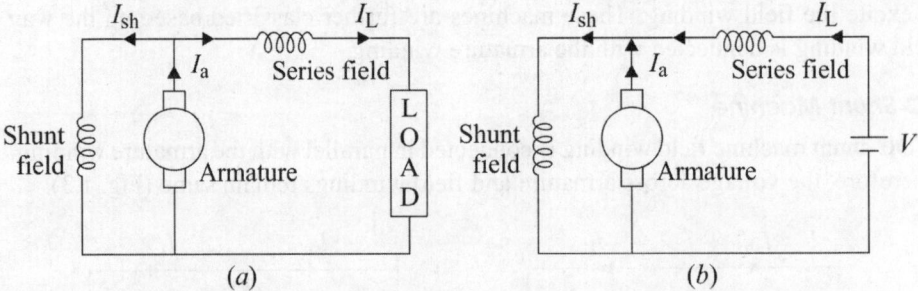

Fig. 1.5: Circuit diagram of short shunt (a) DC generator (b) DC motor

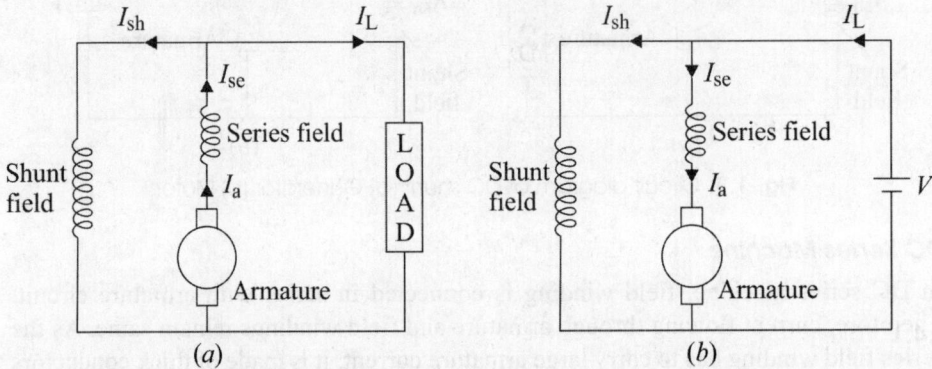

Fig. 1.6: Circuit diagram of long shunt (a) DC generator (b) DC motor

Compound motors can further be classified as cumulative compound and differential compound. In cumulative compound machines, two field windings are arranged in such a way that direction of field current through them is same, then flux due to series field strengthen the flux due to shunt field winding.

In differential compound machines, the direction of field currents in shunt and series field opposes each other, therefore, flux due to series field weakens the flux due to shunt field winding.

1.4 CONSTRUCTION OF DC MACHINE

The construction of a DC machine is shown in Fig. 1.7. A DC machine has following essential parts:

- (i) Yoke or outer enclosure/frame
- (ii) Pole core and pole shoes
- (iii) Field coils
- (iv) Armature core
- (v) Armature winding
- (vi) Commutator
- (vii) Brush assembly and brush gear
- (viii) Bearings

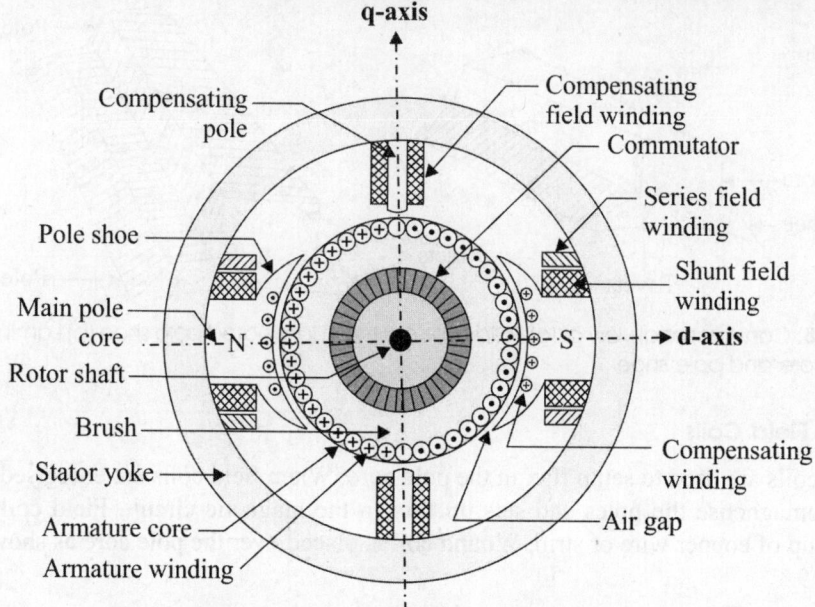

Fig. 1.7: Construction of a DC machine

1.4.1 Yoke

The outer enclosure of a DC machine is known as yoke. It serves following purposes:

(i) It provides protection to the machine from dust, moisture, etc.

(ii) It provides low reluctance path to magnetic flux produced by the pole.

(iii) It provides mechanical support to poles.

As yoke is an essential part of magnetic circuit, it is made up of low reluctance material such as cast iron, cast steel or rolled steel, etc.

1.4.2 Pole Core and Pole Shoes

A pole consists of pole core and pole shoes. Pole core can be a solid iron piece or laminated one. It houses the field winding. Pole shoes are made up of thin laminations. It is an extended part of pole core and due to its typical shape serves following purposes:

(i) It acts as a support to the field coil.

(ii) It reduces the reluctance of magnetic circuit (due to large cross-sectional area).

(iii) It spread out the flux in the airgap

Based on the pole construction, the pole core and pole shoes are shown in Fig. 1.8. Pole core and pole shoes are made of low reluctance material such as cast iron or cast steel.

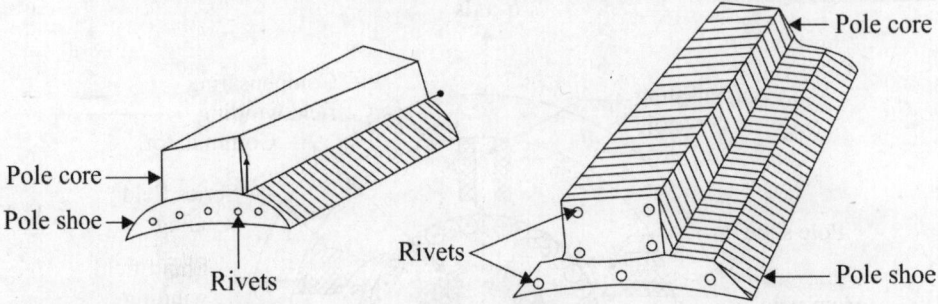

Fig. 1.8: Constructional view of (a) Solid pole core with laminated pole shoe (b) Laminated pole core and pole shoe

1.4.3 Field Coils

Field coils are used to setup flux in the pole core. When field coils are energised they electromagnetise the poles and sets up flux in the magnetic circuit. Field coils are made up of copper wire or strip. Wound coil is placed over the pole core as shown in Fig. 1.9.

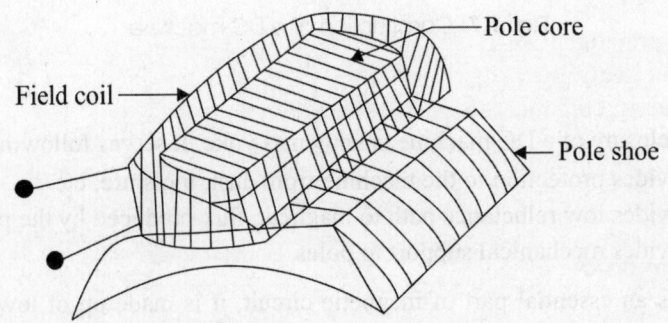

Fig.1.9: Field coil wound over pole

1.4.4 Armature Core

Armature core provides low reluctance path to magnetic lines of flux linking from north pole to south pole. It also houses the armature winding and secure them firmly in armature slots. It is cylindrical in shape and made up of thin laminations to reduce eddy current losses. In DC machines armature is the rotating part of machine. Armature core is made up of high permeability silicon steel laminations. Fig. 1.10(*a*) shows a single armature core lamination,

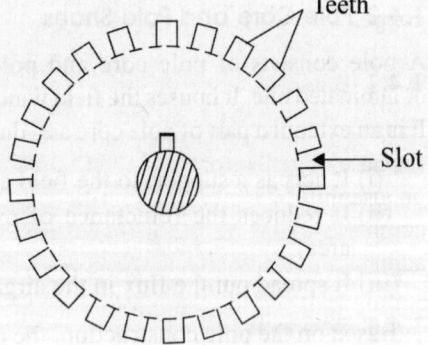

Fig. 1.10(*a*): Armature core lamination

while 1.10 (*b*) represents armature core housing armature winding.

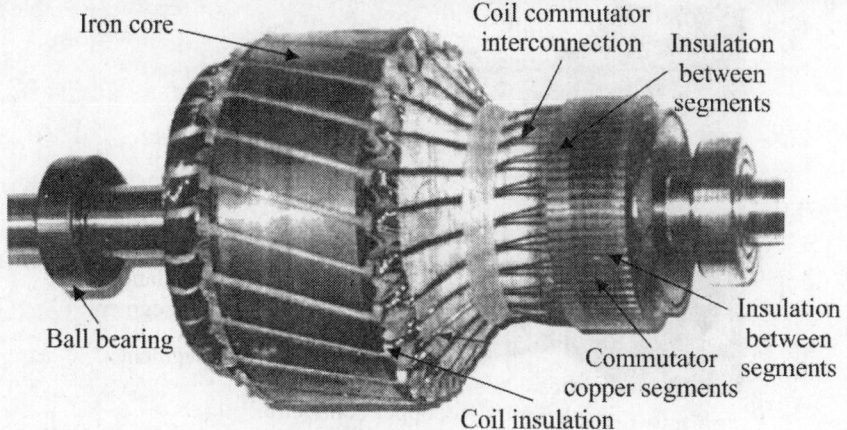

Iron core

Coil commutator interconnection

Insulation between segments

Ball bearing

Insulation between segments

Commutator copper segments

Coil insulation

Fig. 1.10(b): Armature core housing armature winding

1.4.5. Armature Winding

Copper wires housed in armature slots are suitably connected to form armature winding. As per the number of parallel path provided for armature current, armature windings are classified into two categories, i.e. lap winding and wave winding.

In lap winding, conductors are connected such that number of parallel paths are equal to number of poles. In wave winding, the armature conductors provide two parallel path to armature currents, symbol of DC machine is shown in Fig. 1.10 (c).

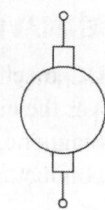

Fig. 1.10(c): Symbol of DC machine

1.4.6 Commutator

Commutator serves the purpose of collecting the alternating current from armature circuit (in case of generator) and supply it to external circuit by making it unidirectional, while its function is *vice versa* in case of motor. It is made up of multiple segments of copper which are insulated by mica as shown in Fig. 1.11.

1.4.7 Brush Assembly and Brush Gear

Brushes are used to collect current from commutator and supply it to external circuit (in case of generator) and *vice versa* in case of a motor. Brushes are made up of carbon or graphite. These brushes are housed in brush holders and are kept pressed on commutator by a spring. Entire arrangement of brush holder, spring and its tension adjusting mechanism is referred as brush gear.

1.4.8 Bearings

Rotor of DC machines is supported on both ends by bearings to provide almost friction less rotation. For low and medium rating machines, ball bearings are used while for heavy duty machines roller bearings are preferred.

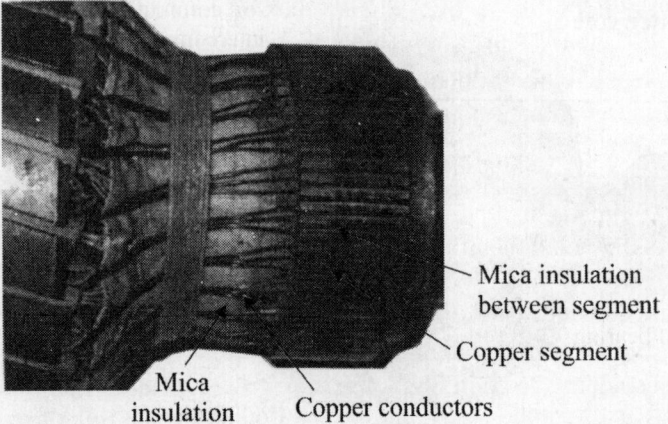

Mica insulation
between segment

Copper segment

Mica
insulation   Copper conductors

Fig. 1.11: Commutator

1.5 GENERAL FEATURES OF DC ARMATURE WINDINGS

(i) A DC machine (generator or motor) generally employs windings distributed in slots over the circumference of the armature core. Each conductor lies at right angles to the magnetic flux and to the direction of its movement. Therefore, the induced emf in the conductor is given by;

$$e = Blv \text{ V}$$

where B = magnetic flux density in Wb/m^2
l = length of the conductor in metres
v = velocity (in m/s) of the conductor

(ii) The armature conductors are connected to form coils. The basic components of all types of armature windings is the armature coil. Figure 1.12(a) shows a single-turn coil. It has two conductors or coil sides connected at the back of the armature. Figure 1.12(b) shows a 4-turn coil which has 8 conductors or coil sides.

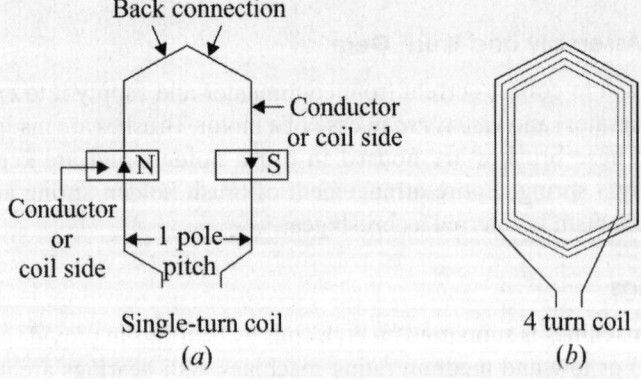

Back connection

Conductor
or coil side

N   S

Conductor
or   ←1 pole→
coil side   pitch

Single-turn coil   4 turn coil
(a)   (b)

Fig. 1.12

The coil sides of a coil are placed a pole span apart, i.e. one coil side of the coil is under N-pole and the other coil side is under the next S-pole at the corresponding position as shown in Fig. 1.12(*a*). Consequently the emf of the coil sides are added together. If the emf induced in one conductor is 2.5 volts, then the emf of a single-turn coil will be = 2 × 2.5 = 5 V.

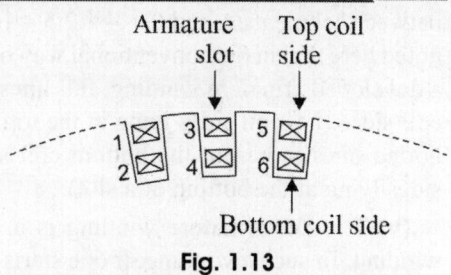

Fig. 1.13

For the same flux and speed, the emf of a 4-turn coil will be = 8 × 2.5 = 20 V.

(iii) Most of DC armature windings are double layer windings, i.e. there are two coil sides per slot as shown in Fig. 1.13. One coil side of a coil lies at the top of a slot and the other coil side lies at the bottom of some other slot. The coil ends will then lie side by side. In two-layer winding, it is desirable to number the coil sides rather than the slots. The coil sides are numbered as indicating in Fig. 1.13. The coil sides at the top of slots are given odd numbers and those at the bottom are given even numbers. The coil sides are numbered in order round the armature.

As discussed above, each coil has one side at the top of a slot and the other side at the bottom of another slot; the coil sides are nearly a pole pitch apart. In connecting the coils, it is ensured that top coil side is joined to the bottom coil side and *vice versa*. This is illustrated in Fig. 1.14. The coil side 1 at the top of a slot is joined to coil side 10 at the bottom of another slot about a pole pitch apart. The coil side 12 at the bottom of a slot is joined to coil side 3 at the top of another slot. How coils are connected at the back of the armature and at the front (commutator end) will be discussed in later sections. It may be noted that as far as connecting the coils is concerned, the number of turns per coil is immaterial. For simplicity, then, the coils in winding diagrams will be represented as having only one turn (i.e. two conductors).

(iv) The coil sides are connected through commutator segments in such a manner so as to form a series-parallel system; a number of conductors are connected in series so as to increase the voltage and two or more such series-connected paths in parallel to share the current Fig. 1.15 shows how the two coils connected through commutator segments (A, B, C, etc.) have their emfs added together. If voltage induced in each conductor is 2.5 V, then voltage

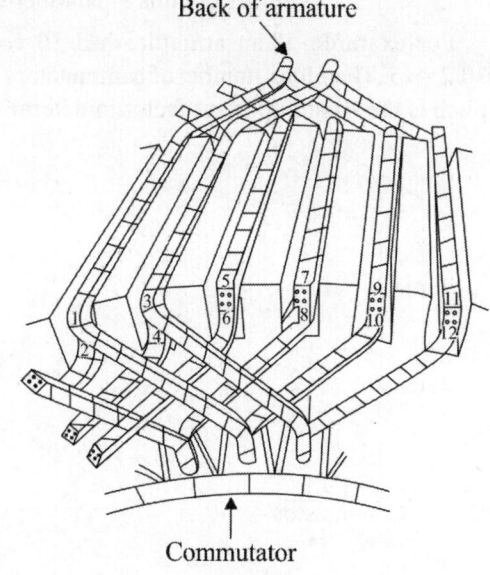

Fig. 1.14

between segments A and C = 4 × 2.5 = 10 V. It may be noted here that in the conventional way of representing a developed armature winding, full lines represent top coil sides (i.e. coil sides lying at the top of a slot) and dotted lines represent the bottom coil sides (i.e. coil sides lying at the bottom of a slot).

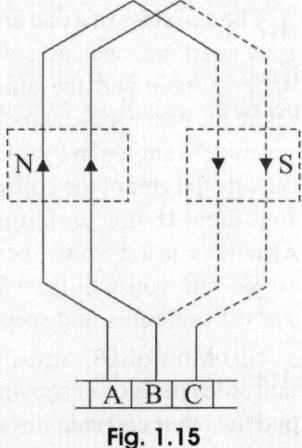

(v) The DC armature winding is a closed circuit winding. In such a winding, if one starts at some point in the winding and traces through the winding, one will come back to the starting point without passing through any external connection. DC armature windings must be of the closed type in order to provide for the commutation of the coils.

Fig. 1.15

1.6 COMMUTATOR PITCH (Y$_C$)

The commutator pitch is the number of commutator segments spanned by each coil of the winding. It is denoted by Y_C.

In Fig. 1.16, one side of the coil is connected to commutator segment 1 and the other side connected to commutator segment 2. Therefore, the number of commutator segments spanned by the coil is 1, i.e. $Y_C = 1$. In Fig. 1.17, one side of the coil is connected to commutator segment 1 and the other side to commutator segment 8. Therefore, the number of commutator segments spanned by the coil = 8 – 1 = 7 segments i.e. $Y_C = 7$. The commutator pitch of a winding is always a whole number. Since each coil has two ends and as two coil connections are joined at each commutator segment,

∴ Number of coils = Number of commutator segments

For example, if an armature has 30 conductors, the number of coils will be 30/2 = 15. Therefore, number of commutators segments is also 15. Note that commutator pitch is the most important factor in determining the type of DC armature winding.

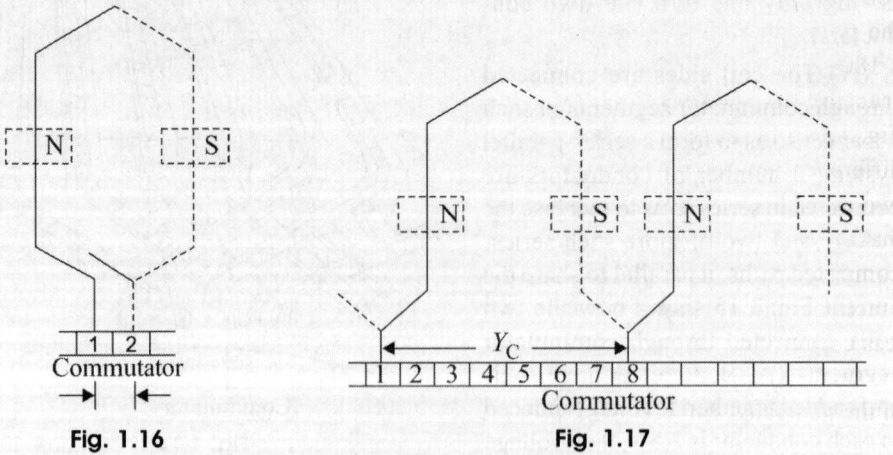

Fig. 1.16   **Fig. 1.17**

1.7 POLE PITCH

It is the distance measured in terms of number of armature slots (or armature conductors) per pole. Thus if a 4-pole generator has 16 coils, then number of slots = 16.

$$\text{Pole pitch} = \frac{16}{4} = 4 \text{ slots}$$

Also $\qquad$ Pole pitch = No. of conductors = $\dfrac{16 \times 2}{4}$ = 8 conductors

The pole pitch is an important factor in the design of armature winding.

1.8 COIL SPAN OR COIL PITCH (Y_S)

It is the distance measured in terms of the number of armature slots (or armature conductors) spanned by a coil. Thus, if the coil span is 9 slots, it means one side of the coil in slot 1 and the other side in slot 10.

1.9 FULL PITCHED COIL AND FRACTIONAL PITCHED COIL

Full pitched coil: If the coil span or coil pitch is equal to pole pitch (i.e. distance between the two sides of a coil is 180° electrical), it is called full-pitched coil. In other words, a full-pitched coil means that when one coil side is under the centre of N-pole, the other coil side must be under the centre of next S-pole as shown in Fig. 1.18. In this case, the emfs in the coil sides are additive and have a phase difference of 0°. If the emf induced in one coil side is 2.0 V, then emf across the coil terminals = 2 × 2.5 = 5 V. The emf induced in a full-pitched coil is maximum. Therefore, coil span or coil pitch should always be one pole pitch unless there is a good reason for making it shorter.

Fractional pitched coil: If the coil span or coil pitch is less than the pole pitch (i.e. coil pitch is < 180° electrical), then it is called fractional pitched coil [see Fig. 1.19]. In this case, the phase difference between the emfs in the two coil sides will not be zero so that the emf in the coil will be less compared to full-pitched coil. Fractional pitch winding requires less copper but if the coil pitch is too small, an appreciable reduction in the generated emf results.

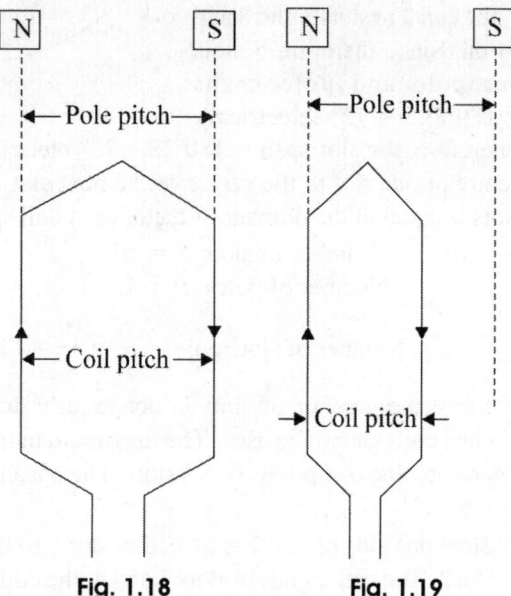

Fig. 1.18 $\qquad$ Fig. 1.19

There are many situations where we cannot use full-pitched coils. For example, for a 2-pole machine, it is very difficult to place full-pitched coils. Similarly, if the number of slots is not exactly divisible by the number of poles, we cannot use full-pitched coils. Under such situations, we have to restore to fractional pitch winding. In this case, the maximum possible pitch may be used as the pitch of the coil.

Example 1.1: The armature of a DC generator has 10 slots. Calculate the coil pitch for a (*i*) 2-pole winding and (*ii*) 4-pole winding. Show the placement of coils in each case.

Solution: A 10-slot armature using double-layer winding requires 10 coils.

(*i*)   Number of slots $S = 10$

Number of poles $P = 2$

$$\text{Number of slots/pole} = \frac{S}{P} = \frac{10}{2} = 5$$

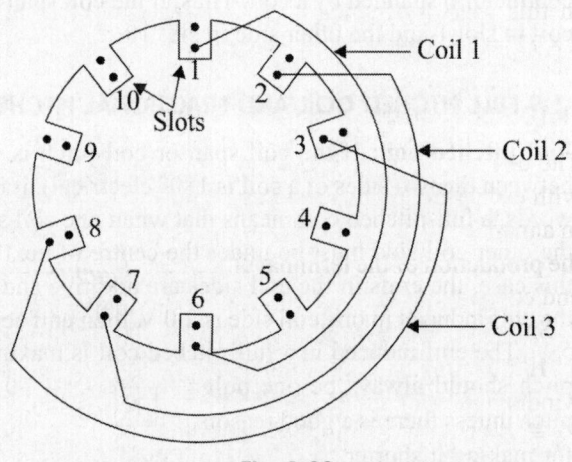

Since the number of slots is exactly divisible by the number of poles, it is possible to use full-pitch coils. Therefore, coil pitch, $Y_S = 5$ slots. The placement of coils in the armature is shown in Fig. 1.20.

Here one side of coil 1 is placed in slot 1 so that the other side of this coil is placed in slot 6. The coil 2 goes in slots 2 and 7, the coil 3 in slots 3 and 8 and so on. Since there are 5 slots per pole and pole spans $= 180°/2.5 = 72°$ (electrical).

Fig. 1.20

Therefore, the slot span $= 180°/5 = 36°$ electrical. It means that the angle from the centre of one slot to the centre of the next slot is 36° electrical. Since the number of slots is equal to the number of teeth, each coil spans have 5 teeth.

(*ii*)   Number of slots $S = 10$

Number of poles $P = 4$

$$\text{Number of slots/pole} = \frac{S}{P} = \frac{10}{4} = 2.5$$

Since the number of slots is not exactly divisible by the number of poles, full-pitched coils cannot be used. The maximum number of slots that the coil can span is 2. Therefore, the coil pitch, $Y_S = 2$ slots. The placement of coils in the armature is shown in Fig. 1.21.

Here one side of coil 1 is placed in slot 1 so that the other side of this coil is placed in slot 3. The coil 2 goes in slots 2 and 4, the coil 3 in slots 3 and 5 and so on. Note that

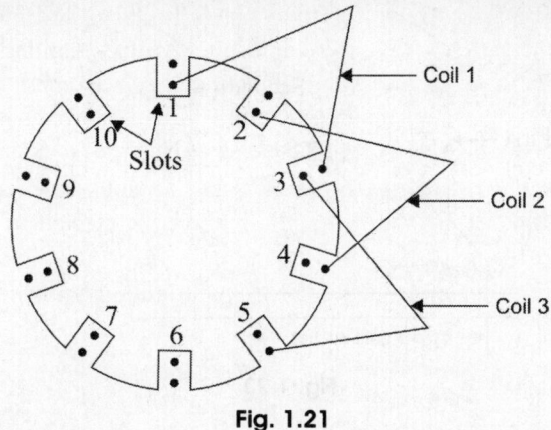

Fig. 1.21

in this case, slot span = 180°/2.5 = 72° (electrical). Therefore, the coil pitch = 2 × 72° = 144° (electrical). Note that for full-pitched coil, the coil pitch is 180° electrical.

1.10 TYPES OF DC ARMATURE WINDINGS

The different armature coils in a DC armature winding must be connected in series with each other by means of end connections (back connection and front connection) in a manner so that the generated voltages of the respective coils will aid each other in the production of the terminal emf of the winding. Two basic methods of making these end connections are: 1. Simplex lap winding 2. Simplex wave winding.

1. Simplex lap winding: For a simplex lap winding, the commutator pitch $Y_C = 1$ and coil span Y_S = pole pitch. Thus the ends of any coil are brought out to adjacent commutator segments and the result of this method of connection is that all the coils of the armature are in sequence with the last coil connected to the first coil. Consequently, closed circuit winding results. This is illustrated in Fig. 1.22 where a part of the lap winding is shown. Only two coils are shown for simplicity. The name lap comes from the way in which successive coils overlap the preceding one.

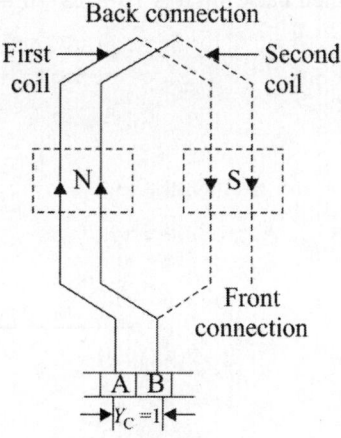

Fig. 1.22

2. Simplex wave winding: For a simplex wave winding, the commutator pitch $Y_C = 2$ pole pitches and coil span = pole pitch. The result is that the coils under consecutive pole pairs will be joined together in series thereby adding together their emfs [see Fig. 1.23]. After passing once around the armature, the winding falls in a slot to the left or right of the starting point and thus connecting to another circuit. Containing in this way, all the

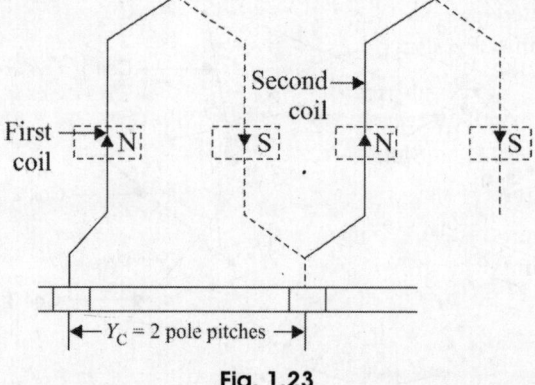

Fig. 1.23

conductors will be connected in a single closed winding. This winding is called wave winding from the appearance (wavy) of the end connections.

1.11 ARMATURE WINDING TERMINOLOGY

Apart from the terms discussed earlier, the following terminology requires discussion:

(i) **Back pitch (Y_B):** It is the distance measured in terms of armature conductors between the two sides of a coil at the back of the armature (see Fig. 1.24). It is denoted by Y_B. For example, if a coil is formed by connecting conductor 1 (upper conductor in a slot) to conductor 12 (bottom conductor in another slot) at the back of the armature, then back pitch is $Y_B = 12 - 1 = 11$ conductors.

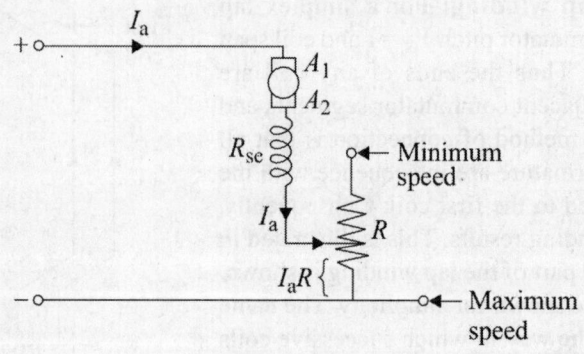

Fig. 1.24

(ii) **Front pitch (Y_F):** It is the distance measured in terms of armature conductors between the coil sides attached to anyone commutator segment [see Fig. 1.24]. It is denoted by Y_F. For example, if coil side 12 and coil side 3 are connected to the same commutator segment, then front pitch is $Y_F = 12 - 3 = 9$ conductors.

(iii) **Resultant pitch (Y_R):** It is the distance (measured in terms of armature conductors) between the beginning of one coil and the beginning of the next coil to which it is connected (see Fig. 1.24). It is denoted by Y_R. Therefore, the resultant pitch is the algebraic sum of the back and front pitches.

(iv) **Commutator pitch (Y_C):** It is the number of commutator segments spanned by each coil of the armature winding.

For simplex lap winding, $Y_C = 1$

For simplex wave winding, $Y_C = 2$ pole pitches (segments)

(v) **Progressive winding:** A progressive winding is one in which, as one traces through the winding, the connections to the commutator will progress around the machine in the same direction as is being traced along the path of each individual coil.

Figure 1.25(*a*) shows progressive lap winding. Note that $Y_B > Y_F$ and $Y_C = +1$.

(vi) **Retrogressive winding:** A retrogressive winding is one in which, as one traces through the winding, the connections to the commutator will progress around the machine in the opposite direction to that which is being traced along the path of each individual coil. Fig. 1.25(*b*) shows retrogressive lap winding. Note that $Y_F > Y_S$ and $Y_C = -1$. A retrogressive winding is seldom used because it requires more copper.

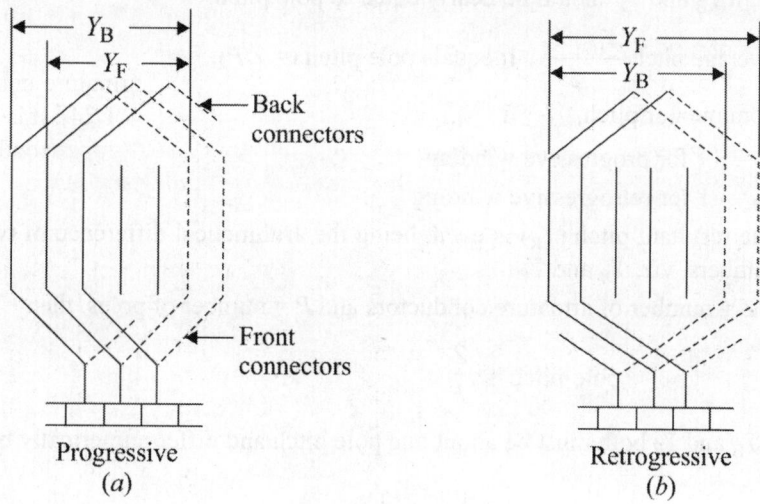

Fig. 1.25

1.12 GENERAL RULES FOR DC ARMATURE WINDINGS

In the design of DC armature winding (lap or wave), the following rules may be followed:

(i) The back pitch (Y_B) as well as front pitch (Y_F) should be nearly equal to pole pitch. This will result in increased emf in the coils.

(ii) Both pitches (Y_B and Y_F) should be odd. This will permit all end connections (back as well as front connections) between a conductor at the top of a slot and one at the bottom of a slot.

(iii) The number of commutator segments is equal to the number of slots or coils (or half the number of conductors).

No. of commutator segments = No. of slots = No. of coils

It is because each coil has two ends and two coil connections are joined at each commutator segment.

(iv) The winding must close upon itself, i.e. it should be a closed circuit winding.

1.13 RELATIONSHIP BETWEEN PITCHES FOR SIMPLEX LAP WINDING

In a simplex lap winding, the various pitches should have the following relation:

(i) The back and front pitches are odd and are of opposite signs. They differ numerically by 2.

$$\therefore \qquad Y_B = Y_F \pm 2 \qquad \text{[for progressive winding]}$$

$$Y_B = Y_F + 2$$

$$Y_B = Y_F - 2$$

(ii) Both Y_B and Y_F should be nearly equal to pole pitch.

(iii) Average pitch $\dfrac{Y_B + Y_F}{2}$. It equals pole pitch ($= Z/P$).

(iv) Commutator pitch, $Y_C \pm 1$

$Y_C = +1$ for progressive winding

$Y_C = -1$ for retrogressive winding

(v) The resultant pitch (Y_R) is even, being the arithmetical difference of two odd numbers, viz. Y_B and Y_F

(vi) If Z = number of armature conductors and P = number of poles, then

$$\text{pole-pitch} = \frac{Z}{P}$$

Since Y_B and Y_F both must be about one pole pitch and differ numerically by 2,

$$\left.\begin{aligned} Y_B &= \frac{Z}{P} + 1 \\[2mm] Y_F &= \frac{Z}{P} - 1 \end{aligned}\right\} \text{For progressive winding}$$

$$\left.\begin{aligned} Y_B &= \frac{Z}{P} - 1 \\[2mm] Y_F &= \frac{Z}{P} + 1 \end{aligned}\right\} \text{For retrogressive winding}$$

It is clear that Z/P must be an even number to make the winding possible.

Example 1.2: A 4-pole, simplex lap-wound armature contains 16 slots and has two coil sides per slot. Find backpitch, front pitch and commutator pitch for (*i*) progressive winding (*ii*) retrogressive winding.

Solution: Pole pitch = $\dfrac{Z}{P} = \dfrac{16 \times 2}{4} = 8$ conductors

(i) For progressive winding

$$\text{Back pitch, } Y_B = \frac{Z}{P} + 1 = 8 + 1 = 9 \text{ conductors}$$

$$\text{Front pitch, } Y_F = \frac{Z}{P} - 1 = 8 - 1 = 7 \text{ conductors}$$

Commutator pitch, $Y_C = -1$ segment

(ii) For retrogressive winding

$$\text{Back pitch, } Y_B = \frac{Z}{P} - 1 = 8 - 1 = 7 \text{ conductors}$$

$$\text{Frontpitch, } Y_F = \frac{Z}{P} + 1 = 8 + 1 = 9 \text{ conductors}$$

Commutator pitch, $Y_C = -1$ segment

Example 1.3: A 4-pole, simplex progressive lap-wound armature contains 21 slots and has two coil sides per slot. Find (*i*) slot span (*ii*) the back pitch (*iii*) commutator pitch (*iv*) the front pitch.

Solution: The slot pitch or slot span is the number of slots per pole. If there are S slots on the armature and the number of poles is P, then

(i) $\qquad$ slot span, $Y_S = \dfrac{S}{P} = \dfrac{21}{4} = 5\dfrac{1}{4}$ Slots

The most convenient way of expressing the span of a coil (or coil pitch) is in terms of the slots.

Thus in this case, the coil pitch should be $5\frac{1}{4}$ slots. Since a coil can span only a whole number of slots, in solving this problem, the fraction 1/4 is dropped.

$$Y_S = 5 \text{ slots}$$

(ii) The back pitch is equal to coil sides per slot (C_S) times the slot span (Y_S) + 1.

∴ $\qquad Y_S = C_S Y_S + 1 = 2 \times 5 + 1 = 11$ conductors

(*iii*) For a progressive simplex lap winding, commutator pitch, $Y_C = +1$ segment

(*iv*) For a progressive simplex lap winding, $Y_F = Y_B - 2 = 11 - 2 = 9$ conductors

Example 1.4: A 4-pole, lap-wound armature has 18 slots and 18 segments. Each slot has two coil sides. If the winding is progressive, determine (*i*) the back pitch (*ii*) the commutator pitch (*iii*) the front pitch.

Solution: Number of armature conductors, $Z = 2 \times 28 = 56$

$$\text{Pole pitch} = \frac{Z}{P} = \frac{56}{4} = 14 \text{ conductors}$$

(i) For a progressive winding,

$$\text{back pitch, } Y_B = \frac{Z}{P} + 1 = 14 + 1 = 15 \text{ conductors}$$

(ii) Commutator pitch, $Y_C = +1$ segment

(iii) Front pitch, $Y_C = \frac{Z}{P} - 1 = 14 - 1 = 13 \text{ conductors}$

Example 1.5: An 8-pole, 240 V, simplex lap-wound armature delivers 125 A. Determine (*i*) the voltage per path (*ii*) the current per path and (*iii*) the current per brush.

Solution.

(i) Voltage per path = Generator terminal voltage = 240 V

(ii) Current per path = $\dfrac{I_a}{A} = \dfrac{I_a}{P} = \dfrac{125}{8} = 15.62$ A

(iii) Current per brush = $\dfrac{125}{2} = 62.5$ A

1.14 SIMPLEX WAVE WINDING

The essential difference between a lap winding and a wave winding is in the commutator connections. In a simplex lap winding, the coils approximately pole pitch apart are connected in series and the commutator pitch $Y_C = \pm 1$ segment. As a result, the coil voltages gets added. This is illustrated in Fig. 1.26. In a simplex wave winding, the coils approximately pole pitch apart are connected in series and the commutator pitch $Y_C = 2$ pole pitches (segments). Thus in a wave winding, successive coils "wave" forward under successive poles instead of "lapping" back on themselves as in the lap winding. This is illustrated in Fig. 1.27 and 1.28.

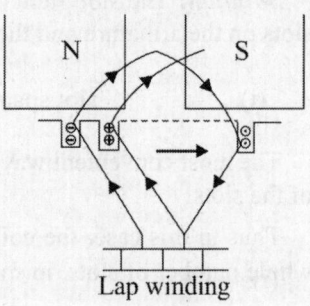

Lap winding

Fig. 1.26

The simplex wave winding must not close after it passes once around the armature but it must connect to a commutator segment adjacent to the first and the next coil must be adjacent to the first as indicated in Figs 1.27 and 1.28. This is repeated each time around until connections are made to all the commutator segments and all the slots are occupied after which the winding automatically returns to the starting point. If, after passing once around the armature, the winding connects to a segment to the left of the starting point, the winding is retrogressive [see Fig. 1.27]. If it connects to a segment to the right of the starting point, it is progressive [see Fig. 1.28]. This type of winding is called wave winding because it passes around the armature in a wave-like form.

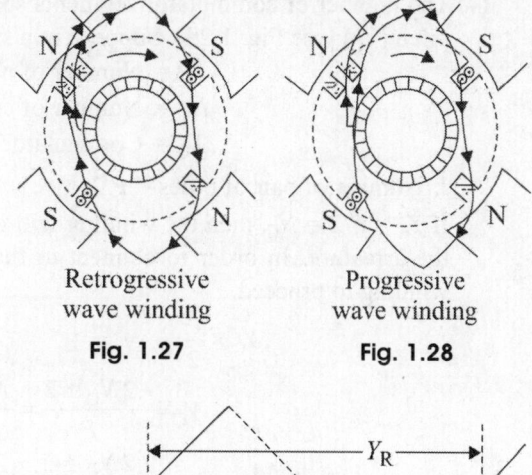

Retrogressive   Progressive
wave winding   wave winding

Fig. 1.27   **Fig. 1.28**

Various pitches: The various pitches retrogressive in a wave winding are defined in a manner similar to lap winding.

(i) The distance measured in terms of armature conductors between the two sides of a coil at the back of the armature is called back pitch Y_B [see Fig. 1.29]. The Y_B must be an odd integer so that a top conductor and a bottom conductor will be joined.

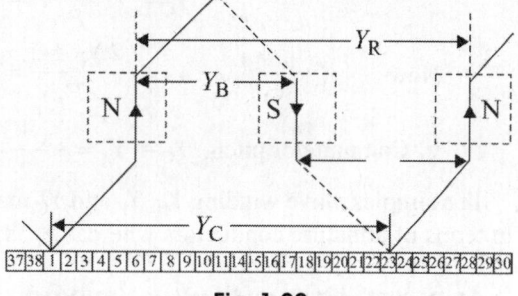

Fig. 1.29

(ii) The distance measured in terms of armature conductors between the coil sides attached to any one commutator segment is called front pitch Y_F [see Fig. 1.29]. The Y_F must be an odd integer so that a top conductor and a bottom conductor will be joined.

(iii) Resultant pitch, $Y_R = Y_B + Y_F$ [see Fig. 1.29]

The resultant pitch must be an even integer since Y_B and Y_F are odd, Further Y_F is approximately two pole pitches because Y_B as well as Y_F is approximately one pole pitch.

(iv) Average pitch, $Y_A = \dfrac{Y_B + Y_F}{2}$

When one turn of armature has been completed, the winding should connect to the next top conductor (progressive) or to the preceding top conductor (retrogressive). In either case, the difference will be of 2 conductors or one slot. If P is the number of poles and Z is the total number of armature conductors, then,

$$P \times Y_A = Z \pm 2$$

or $$Y_A = \dfrac{Z \pm 2}{P} \qquad \qquad ...(1.1)$$

Since P is always even and $Z = P_{YA} \pm 2$, Z must be even. It means that $Z \pm 2/P$ must be an integer. In Eq. (1.1), plus + sign will give progressive winding and the − sign gives retrogressive winding.

(v) The number of commutator segments spanned by a coil is called commutator pitch (Y_C) [see Fig. 1.29]. Suppose in a simplex wave winding,

$$P = \text{Number of poles};$$
$$N_C = \text{Number of commutator segments};$$
$$Y_C = \text{Commutator pitch}$$

$\therefore$ Number of pair of poles $= P/2$

If $Y_C \times P/2 = N_C$, then the winding will close on itself in passing once around the armature. In order to connect to the adjacent conductor and permit the winding to proceed.

$$Y_C \times \frac{P}{2} = N_C \pm 1$$

$$Y_C = \frac{2N_C \pm 2}{P} = \frac{N_C \pm 1}{P/2} = \frac{\text{No. of commutator seg.} \pm 1}{\text{Number of pair of poles}}$$

Now
$$Y_C = \frac{2N_C \pm 2}{P} = \frac{Z \pm 2}{P} = Y_A \qquad (\ 2N_C = Z)$$

$\therefore$ Commutator pitch, $Y_C = Y_A = \dfrac{Y_B + Y_F}{2}$

In a simplex wave winding Y_B, Y_F and Y_C may be equal. Note that Y_B, Y_F and Y_A are in terms of armature conductors whereas Y_C is in terms of commutator segments.

1.15 DESIGN OF SIMPLEX WAVE WINDING

In the design of simplex wave winding, the following points may be kept in mind:

(i) Both pitches Y_B and Y_F are odd and are of the same sign.

(ii) Average pitch, $Y_A = \dfrac{Z \pm 2}{P}$

(iii) Both Y_B and Y_F are nearly equal to pole pitch and may be equal or differ by 2. If they differ by 2, they are one more and one less than Y_A.

(iv) Commutator pitch is given by

$$Y_C = Y_A = \frac{\text{Number of commutator segments} \pm 1}{\text{Number of pair of poles}}$$

The plus sign for progressive winding and negative for retrogressive winding.

(v) $$Y_A = \frac{Z \pm 2}{P}$$

Since Y_A must be a whole number, there is a restriction on the value of Z, with $Z = 180$, this winding is impossible for a 4-pole machine because Y_A is not a whole number.

(vi) $$Z = PY_A \pm 2$$

$\therefore$ Number of coils $= \dfrac{Z}{2} = \dfrac{PY_A \pm 2}{2}$

Example 1.6: A 4-pole armature has 30 armature conductors. The armature is to be simplex wave-wound with single-turn coils. Determine for a retrogressive winding (*i*) back pitch (*ii*) front pitch (*iii*) commutator pitch.

Solution:

(*i*) Average pitch, $Y_A = \dfrac{Z \pm 2}{P}$

For a retrogressive winding, $Y_A = \dfrac{Z-2}{P} = \dfrac{30-2}{4} = 7$ conductors

$\therefore$ $Y_B = 7$ conductors

(*ii*) $Y_A = \dfrac{Y_B + Y_F}{2}$

or $7 = \dfrac{7 + Y_F}{2}$ $\therefore Y_F = 1$ conductors

(*iii*) $Y_C = Y_A = 7$ commutator segments

Alternatively, for a retrogressive winding,

$$Y_C = \frac{\text{No. of commutator segments} - 1}{\text{No. of pair of poles}} = \frac{15-1}{2}$$
$$= 7 \text{ segments}$$

1.16 DUMMY COILS

In a simplex wave winding, the average pitch Y_A (or commutator pitch Y_C) should be a whole number. Sometimes the standard armature punchings available in the market have slots that do not satisfy the above requirement so that more coils (usually only one more) are provided that can be utilised. These extra coils are called dummy or dead coils. The dummy coil is inserted into the slots in the same way as the others to make the armature dynamically balanced but it is not a part of the armature winding.

Let us illustrate the use of dummy coils with a numerical example. Suppose the number of slots is 22 and each slot contains 2 conductors. The number of poles is 4. For simplex wave wound armature,

$$Y_A = \frac{Z \pm 2}{P} = \frac{2 \times 2 \pm 2}{4} = \frac{44 \pm 2}{4} = 11\frac{1}{2} \text{ or } 10\frac{1}{2}$$

Since the results are not whole numbers, the number of coils (and hence segments) must be reduced. If we make one coil dummy, we have 42 conductors and

$$Y_A = \frac{42 \pm 2}{4} \quad 11 \text{ or } 10$$

This means that armature can be wound only if we use 21 coils and 21 segments. The extra coil or dummy coil is put in the slot. One end of this coil is taped and the other end connected to the unused commutator segment (segment 22) for the sake of appearance. Since only 21 segments are required, the two (21 and 22 segments) are connected together and considered as one.

1.17 APPLICATIONS OF LAP- AND WAVE-WINDINGS

In multipolar machines, for a given number of poles (P) and armature conductors (Z), a wave winding has a higher terminal voltage than a lap winding because it has more conductors in series. On the other hand, the lap winding carries more current than a wave winding because it has more parallel paths.

In small machines, the current-carrying capacity of the armature conductors is not critical and in order to achieve suitable voltages, wave windings are used. On the other hand, in large machines suitable voltages are easily obtained because of the availability of large number of armature conductors and the current carrying capacity is more critical. Hence in large machines, lap windings are used.

Note: In general, a high-current armature is lap-wound to provide a large number of parallel paths and a low-current armature is wave-wound to provide a small number of parallel paths.

1.18 MULTIPLEX WINDINGS

A simplex lap-wound armature has as many parallel paths as the number of poles. A simplex wave-wound armature has two parallel paths irrespective of the number of poles. In case of a 10-pole machine, using simplex windings, the designer is restricted to either two parallel circuits (wave) or ten parallel circuits (lap). Sometimes it is desirable to increase the number of parallel paths. For this purpose, multiplex windings are used. The main purpose of multiplex windings is to increase the number of parallel paths enabling the armature to carry a large amount of current. The degree of multiplicity or plex determines the number of parallel paths in the following manner:

(*i*) A lap winding has pole times the degree of plex parallel paths.

Number of parallel paths, $A = P \times$ plex

Thus a duplex lap winding has $2P$ parallel paths, triplex lap winding has $3P$ parallel paths and so on. If an armature is changed from simplex lap to duplex lap without making any other change, the number of parallel paths is doubled and each path has half as many coils. The armature winding then supply twice as much current at half the voltage.

(*ii*) A wave winding has two times the degree of plex parallel paths.

Number of parallel paths, $A = 2 \times$ plex

Note that the number of parallel paths in a multiplex wave winding depends upon the degree of plex and not on the number of poles. Thus a duplex wave winding has 4 parallel paths, triplex wave winding has 6 parallel paths and so on.

1.19 WORKING OF DC GENERATOR

For generating action supply (DC) is given to the field winding and rotating the rotor in the field by some prime mover. The armature conductor cuts the flux, hence an emf induce in the armature according to Faraday's law of electromagnetic induction and current produces.

The magnitude of this induced emf is directly proportional to rate of change of flux and it direction is given by Fleming's right hand rule.

Working: To understand the working of DC generator, let us consider a single turn rectangular copper coil ABCD as shown in Fig. 1.30. Let this coil rotates in magnetic field, two ends of this coil are joined to two slip rings a and b. The current is collected from slip rings with the help of brushes 1 and 2.

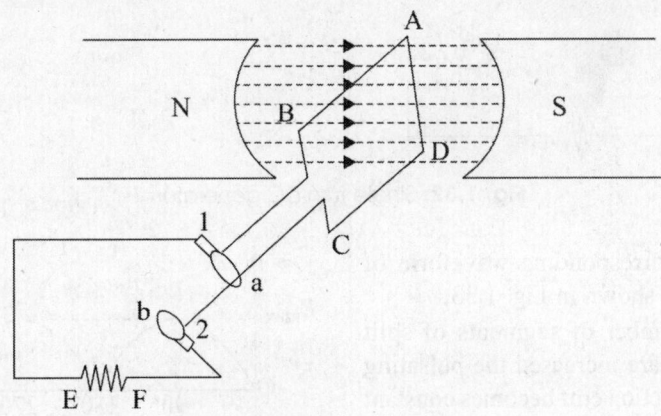

Fig. 1.30: Single turn generator

When the coil is perpendicular to the magnetic line of fluxes as shown in Fig. 1.30. The flux linking with the coil is maximum but its rate of change is zero. Hence induced emf in this position is zero.

Let the coil is rotating in clockwise direction, as the coil sides start taking successive positions the rate of flux linking with coil sides increases and reaches maximum value when coil is parallel to the magnetic lines of fluxes.

As the coil turns further, the rate of change of flux starts decreasing and reaches to zero as the coil rotates by 180°.

When coil rotates further, rate of change of flux starts increasing in negative direction upto negative maximum at 270°. On further rotation the rate of change of flux decreases towards zero as the coil complete one turn at 360°. Therefore, the induce emf has alternating nature as shown in Fig. 1.31.

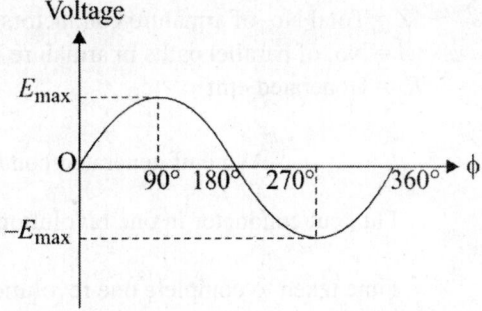

Fig. 1.31: Waveform of induced emf of a single turn generator

To make this emf unidirectional, split rings are used in place of slip rings, as shown in Fig. 1.32.

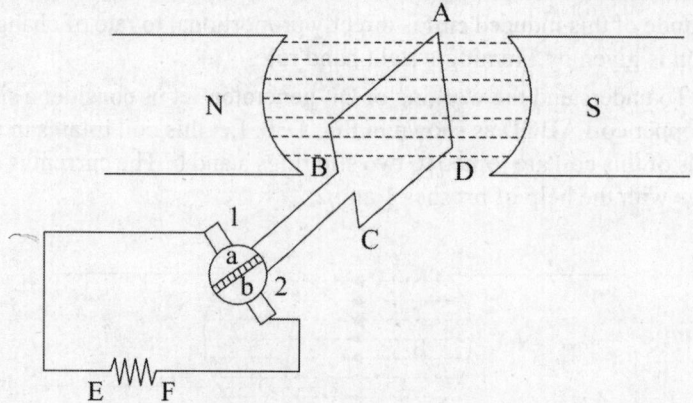

Fig. 1.32: Single turn DC generator

The corresponding waveform of emf is shown in Fig. 1.33.

As number of segments of split rings, are increased the pulsating undirection emf becomes constant DC. A multiple segment split ring arrangement is called commutator, which makes alternating emf of armature as a constant DC for external circuit.

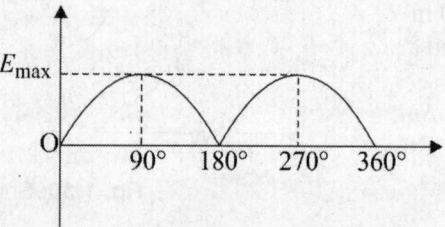

Fig. 1.33: Waveform of induced emf with split ring and no. of turns

1.20 EMF EQUATION OF DC GENERATOR

Let ϕ = Flux/pole (weber)

P = No. of poles

N = Number of revolution/minute made by armature (rpm)

Z = Total No. of armature conductors = No. of slots × No. of conductors/slot

A = No. of parallel paths in armature

E_g = Generated emf

$$\text{Avg emf generated/conductor} = \frac{d\phi}{dt}\ \text{volt}$$

Flux cut/conductor in one revolution, $d\phi = P.\phi$ Wb

Time taken to complete one revolution, $dt = \frac{60}{N}$ sec

(as N revolutions are completed in 60 sec)

$$\text{Avg emf generated/conductor} = \frac{d\phi}{dt} = \frac{p.\phi}{60/N} = \frac{\phi \cdot PN}{60}$$

No. of armature conductors/parallel path $= \dfrac{Z}{A}$

Generator emf, $E_g =$ Avg. emf generator/parallel path

$$\boxed{E_g = \dfrac{\phi PN}{60} \cdot \dfrac{Z}{A}}$$

or $\qquad\qquad\qquad\qquad\qquad E_g = \dfrac{\phi ZPN}{60A}\,\text{V}$

For wave winding $\qquad\qquad\qquad\qquad E_g \propto \phi N$

SOLVED EXAMPLES

Example 1.7: A DC generator has an emf of 100 V, when the useful flux per pole is 20 mWb and the speed is 800 rpm. Calculate the generated emf (*i*) with the same flux and a speed of 1000 rpm (*ii*) with a flux per pole of 24 mWb and a speed of 900 rpm.

[UPTU, 2003]

Solution: $\qquad\qquad\qquad E_{g1} = 100\text{ V}$

$$\phi_1 = 20 \times 10^{-3}\text{ Wb}$$

$$N_1 = 800\text{ rpm}$$

As $\qquad\qquad\qquad\qquad E \propto \phi N$

So, $\qquad\qquad\qquad\qquad \dfrac{E_{g2}}{E_{g1}} = \dfrac{\phi_2}{\phi_1} \times \dfrac{N_2}{N_1}$

(*i*) $\qquad\qquad\qquad\qquad \phi_2 = \phi_1,\ N_2 = 1000\text{ rpm}$

$$\dfrac{E_{g2}}{E_{g1}} = \dfrac{\phi_1}{\phi_1} \times \dfrac{1000}{800}$$

or $\qquad E_{g2} = E_{g1} \times \dfrac{1000}{800} = 100 \times \dfrac{1000}{800} = 125\text{ V}$

(*ii*) $\qquad\qquad \phi_2 = 24 \times 10^{-3}\text{ Wb},\ N_2 = 900\text{ rpm}$

or $\quad E_{g2} = E_{g1} \times \dfrac{\phi_2}{\phi_1} \times \dfrac{N_2}{N_1} = 100 \times \dfrac{24 \times 10^{-3}}{20 \times 10^{-3}} \times \dfrac{900}{800} = 135\text{ V}$ **Ans.**

Example 1.8: A four-pole, wave wound generator has 51 slots, each slot containing 20 conductors. Calculate generated voltage, when it is running at 1500 rpm. Assume flux per pole as 7 mWb. **[Allahabad University, 1993]**

Solution: $\qquad\qquad E_g = \dfrac{\phi ZN}{60}\left(\dfrac{P}{A}\right)$

$$\phi = 7\text{ mWb} = 7 \times 10^{-3}\text{ Wb}$$

$$Z = 51 \times 20 = 1020 \text{ conductors}$$
$$N = 1500 \text{ rpm}$$
$$P = 4$$
$$A = 2$$

$$E_g = \frac{7 \times 10^{-3} \times 1020 \times 1500}{60} \left(\frac{4}{2}\right)$$

$$= 357 \text{ volt } \textbf{Ans.}$$

Example 1.9: In a DC generator, what will be the change in induced emf, if flux is reduced by 20% and speed is increased by 20%? **[UPTU 2005-06]**

Solution:
$$\frac{E_{g2}}{E_1} = \frac{\phi_2}{\phi_1} \times \frac{N_2}{N_1}$$

$$\phi_2 = 0.8\phi_1$$
$$N_2 = 1.2N_1$$

$$E_{g2} = E_{g1} \times \frac{0.8\phi_1}{\phi_1} \times \frac{1.2N_1}{N_1}$$

$$= 0.96 E_{g1}$$

$\therefore$ Change in induced emf $= E_{g1} - 0.96 E_{g1}$
$$= 0.04 \, E_{g1}$$

or induced emf is reduced by 4%. **Ans.**

1.21 IMPORTANT RELATIONS FOR VARIOUS TYPES OF DC GENERATORS

1.21.1 Separately Excited DC Generator

For separately excited DC generator, shown in Fig. 1.34.

Let V_f = Field voltage provided by external source

I_f = Field current

R_f = Field resistance

I_a = Armature current

I_L = Load current

R_a = Armature resistance

E_g = Generated emf

V = Output voltage

$I_a = I_L$

$$I_f = \frac{V_f}{R_f}$$

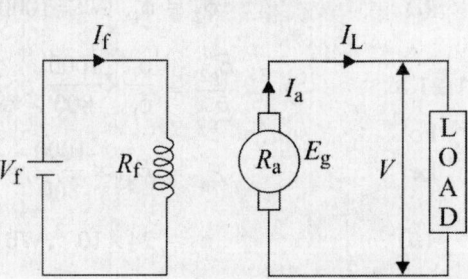

Fig. 1.34: Separately excited DC generator

$$V = E_g - I_a R_a - \text{brush contact drop (if any)}$$

Power developed $\qquad \boxed{P_g = E_g \cdot I_a}$

Power delivered $\qquad \boxed{P_L = V \cdot I_L}$

1.21.2 Shunt Wound DC Generator

For shunt wound DC generator as shown in Fig. 1.35.

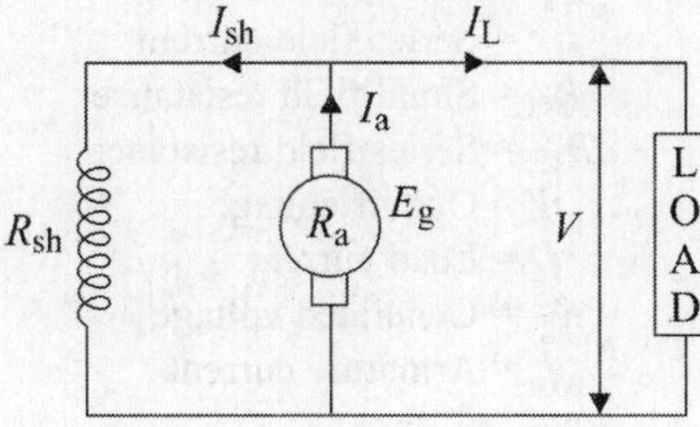

Let I_{sh} = Shunt field current
$\quad R_{sh}$ = Shunt field resistance
$\quad\; V$ = Output voltage
$\quad E_g$ = Generated voltage
$\quad I_a$ = Armature current
$\quad I_L$ = Load current
$\quad R_a$ = Armature resistance

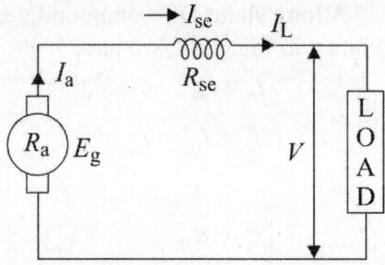

Fig. 1.35: DC shunt generator

$$\boxed{I_{sh} = \frac{V}{R_{sh}}}$$

$I_a = I_L + I_{sh}$

$V = E_g - I_a R_a - $ brush contact drop (if any)

Power developed $\qquad \boxed{P_g = E_g \cdot I_a}$

Power delivered $\qquad \boxed{P_L = V \cdot I_L}$

1.21.3 Series Wound DC Generator

For DC series generator shown in Fig. 1.36.

Let I_{se} = Series field current
$\quad R_{se}$ = Series field resistance
$\quad\; V$ = Output voltage
$\quad E_g$ = Generated voltage
$\quad I_a$ = Armature current
$\quad I_L$ = Load current
$\quad R_a$ = Armature resistance
$\quad I_{se} = I_a = I_L$

Fig. 1.36: DC series generator

$$V = E_g - I_a \left(R_a + R_{se} \right) - \text{brush contact drop (if any)}$$

Power developed $\qquad \boxed{P_g = E_g \cdot I_a}$

Power delivered $\qquad \boxed{P_L = V \cdot I_L}$

1.21.4 Compound Wound DC Generator

Case 1: Short Shunt DC Compound Generator

A short shunt DC compound generator is shown in Fig. 1.37.

Let I_{sh} = Shunt field current

I_{se} = Series field current

R_{sh} = Shunt field resistance

R_{se} = Series field resistance

V = Output voltage

I_L = Load current

E_g = Generated voltage

I_a = Armature current

$\quad = I_{sh} + I_{se}$

$I_{se} = I_L$

$I_{sh} = \dfrac{V + I_{se}\,R_{se}}{R_{sh}}$

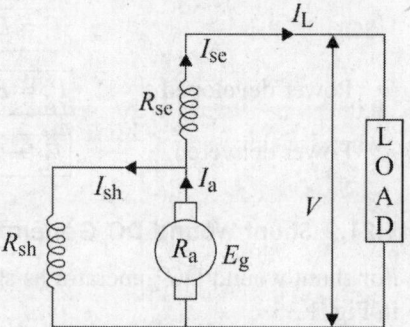

Fig. 1.37: Short shunt DC compound generator

$V = E_g - I_a R_a - I_{se}\,R_{se} -$ brush contact drop (if any)

Power developed $\quad \boxed{P_g = E_g \cdot I_a}$

Power delivered $\quad \boxed{P_L = V \cdot I_L}$

Case 2: Long Shunt DC Compound Generator

A long shunt DC compound generator is shown in Fig. 1.38, we have

$I_a = I_{se}$

$I_a = I_{sh} + I_L$

$I_{sh} = \dfrac{V}{R_{sh}}$

$V = E_g - I_a R_a - I_{se}\,R_{se} -$ brush contact drop (if any)

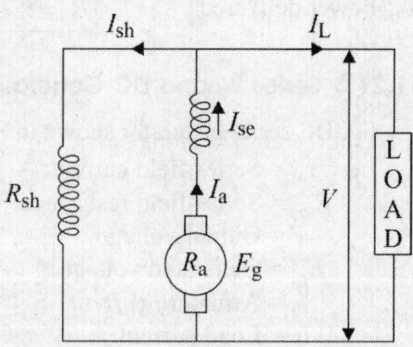

Fig. 1.38: Long shunt DC compound generator

or $\quad V = E_g - I_a(R_a + R_{se}) -$ brush contact drop (if any)

Power developed $\quad \boxed{P_g = E_g \cdot I_a}$

Power delivered $\quad \boxed{P_L = V \cdot I_L}$

SOLVED EXAMPLES

Example 1.10: A 20 kW, 200 V shunt generator has an armature resistance of 0.05 W, and a shunt field resistance of 200 W. Calculate power developed in armature, when it delivers rated output. **[UPTU 2006–07]**

Solution: Given $P_L = 20$ kW, $V = 200$ V, $R_{sh} = 200\ \Omega$, $R_a = 0.05\ \Omega$

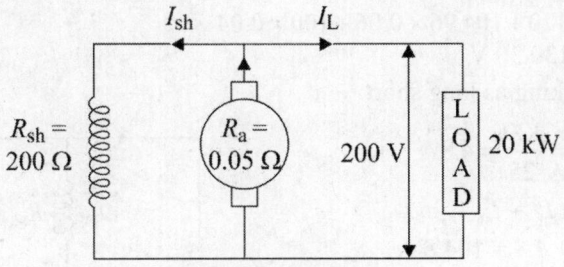

Fig. 1.39

$$I_L = \frac{P_L}{V} = \frac{20 \times 1000}{200} = 100\ A$$

$$I_{sh} = \frac{V}{R_{sh}} = \frac{200}{200} = 1\ A$$

$$I_a = I_L + I_{sh} = 100 + 1 = 101\ A$$

$$E_g = V + I_a R_a = 200 + 101 \times 0.05 = 205.05\ V$$

$$P_g = E_g \cdot I_a = 205.05 \times 101 = 20710\ W = 20.71\ kW\ \textbf{Ans.}$$

Example 1.11: In a 120 V compound generator, the resistances of armature, shunt and series field windings are 0.06 W, 25 W and 0.04 W, respectively. Load current is 100 A. Find the induced emf and armature current, when the machine is connected as (*i*) short shunt, (*ii*) long shunt.

Solution: Given, $V = 120$ V, $R_a = 0.06\ \Omega$, $R_{sh} = 25\ \Omega$, $R_{se} = 0.04\Omega$, $I_L = 100A$

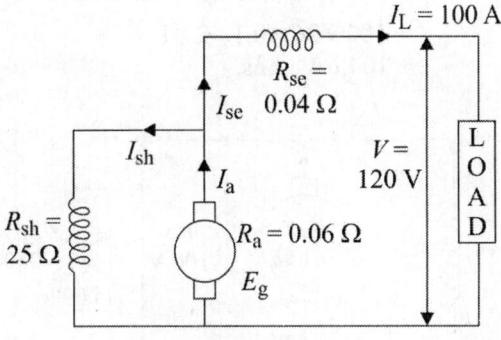

Fig. 1.40

(*i*) When working as short shunt

$$I_{se} = I_L = 100 \text{ A}$$

$$I_{sh} = \frac{V + I_{se}R_{se}}{R_{sh}} = \frac{120 + 100 \times 0.04}{25} = 4.96 \text{ A}$$

$$I_a = I_{sh} + I_{se} = 104.96 \text{ A}$$

$$E_g = V + I_a R_a + I_{se} R_{se}$$

$$= 120 + 104.96 \times 0.06 + 100 \times 0.04$$

$$E_g = 130.30 \text{ V} \qquad \qquad \textbf{Ans.}$$

(*ii*) When working as long short

$$I_{sh} = \frac{V}{R_{sh}} = \frac{120}{25} = 4.8 \text{ A}$$

$$I_a = I_{se} = I_L + I_{sh}$$

$$I_a = 100 + 4.8 = 104.8 \text{ A}$$

$$E_g = V + I_a(R_a + R_{se})$$

$$= 120 + 104.96 (0.06 + 0.04)$$

$$E_g = 130.48 \text{ V}$$

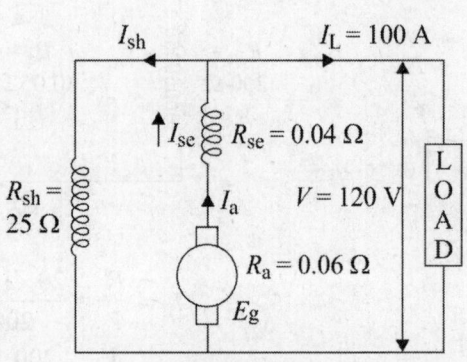

Fig. 1.41

Example 1.12: A DC shunt generator having armature and field resistance of $0.1\ \Omega$ and $0.1\ \Omega$, respectively. It supplies power to fifty 100 V, 50 W bulb. Calculate armature current and generated emf. Consider brush contact drop of 1 V/brush.

Solution:

$$P_L = 50 \times 50 \text{ W}$$

$$= 2500 \text{ W}$$

$$I_L = \frac{P_L}{V} = \frac{2500}{100} = 25 \text{ A}$$

$$I_{sh} = \frac{V}{R_{sh}} = \frac{100}{100} = 1 \text{ A}$$

$$I_a = I_L + I_{sh} = 25 + 1 = 26 \text{ A}$$

$$E_g = V + I_a R_a + \text{ brush contact drop}$$

$$= 100 + 26 \times 0.1 + 2 \times 1$$

$$= 104.6 \text{ V } \textbf{Ans.}$$

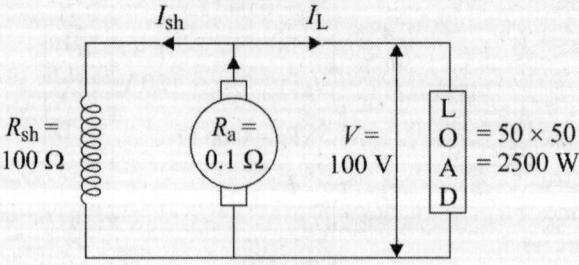

Fig. 1.42

1.22 ARMATURE REACTION

So far we have assumed that the only flux acting in a DC machine is due to the main poles called main flux. However, current flowing through armature conductors also creates a magnetic flux (called armature flux) that distorts and weakens the flux coming from the poles. This distortion and field weakening takes place in both generators and motors. The action of armature flux on the main flux is known as armature reaction.

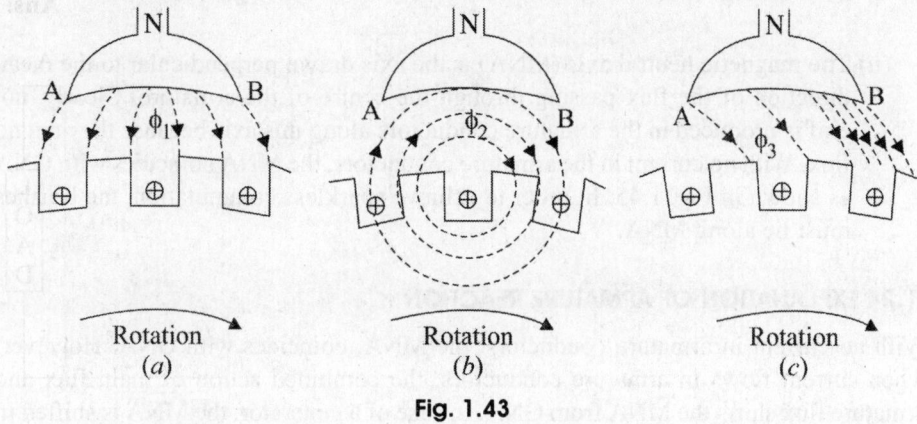

Fig. 1.43

The phenomenon of armature reaction in a DC generator is shown in Fig. 1.43. Only one pole is shown for clarity. When the generator is on no-load, a small current flowing in the armature does not appreciably affect the main flux ϕ_1, coming from the pole [See Fig. 1.43(a)]. When the generator is loaded, the current flowing through armature conductors sets up flux ϕ_2. Figure 1.43(b) shows flux due to armature current alone. By superimposing ϕ_1 and ϕ_2, we obtain the resulting flux ϕ_3 as shown in Fig. 1.43(c). Referring to Fig. 1.43(c), it is clear that flux density at the trailing pole tip (point B) is increased what at the leading pole tip (point A), it is decreased. This unequal field distribution produces the following two effects:

(i) The main flux is distorted.

(ii) Due to higher flux density at pole tip B, saturation builds up. Consequently, the increase in flux at pole tip B is less than the decrease in flux under pole tip A. Flux ϕ_3 at full load is, therfore, less than flux ϕ_1 at no load. As we shall see, the weaking of flux due to armature reaction depends upon the position of brushes.

1.23 GEOMETRICAL AND MAGNETIC NEUTRAL AXES

(i) The geometrical neutral axis (GNA) is the axis that bisects the angle between the centre line of adjacent poles [see Fig. 1.44]. Clearly, it is the axis of symmetry between two adjacent poles.

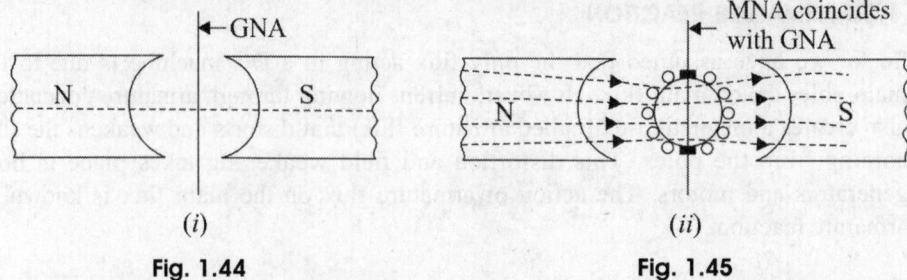

(*i*)

Fig. 1.44

(*ii*)

Fig. 1.45

(*ii*) The magnetic neutral axis (MNA) is the axis drawn perpendicular to the mean direction of the flux passing through the centre of the armature. Clearly, no. emf is produced in the armature conductors along this axis because they cut no flux. With no current in the armature conductors, the MNA coincides with GNA as shown in Fig. 1.45. In order to achieve sparkless commutation, the brushes must lie along MNA.

1.24 EXPLANATION OF ARMATURE REACTION

With no current in armature conductors, the MNA, coincides with GNA. However, when current flows in armature conductors, the combined action of main flux and armature flux shifts the MNA from GNA. In case of a generator, the MNA is shifted in the direction of rotation of the machine. In order to achieve sparkless commutation, the brushes have to be moved along the new MNA. Under such a condition, the armature reaction produces the following two effects:

1. It demagnestises or weakens the main flux
2. It cross-magnetises or distorts the main flux

Let us discuss these effects of armature reaction of considering a 2-pole generator (through the following remarks also hold good for a multipolar generator).

(*i*) Figure 1.46 shows the flux due to (main flux) when the armature conductors carry no current. The flux across the air gap is uniform. The mmf producing the main flux is represented in magnitude and direction by the vector OF_m in Fig. 1.46. Note that OF_m is perpendicular to GNA.

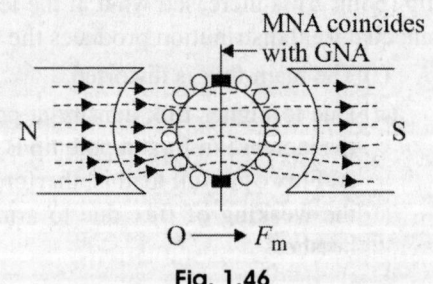

Fig. 1.46

As conductors, the pole tip they first meet is the leading pole tip and the other trailing pole tip. If the brushes remain along GNA (i.e. they are not shifted along new MNA), only cross–magnetising effect (i.e. distortion of main flux) of armature reaction occurs. In case of a generator, the flux is weakened at the leading pole tip and strengthened at the trailing pole tip. Clearly, it is the distortion of the main flux. Since the decrease of flux at one pole tip is equal to the increase in flux at the other tip, the flux per pole remains the same.

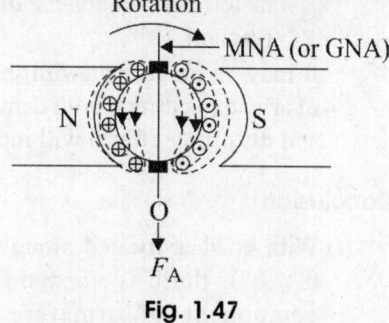

Fig. 1.47

(*ii*) Figure 1.47 shows the flux due to current flowing in armature conductors alone (main poles unexcited). The armature conductors to the left of GNA carry current 'in' (×) and those to the right carry current 'out' (•). The direction of magnetic lines of force can be found by cork screw rule. It is clear that armature flux is directed downward parallel to the brush axis. The mmf producing the armature flux is represented in magnitude and direction by the vector OF_A in Fig. 1.47.

(*iii*) Figure 1.48 shows the flux due to the main poles and that due to current in armature conductors acting together. The resultant mmf OF is the vector sum of OF_m and OF_A as shown in Fig. 1.48. Since MNA is always perpendicular to the resultant mmf, the MNA is shifted through an angle θ. Note that MNA is shifted in the direction of rotation of the generator.

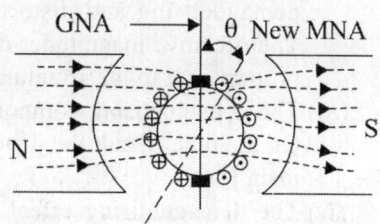

(*iv*) In order to achieve sparkless commutation, the brushes must lie along the MNA. Consequently, the brushes are shifted through an angle θ so as to lie along the new MNA as shown in Fig. 1.48. Due to brush shift, the mmf F_A

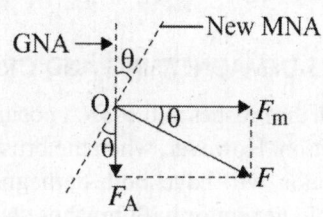

Fig. 1.48

of the armature is also rotated through the same angle θ. It is because some of the conductors which were earlier under N-pole now come under S-pole and *vice versa*. The result is that armature mmf F_A will no longer be vertically downward but will be rotated in the direction of rotation through an angle θ as shown in Fig. 1.49. Now F_A can be resolved into rectangular components F_c and F_d.

(*a*) The component F_d is in direct opposition to the mmf OF_m due to main poles. It has a demagnetising effect on the flux due to main poles. For this reason, it is called the demagnetising or weakening component of armature reaction.

(*b*) The component F_C is at right angles to the mmf OF_m due to main poles. It distorts the main field. For this reason, it is called the cross-magnetising or distorting component of armature reaction.

It may be noted that with the increase of armature current, both demagnetising and distorting effects will increase.

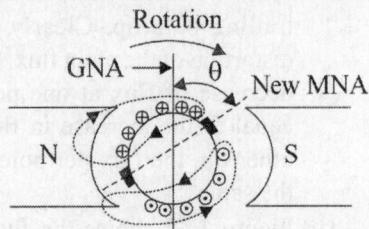

Conclusion

(i) With brushes located along GNA (i.e. $\theta = 0°$), there is no demagnetising component of armature reaction ($F_d = 0$). There is only distorting or cross-magnetising effect of armature reaction.

(ii) With the brushes shifted from GNA, armature reaction will have both demagnetising and distorting effects.

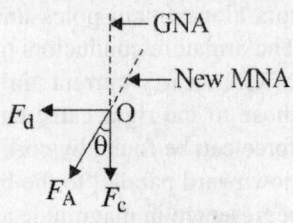

Fig. 1.49

Their relative magnitudes depend on the amount of shift. This shift is directly proportional to the armature current.

(iii) The demagnetising component of armature reaction weakens the main flux. On the other hand, the distorting component of armature reaction distorts the main flux.

(iv) The demagnetising effect leads to reduced generated voltage while cross-magnetising effect leads to sparking at the brushes.

1.25 DEMAGNETISING AND CROSS-MAGNETISING CONDUCTORS

With the brushes in the GNA position, there is only cross-magnetising effect of armature reaction. However, when the brushes are shifted from the GNA position, the armature reaction will have both demagnetising and cross-magnetising effects. Consider a 2-pole generator with brushes shifted (lead) and mechanical degrees from GNA. We shall identify the armature conductors that produce demagnetising effect and those that produce cross-magnetising effect.

(i) The armature conductors $\theta_m^°$ on either side of G.N.A. produce flux in direct opposition to main flux as shown in Fig. 1.50. Thus the conductors lying within angles AOC = BOD = $2\theta_m$ at the top and bottom of the armature produce demagnetising effect. These are called demagnetising armature conductors and constitute the demagnetising ampere-turns of armature reaction (Remember two conductors constitute a turn).

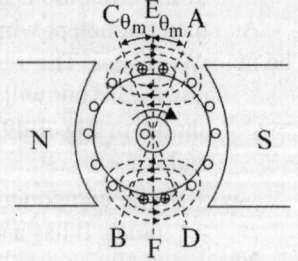

Fig. 1.50

(ii) The axis of magnetisation of the remaining armature conductors lying between angles AOD and COB is at right angles to the main flux as shown in Fig. 1.51.

These conductors produce the cross-magnetising (or distorting) effect, i.e. they produce uneven flux distribution on each pole. Therefore, they are called cross-magnetising conductors and constitute the cross-magnetising ampere-turns of armature reaction.

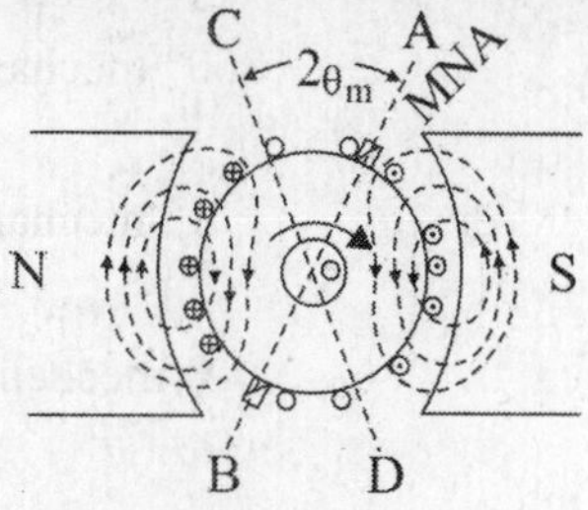

Fig. 1.51

1.26 CALCULATION OF DEMAGNETISING AMPERE-TURNS PER POLE (AT$_d$/POLE)

It is sometimes desirable to neutralise the demagnetising ampere-turns of armature reaction. This is achieved by adding extra ampere-turns to the main field winding. We shall now calculate the demagnetising ampere-turns per pole $(AT_d/Pole)$.

Let Z = total number of armature conductors

I = current in each armature conductor

$= I_a/2$ for simplex wave winding

$= I_a/P$ for simplex lap winding

θ_m = forward lead in mechanical degrees

Referring to Fig. 1.50, we have,

total demagnetising armature conductors = conductors in angles AOC and BOD

$$= \frac{4\theta_m}{360} \times Z$$

Since two conductors constitute one turn

$\therefore$ total demagnetising ampere-turns $= \frac{1}{2}\left[\frac{4\theta_m}{360} \times Z\right] \times I = \frac{2\theta_m}{360} \times ZI$

These demagnetising ampere-turns are due to a pair of poles.

Demagnetising ampere-turns/pole $= \frac{\theta_m}{360} \times ZI$

i.e. $\qquad AT_d/\text{pole} = \frac{\theta_m}{360} \times ZI$

As mentioned above, those demagnetising ampere-turns of armature reaction can be neutralised by putting extra turns on each pole of the generator.

$\therefore \qquad$ No. of extra turns/pole $= \dfrac{AT_d}{I_{sh}}$

$$= \dfrac{AT_d}{I_d}$$

Note: When a conductor passes a pair of poles, one cycle of voltage is generated. We say one cycle contains 360 electrical degrees. Suppose there are P poles in a generator. In one revolution there are 360 mechanical degrees and $360 \times P/2$ electrical degrees.

$$\therefore \qquad 360° \text{ mechanical} = 360 \times \frac{P}{2} \text{ electrical degrees}$$

or $\qquad 1° \text{ mechanical} = \dfrac{P}{2} \text{ electrical degrees}$

$$\therefore \qquad \theta(\text{mechanical}) = \frac{\theta(\text{electrical})}{\text{Pair of poles}}$$

or $\qquad \theta_m = \dfrac{\theta_e}{P/2} \qquad \therefore \ \theta_m = \dfrac{2\theta_e}{P}$

1.27 CROSS-MAGNETISING AMPERE–TURNS PER POLE (AT$_C$/POLE)

We now calculate the cross-magnetising ampere-turns per pole (AT/pole).

Total armature reaction ampere-turns per pole

$$= \frac{Z/2}{P} \times I = \frac{Z}{2P} \times I \quad (\therefore \text{ two conductors make one turn})$$

Demagnetising ampere-turns per pole is given by;

$$AT_d/\text{pole} = \frac{\theta_m}{360} \times ZI$$

$\therefore$ Cross-magnetising ampere-turns/pole are

$$AT_c/\text{pole} = \frac{Z}{2P} \times I - \frac{\theta_m}{360} \times ZI$$

$$= ZI\left(\frac{1}{2P} - \frac{\theta_m}{360}\right)$$

$$\therefore \qquad AT_c/\text{pole} = ZI\left(\frac{1}{2P} - \frac{\theta_m}{360}\right)$$

1.28 COMPENSATING WINDINGS

The cross-magnetising effect of armature reaction may cause trouble in DC machines subjected to large fluctuations in load. In order to neutralise the cross-magnetising effect of armature reaction, a compensating winding is used.

A compensating winding is an auxiliary winding embedded in slots in the pole faces as shown in Fig. 1.52. It is connected in series with armature in a manner so that the direction of current through the conductors in anyone pole face will

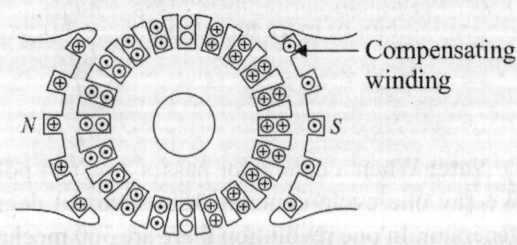

Fig. 1.52

be opposite to the direction of the current through the adjacent armature conductors [see Fig. 1.52]. Let us now calculate the number of compensating conductors/pole face.

In calculating the conductors per pole face required for the compensating winding, it should be remembered that the current in the compensating conductors is the armature current I_a whereas the current in armature conductors is I_a/A, where A is the number of parallel paths.

Let Z_C = No. of compensating conductors/pole face

   Z_a = No. of active armature conductors

   I_a = Total armature current

   I_a/A = Current in each armature conductor

$\therefore$ $Z_C I_a = Z_a \times \dfrac{I_0}{A}$

or $Z_C = \dfrac{Z_0}{A}$

The use of a compensating winding considerably increases the cost of a machine and is justified only for machines intended for severe service, e.g. for high speed and high voltage machines.

1.29 AT$_C$ /POLE FOR COMPENSATING WINDING

Only the cross-magnetising ampere-turns produced by conductors under the pole face are effective in producing the distortion in the pole cores. If Z is the total number of armature conductors and P is the number of poles, then,

No. of armature conductors/pole = $\dfrac{Z}{P}$

No. of armature turns/pole = $\dfrac{Z}{2P}$

No. of armature turns under pole face = $\dfrac{Z}{2P} \times \dfrac{\text{Pole arc}}{\text{Pole pitch}}$

If I is the current through each armature conductor, then,
AT/pole required for compensating winding

$$= \frac{ZI}{2P} \times \frac{\text{Pole arc}}{\text{Pole pitch}}$$

$$= \text{Armature AT}_c / \text{pole} \times \frac{\text{Pole arc}}{\text{Pole pitch}}$$

Example 1.13: A 4-pole generator has a wave-wound armature with 722 conductors and it delivers 100 A on full-load. If the brush lead is 8°, calculate the armature demagnetising and cross-magnetising ampere-turns per pole.

Solution: $I = I_a/2 = 100/2 = 50$ A; $Z = 722$; $Q_m = 8°$

$$\text{AT}_C/\text{pole} = ZI \times \frac{Q_m}{360} = 722 \times 50 \times \frac{8}{360} = 802$$

$$\text{AT}_C/\text{pole} = ZI\left(\frac{1}{2P} - \frac{Q_m}{360}\right) = 722 \times 50 \times \left(\frac{1}{2 \times 4} - \frac{8}{360}\right) = 3710$$

Example 1.14: A 4-pole generator supplies a current of 143 A. It has 492 conductors (*i*) wave connected (*ii*) lap connected. When delivering full load, the brushes are given an actual angle of 10°. Calculate the demagnetising ampere-turns/pole. The field winding is shunt connected and takes 10 A. Calculate the number of extra shunt field turns necessary to neutralise this demagnetisation.

Solution: $Z = 492$; $\theta_m = 10°$; $I_a = 143 + 10 = 153$ A

(*i*) Wave-connected $\qquad I = \dfrac{I_0}{2} = \dfrac{153}{2}$ A

$$\text{AT}_d/\text{pole} = ZI \times \frac{\theta_m}{360} = 492 \times \frac{153}{2} \times \frac{10}{360} = 1046\text{AT}$$

$$\text{Extra shunt field turns/pole} = \frac{\text{AT}_d/\text{pole}}{I_{Sh}} = \frac{1046}{10} = 105$$

(*ii*) Lap-connected $\qquad I = \dfrac{I_P}{P} = \dfrac{153}{4}$ A

$$\text{AT}_d/\text{pole} = Z_1 \times \frac{\theta_m}{360} = 492 \times \frac{153}{4} \times \frac{10}{360} = 523\text{AT}$$

$$\text{Extra shunt field turns/pole} = \frac{523}{10} = 52$$

Example 1.15: A 500 kW, 500 Y 10 pole DC generator has a lap wound armature with 800 conductors. Calculate the number of pole-face conductors in each pole of a compensating winding if the pole face covers 75% of pole pitch.

Solution:

$$\text{Load current supplied} = \frac{500 \times 10^3}{500} = 100\,\text{A}$$

Current in each armature conductor $I = 1000/10 = 100$ A

$$\text{AT/pole for compensating winding} = \text{Armature AT/pole} \times \text{pole arc} \times \frac{\text{pole arc}}{\text{pole pitch}}$$

$$= \frac{ZI}{2P} \times 0.75 = \frac{800 \times 100}{2 \times 10} \times 0.75 = 300\text{AT}$$

$$\text{Turns per pole} = \frac{3000}{I_a} = \frac{3000}{1000} = 3$$

Example 1.16: A 400 V, 1000 A, lap-wound DC machine has 10 poles and 860 armature conductors. Calculate the number of conductors in the pole face to give full compensation if the pole face covers 70% of pole span.

Solution: The current in each armature conductor is

$$I = \frac{1000}{10} = 100\,\text{A}$$

$$\text{AT/pole for compensating winding} = \text{Armature AT/pole} \times \frac{\text{pole arc}}{\text{pole pitch}}$$

$$= \frac{ZI}{2P} \times 0.7 = \frac{860 \times 100}{2 \times 10} \times 0.7 = 3010$$

1.30 COMMUTATION

Figure 1.53 shows the schematic diagram of 2-pole lap-wound generator. There are two parallel paths between the brushes. Therefore, each coil of the winding carries one half ($I_a/2$ in this case) of the total current (I_a) entering or leaving the armature.

Note that the currents in the coils connected to a brush are either all towards the brush (positive brush) or all directed away from the brush (negative brush). Therefore, current in a coil will reverse as the coil passes a brush. This reversal of current as the coil passes a brush is called commutation.

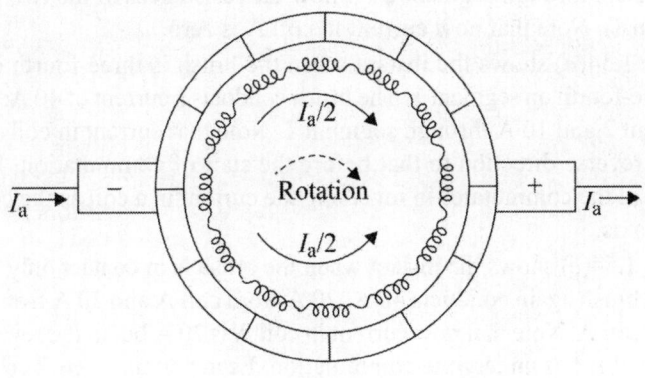

Fig. 1.53

The reversal of current in a coil as the coil passes the brush axis is called commutation.

When commutation takes place, the coil undergoing commutation is short-circuited by the brush. The brief period during which the coil remains short-circuited is known as commutation period T_C. If the current reversal is completed by the end of commutation period. it is called ideal commutation. If the current reversal is not completed by that time, then sparking occurs between the brush and the commutator which results in progressive damage to both.

Ideal commutation: Let us discuss the phenomenon of ideal commutation (i.e. coil has no inductance) in one coil in the armature winding shown in Fig. 1.53. For this purpose. we consider the coil A. The brush width is equal to the width of one commutator segment and one mica insulation. Suppose the total armature current is 40 A. Since there are two parallel paths, each coil carries a current of 20 A.

(i) In Fig. 1.54(a), the brush is in contact with segment 1 of the commutator. The commutator segment 1 conducts a current of 40 A to the brush; 20 A from coil A and 20 A from the adjacent coil as shown. The coil A has yet to undergo commutation.

(ii) As the armature rotates, the brush will make contact with segment 2 and thus short-circuits the coil A as shown in Fig. 1.54(b). There are now two parallel paths into the brush as long as the short–circuit of coil A exists. Figure 1.54(b) shows the instant when the brush is one-fourth on segment 2 and three-fourth on segment 1. For this condition, the resistance of the path through segment 2 is three times the resistance of the path through segment 1 (∵ contact resistance varies inversely as the area of contact of brush with the segment). The brush again conducts a current of 40 A; 30 A through segment 1 and 10 A through segment 2. Note that current in coil A (the coil undergoing commutation) is reduced from 20 A to 10 A.

(iii) Figure 1.54(c) shows the instant when the brush is one-half on segment 2 and one-half on segment 1. The brush again conducts 40 A; 20 A through segment 1 and 20 A through segment 2 (now the resistances of the two parallel paths are equal). Note that now current in coil A is zero.

(iv) Figure 1.54(d) shows the instant when the brush is three-fourth on segment 2 and one-fourth on segment 1. The brush conducts a current of 40 A; 30 A through segment 2 and 10 A through segment 1. Note that current in coil A is 10 A but in the reverse direction to that before the start of commutation. However, the action of the commutator in reversing the current in a coil as the coil passes the brush axis.

(v) Figure 1.54(e) shows the instant when the brush is in contact only with segment 2. The brush again conducts 40 A; 20 A from coil A and 20 A from the adjacent coil to coil A. Note that now current in coil A is 20 A but in the reverse direction. Thus coil A has undergone commutation. Each coil undergoes commutation in this way as it passes the brush axis. Note that during commutation, the coil under consideration remains short-circuited by the brush.

Figure 1.55 shows the current-time graph for the coil A undergoing commutation. The horizontal line AB represents a constant current of 20 A upto the beginning of commutation. From A the finish of commutation, it is represented by another horizontal line CD on the opposite side of the zero line and at the same distance from it as AB i.e., the current has exactly reversed quantum (–20 A). The way in which current changes from B to C depends upon the conditions under which the coil undergoes commutation. If the current changes at a uniform rate (i.e. BC is a straight line), then it is called ideal commutation as shown in Fig. 1.55.

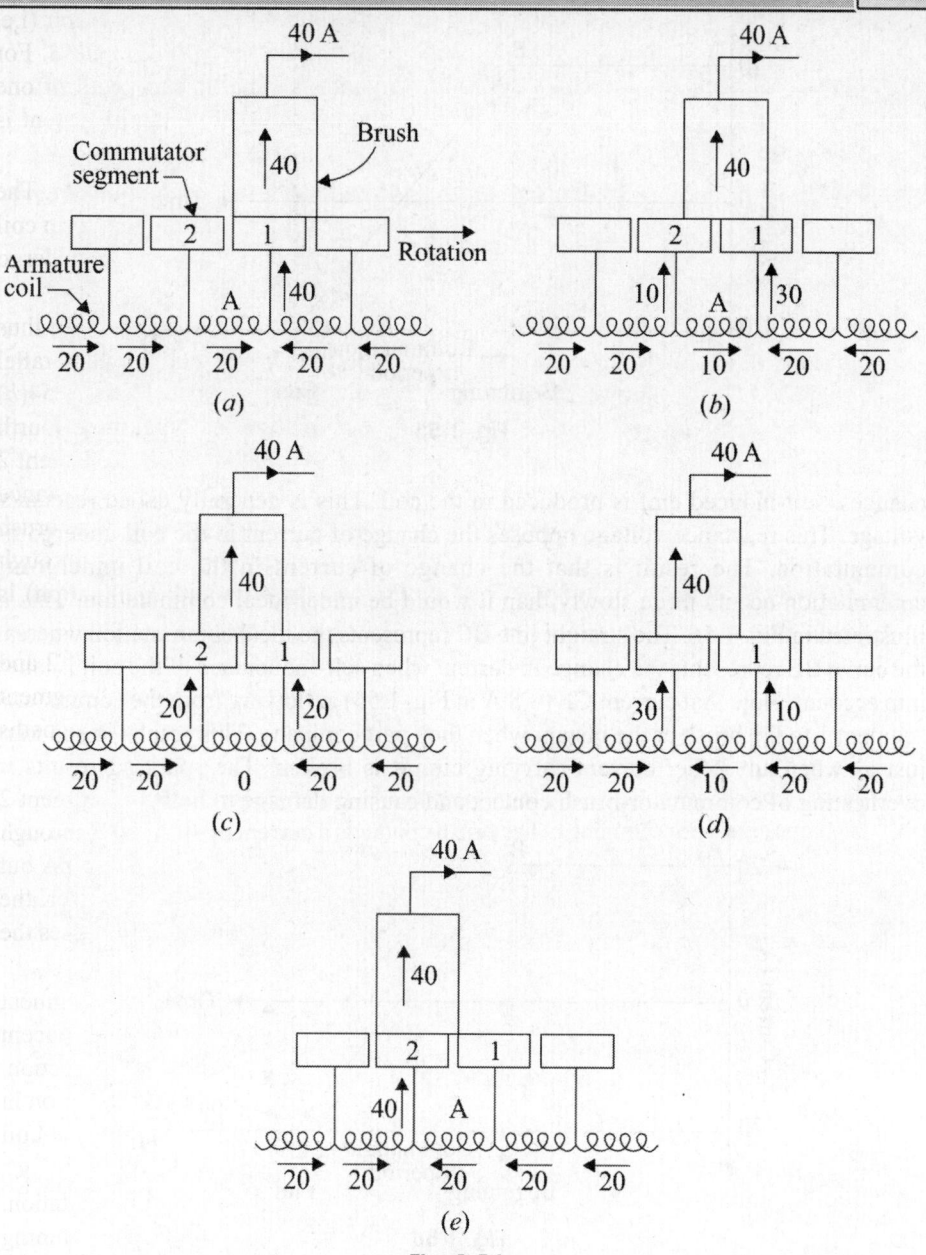

Fig. 1.54

Under such conditions, no sparking will take place between the brush and the commutator.

Practical difficulties: The ideal commutation (i.e. straight line change of current) cannot be attained in practice. This is mainly due to the fact that the armature coils have appreciable inductance. When the current in the coil undergoing commutation

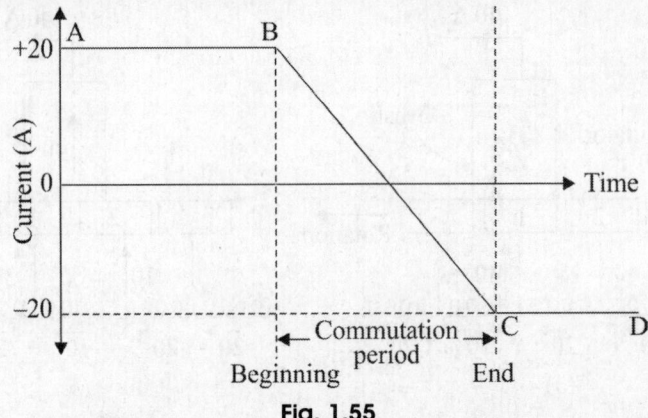

Fig. 1.55

changes, self-induced emf is produced in the coil. This is generally called reactance voltage. This reactance voltage opposes the change of current in the coil undergoing commutation. The result is that the change of current in the coil undergoing commutation occurs more slowly than it would be under ideal commutation. This is illustrated in Fig. 1.56. The straight line BC represents the ideal commutation whereas the curve BE represents the change in current when self-inductance of the coil is taken into account. Note that current CE (= 8 A in Fig. 1.56) is flowing from the commutator segment 1 to the brush at the instant when they part company. This results in sparking just as when any other current-carrying circuit is broken. The sparking results in overheating of commutator-brush contact and causing damage to both.

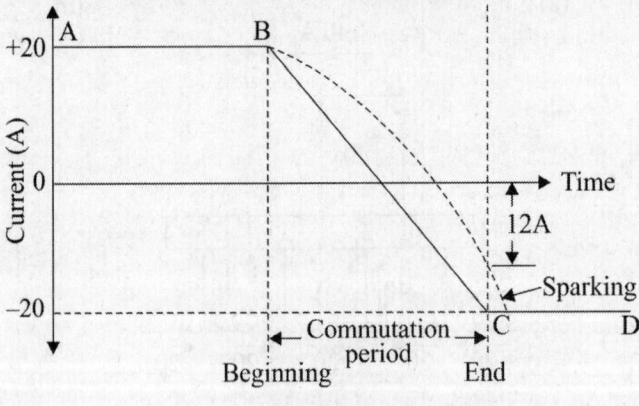

Fig. 1.56

Figure 1.57 illustrates how sparking takes place between the commutator segment and the brush. At the end of commutation or short–circuit period, the current in coil A is reversed to a value of 12 A (instead of 20 A) due to inductance of the coil. When the brush breaks contact with segment 1, the remaining 8A current jumps from segment 1 to the brush through air causing sparking between segment 1 and the brush.

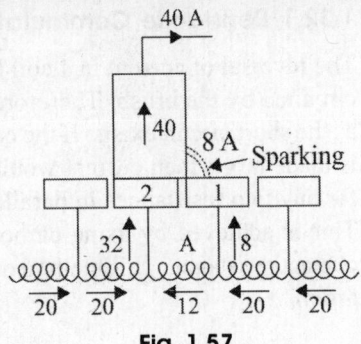

Fig. 1.57

1.31 CALCULATION OF REACTANCE VOLTAGE

Reactance voltage = Coefficient of self–inductance × Rate of change of current.

When a coil undergoes commutation, two commutator segments remain short-circuited by the brush. Therefore, the time of short circuit (or commutation period T_C) is equal to the time required by the commutator to move a distance equal to the circumferential thickness of the brush minus the thickness of one insulating strip of mica.

Let W_b = brush width in cm;

W_m = mica thickness in cm

v = peripheral speed of commutator in cm/s

T_C in seconds

The commutation period is very small, say of the order of 1/500 second.

Let the current in the coil undergoing commutation change from $+I$ to $-I$ (ampere) during the commutation. If L is the inductance of the coil, then reactance voltage is given by

$$E_R = L \times \frac{2I}{T_C} \quad \text{(for linear commutation)}$$

Example 1.17: Calculate the reactance voltage for a machine with the following criteria:

Revolution per minute = 900
Number of commutator segments = 55
Brush width in commutator segments = 1.74
Coefficient of self induction = 153 × 10⁻⁶ H

Assume linear commutation and neglect mica thickness.

Solution:

Peripheral speed of the commutator,

$$v = 55 \times 60 = 825 \text{ segments}$$

Commutation period, $T_C = \dfrac{W_b - W_m}{v} = \dfrac{1.74 - 0}{825} = 2.11 \times 10^{-3} \text{ s}$

Reactance voltage, $E_R = L \times \dfrac{2I}{T_C} = 153 \times 10^{-6} \times \dfrac{2 \times 27}{2.11 \times 10^{-3}} = 3.91 \text{ V}$

1.32 METHODS OF IMPROVING COMMUTATION

Improving commutation means to make current reversal in the short-circuited coil as sparkless as possible. The following are the two principal methods of improving commutation: (*i*) Resistance commutation (*ii*) emf commutation.

We shall discuss each method in details.

1.32.1 Resistance Commutation

The reversal of current in a coil (i.e. commutation) takes place when the coil is short-circuited by the brush. Therefore, there are two parallel paths for the current as long as the short circuit exists. If the contact resistance between the brush and the conductor is made large, then current would divide in the inverse ratio of contact resitances (as for any two resistances in parallel). This is the key point in improving commutation. This is achieved by using carbon brushes (instead of Cu brushes) which have high contact resistance. This method of improving commutation is called *resistance commutation.*

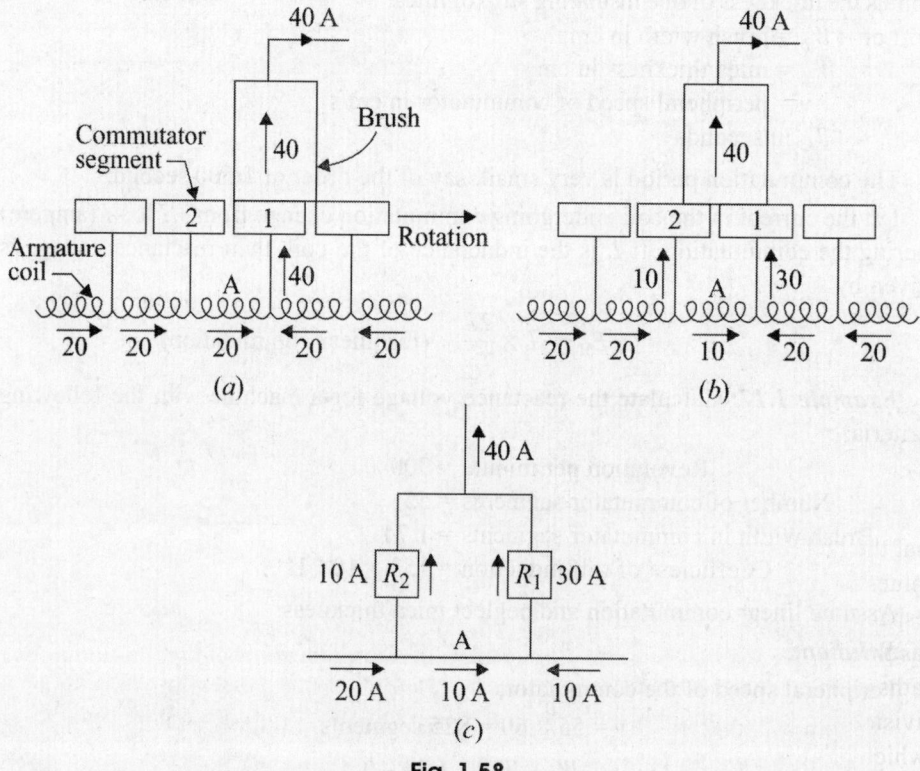

Fig. 1.58

Figures 1.58 and 1.59 illustrate how high contact resistance of carbon brush improves commutation (i.e. reversal of current) in coil A. In Fig. 1.58(*a*), the brush is entirely on segment 1 and, therefore, the current in coil A is 20 A. The coil A is yet to undergo commutation. As the armature rotates, the brush short–circuits the coil A and there are two parallel paths for the current into the brush. Figure 1.58(*b*) shows the instant when the brush is one-fourth on segment 2 and three-fourth on segment 1. The equivalent electric circuit is shown in Fig. 1.58(*c*) where R_1 and R_2 represent the brush contact resistances on segments 1 and 2. A resistor is not shown for coil A since it is assumed

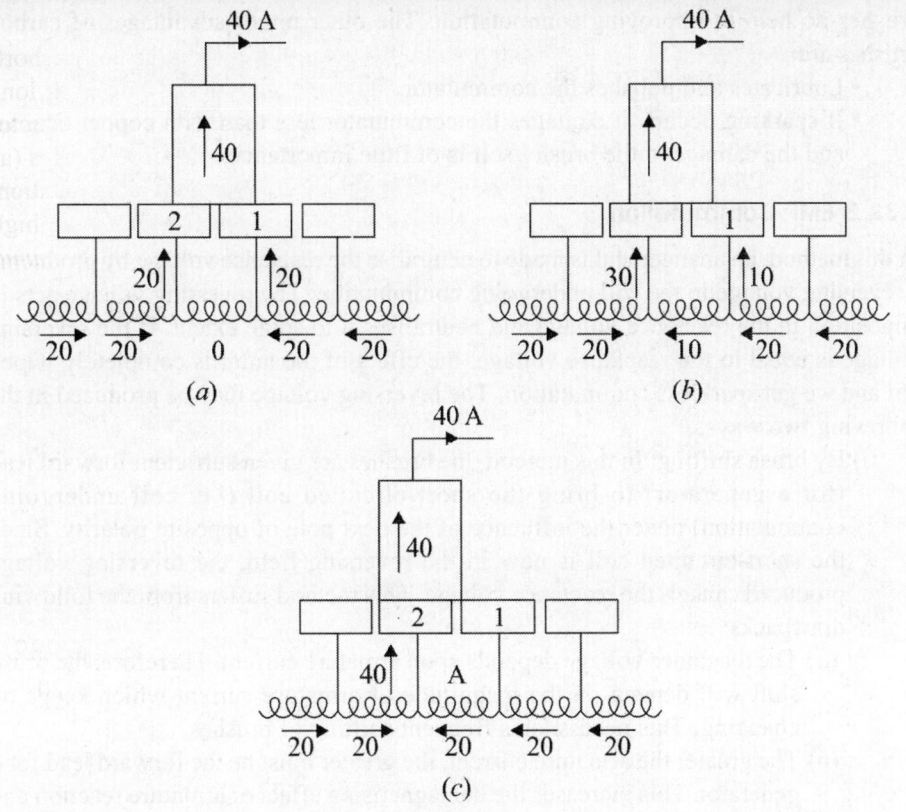

Fig. 1.59

that the coil resistance is negligible as compared to the brush contact resistance. The values of current in the parallel paths of the equivalent circuit are determined by the respective resistances of the paths. For the condition shown in Fig. 1.58(b), resistor R_2 has three times the resistance of resistor R_1. Therefore, the current distribution in the paths will be as shown. Note that current in coil A is reduced from 20 A to 10 A due to division of current in the inverse ratio of contact resistances. If the Cu brush is used (which has low contact resistance), $R_1 = R_2$ and the current in coil A would not have reduced to 10 A.

As the carbon brush passes over the commutator, the contact area with segment 2 increases and that with segment 1 decreases, i.e. R_2 decreases and R_1 increases. Therefore, more and more current passes to the brush through segment 2. This is illustrated in Figs 1.59(a) and 1.59(b). When the break between the brush and the segment 1 finally occurs [See Fig. 1.59(c)], the current in the coil is reversed and commutation is achieved.

It may be noted that the main cause of sparking during commutation is the production of reactance voltage and carbon brushes cannot prevent it. Nevertheless, the carbon

brushes do help in improving commutation. The other minor advantages of carbon brushes are:
- Lubricates and polishes the commutator.
- If sparking occurs, it damages the commutator less than with copper brushes and the damage to the brush itself is of little importance.

1.32.2 EMF Commutation

In this method, an anangement is made to neutralise the reactance voltage by producing a reversing voltage in the coil undergoing commutation. The reversing voltage acts in opposition to the reactance voltage and neutralises it to some extent. If the reversing voltage is equal to the reactance voltage, the effect of the latter is completely wiped out and we get sparkless commutation. The reversing voltage may be produced in the following two ways:

(*i*) By brush shifting: In this method, the brushes are given sufficient forward lead (for a generator) to bring the short-circuited coil (i.e. coil undergoing commutation) under the influence of the next pole of opposite polarity. Since the short-circuited coil is now in the reversing field, the reversing voltage produced cancels the reactance voltage. This method suffers from the following drawbacks:

(*a*) The reactance voltage depends upon armature current. Therefore, the brush shift will depend on the magnitude of armature current which keeps on changing. This necessitates frequent shifting of brushes.

(*b*) The greater the armature current, the greater must be the forward lead for a generator. This increases the demagnetising effect of armature reaction and further weakens the main field.

(*ii*) By using interpoles or compoles: The best method of neutralising reactance voltage is by using interpoles or compoles. This method is discussed as under.

1.33 INTERPOLES OR COMPOLES

The best way to produce reversing voltage to neutralise the reactance voltage is by using interpoles or compoles. These are small poles fixed to the yoke and spaced midway between the main poles [see Fig. 1.60]. They are wound with comparatively few turns and connected in series with the armature so that they carry armature current. Their polarity is the same as the next main pole ahead in the direction of rotation for a generator [see Fig. 1.60], connections for a d.c. generator with interpoles is shown in Fig. 1.61.

Functions of interpoles: The machines fitted with interpoles have their brushes set on geometrical neutral axis (no lead). The interpoles perform the following two functions:

(*i*) As their polarity is the same as the main pole ahead (for a generator), they induce an emf in the coil (undergoing commutation) which opposes reactance voltage. This leads to sparkless commutation. The emf induced by compoles is known as commutating or reversing emf. Since the interpoles carry the armature

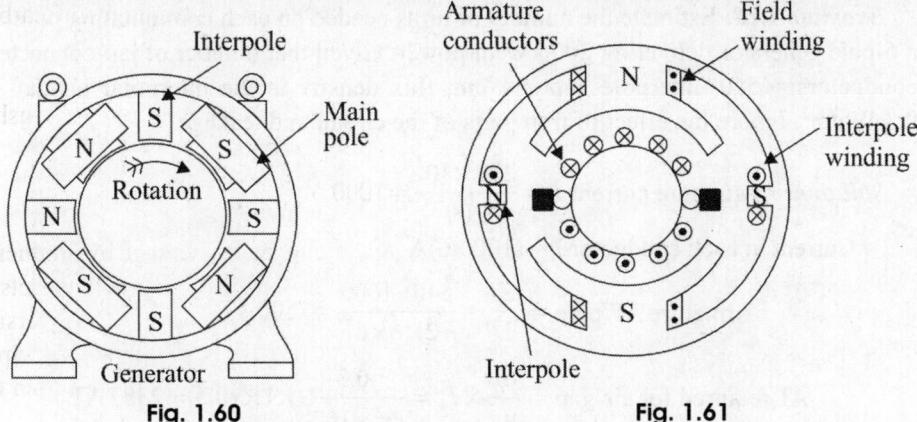

Fig. 1.60   Fig. 1.61

current and the reactance voltage is also proportional to armature current, the neutralisation of reactance voltage is automatic.

(*ii*) The mmf of the compoles neutralises the cross-magnetising effect of armature reaction in small region in the space between the main poles. It is because the two mmfs oppose each other in this region.

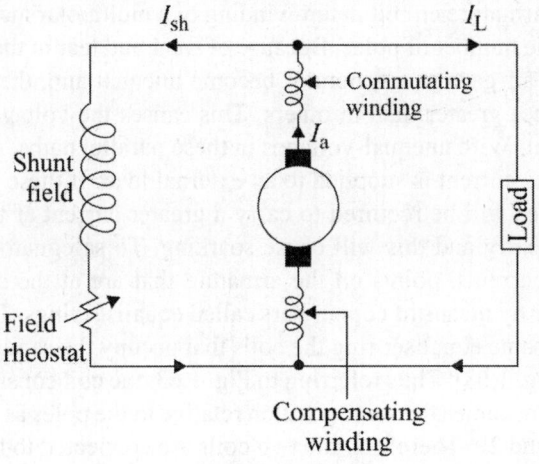

Fig. 1.62

Figure 1.62 shows the circuit diagram of a shunt generator with commutating winding and compensating winding. Both of these windings are connected in series with the armature and so they carry the armature current. However, the functions they perform must be understood clearly. The main function of commutating winding is to produce reversing (or commutating) emf in order to cancel the reactance voltage. In addition to this, the mmf of the commutating winding neutralises the cross-magnetising ampere-turns in the space between the main poles. The compensating winding neutralises the cross-magnetising effect of armature reaction under the pole faces.

Example 1.18: Estimate the number of turns needed on each commutating pole of a 6-pole generator delivering 200 kW of power. Given that number of lap-connected conductors = 540, interpole gap = 1 cm, flux density in the interpolar air-gap = 0.3 Wb/m². Ignore the effect of iron parts of the circuit and leakage.

Solution: Armature current $I = \dfrac{200 \times 10^3}{200} = 1000 \, \text{A}$

Current in each conductor $I = (1000/6)\text{A}$

Armature AT/pole $= \dfrac{ZI}{2P} = \dfrac{540 \times 1000}{2 \times 6 \times 6} = 7500 \, \text{AT}$

AT required for air-gap $= \dfrac{B}{\mu_0} \times I_g = \dfrac{0.3}{4\pi \times 10^{-7}} \times 1 \times 10^{-2} = 2387 \, \text{AT}$

AT for each commutating pole = 7500 + 2387 = 9887 AT

∴ Number of turns on each commutating pole

$$= 9887/I_a = 9887/1000 = 10$$

1.34 EQUALISING CONNECTIONS

We know that the armature circuit in lap-winding of a multipolar machine has as many parallel paths as the number of poles. Because of wear and tear in the bearings, and for other reasons, the air gaps in a generator become unequal and, therefore, the flux in some poles becomes greater than in others. This causes the voltages of the different paths to be unequal. With unequal-voltages in these parallel paths, circulating current will flow even if no current is supplied to an external load. If these currents are large, some of the brushes will be required to carry a greater current at full load than they were designed to carry and this will cause sparking. To safeguard the brushes from these circulating currents, points on the armature that are at the same potential are connected together by means of copper bars called equaliser rings. This is achieved by connecting to the same equaliser ring the coils that occupy the same positions relative to the poles [see Fig. 1.63]. Thus referring to Fig. 1.63, the coil consisting of conductor 1 and conductor-8 occupies the same position relative to the poles as the coil consisting of conductors 13 and 20. Therefore, the two coils are connected to the same equaliser ring. The equalisers provide a low resistance path for the circulating current. As a result, the circulating current due to the slight differences in the voltages of the various parallel paths passes through the equaliser rings instead of passing through the brushes. This reduces sparking.

Equaliser rings should be used only on windings in which the number of coils is a multiple of the number of poles. For best results, each coil should be connected to an equaliser ring but this is seldom done. Satisfactory results are obtained by connecting about every third coil to an equaliser ring. In order to distribute the connections to the equaliser rings equally, the number of coils per pole must be divisible by the connection pitch.

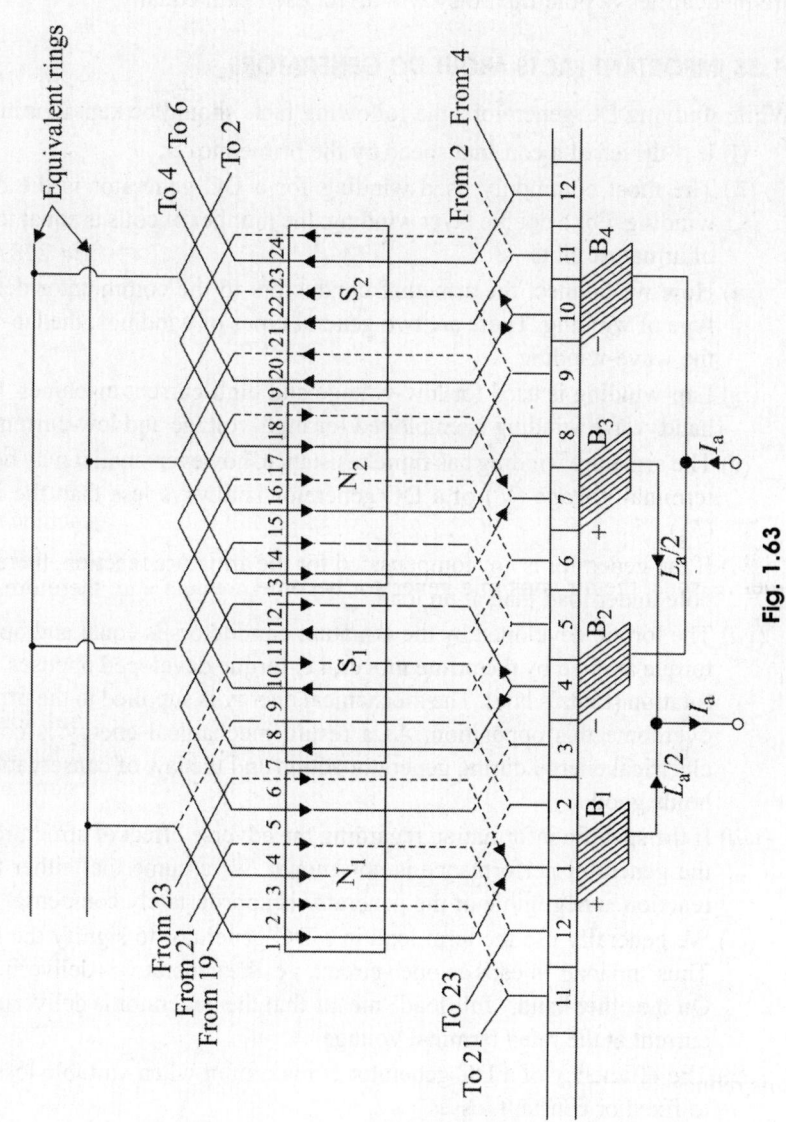

Fig. 1.63

Note: Equaliser rings are not used in wave-winding because there is no imbalance in the voltages of the two parallel paths. This is due to the fact that conductors in each of the two paths pass under all N and S poles successively (unlike a lap-winding where all conductors in any parallel path lie under one pair of poles). Therefore, even if there are inequalities in pole flux, they will affect each path equally.

1.35 IMPORTANT FACTS ABOUT DC GENERATORS

While studying DC generators, the following facts should be kept in mind:

(*i*) It is driven at a constant speed by the prime mover.

(*ii*) The most commonly used winding for a DC generator is the double layer winding. For a double layer winding, the number of coils is equal to the number of armature slots.

(*iii*) How we connect the armature conductors to the commutator determines the type of winding. There are two general types of windings; the lap-winding and the wave-winding.

(*iv*) Lap-winding is used for low-voltage and high-current machines. On the other hand, wave winding is employed for high-voltage and low-current machines.

(*v*) The armature winding has finite resistance, however small it may be. Therefore, terminal voltage (V) of a DC generator is always less than the induced emf (E_g).

(*vi*) If the generator is not compensated for the armature reaction, there is less flux/pole under load than at no load.

(*vii*) The torque developed by the armature conductors is equal and opposite to the torque applied by the prime mover, i.e. torque developed opposes the armature rotation (Lenz's law). The mechanical energy is supplied to the prime mover to overcome this opposition. As a result, mechanical energy is converted into electrical energy during generator action and the law of conservation of energy holds good.

(*viii*) If the specific information regarding the adverse effect of armature reaction on the generator performance is not known, we assume that either the armature reaction is negligible or the generator is appropriately compensated for it.

(*ix*) We generally use the term load in a DC generator to signify the load current. Thus "no load" means an open circuit, i.e. the generator is delivering no current. On the other hand, "full-load" means that the generator is delivering rated load current at the rated terminal voltage.

(*x*) The efficiency of a DC generator is maximum when variable losses are equal to fixed or constant losses.

1.36 DC GENERATOR CHARACTERISTICS

The following are the three most important characteristics of a DC generator:

1. **Open circuit characteristic (OCC):** This curve shows the relation between the generated emf at no-load (E_0) and the field current (I_f) at constant speed. It

is also known as magnetic characteristics or no-load saturation curve. Its shape is practically the same for all generators wheather separately- or self-excited. The data for OCC curve are obtained experimentally by operating the generator at no load and constant speed and recording the change in terminal voltage as the field current is varied.

2. **Internal or total characteristic (E/I_a):** This curve shows the relation between the generated emf on load (E) and the armature current (I_a). The emf E is less than E_0 due to demagnetising effect of armature reaction. Therefore, this curve will lie below the open circuit characteristics (OCC). The internal characteristic is of interest chiefly to the designer. It cannot be obtained directly by experiment. It is because a voltmeter cannot read the emf generated on load due to the voltage drop in armature resistance. The internal characteristic can be obtained from external characteristic if winding resistance are known because armature reaction effect is included in both characteristics.

3. **External characteristic (V/I_L):** This curve shows the relation between the terminal voltage (V) and load current (I_L). The terminal voltage V will be less than E due to voltage drop in the armature circuit. Therefore, this curve will lie below the internal characteristic. This characteristics is very important in determining the suitability of a generator for a given purpose. It can be obtained by making simultaneous measurements of terminal voltage and load current (with voltmeter and ammeter) of a loaded generator.

1.37 OPEN CIRCUIT CHARACTERISTIC OF A DC GENERATOR

The OCC for a DC generator is determined as follows (Fig. 1.64). The field winding of the DC generator (series or shunt) is disconnected from the machine and is separately excited from an extemal DC source as shown in Fig. 1.64(b). The generator is run at fixed speed (i.e. normal speed). The field current (I_f) is increased from zero in steps and the corresponding values of generated emf (E_0) read off on a voltmeter connected across the armature terminals. On plotting the relation between E_0 and I_a we get the open circuit characteristic as shown in Fig. 1.64(a).

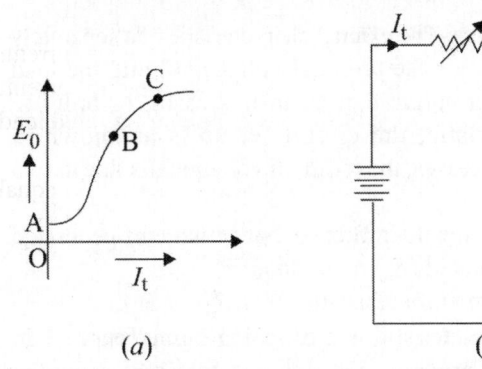

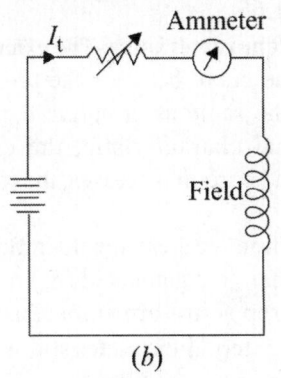

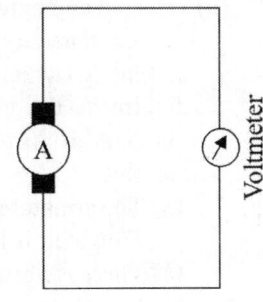

(a) (b)

Fig. 1.64

The following points may be noted from OCC:

(*i*) When the field cutrent is zero, there is some generated emf OA. This is due to the residual magnetism the field poles.

(*ii*) Over a fairly wide range of field current (upto point *B* in the curve), the curve is linear. It is because in this range; reluctance of iron is negligible as compared to that of air gap. The air gap reluctance is constant and hence linear relationship.

(*iii*) After point B on the curve, the reluctance of iron also comes into picture. It is because at higher flux densities, m_1 for iron decreases and reluctance of iron is no longer negligible. Consequently, the curve deviates from linear relationship.

(*iv*) After point *C* on the curve, the magnetic saturation of poles begins and E_0 tends to level off.

The reader may note that the OCC of even self-excited generator is obtained by running it as a separately excited generator.

1.38 CHARACTERISTICS OF A SEPARATELY EXCITED DC GENERATOR

The obvious disadvantage of a separately excited DC generator is that we require an external DC source for excitation. But since the output voltage may be controlled more easily and over a wide range (from zero to a maximum), this type of excitation finds many applications.

(*i*) **Open circuit characteristic:** The OCC of a separately excited generator is determined in a manner described in Section 1.37. Figure 1.65 shows emf and the variation of generated emf on no load with field current for various fixed speeds. Note that if the value of constant speed is increased, the steepness of the curve also increases. When the field current is zero, the residual magnetism in the poles will give rise to the small initial emf as shown in Fig. 1.65.

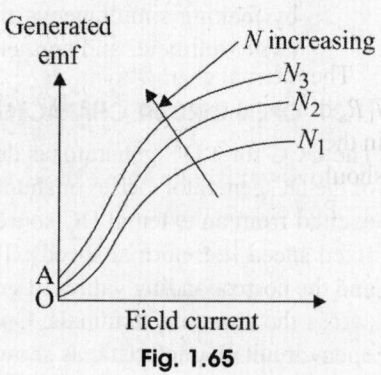

Fig. 1.65

(*ii*) **Internal and external characteristics:** The extenal characteristic of a separately excited generator is the curve between the terminal voltage (*V*) and the load current I_L (which is the same as armature current in this case). In order to determine the external characteristic, the circuit set up is as shown in Fig. 1.66(*a*). As the load current increases, the terminal voltage falls due to two reasons:

(*a*) The armature reaction weakens the main flux so that actual emf generated *E* on load is less than that generated (E_0) on no load.

(*b*) There is voltage drop across armature resistance ($= I_L R_a = I_a R_a$).

Due to these reasons, the external characteristic is a drooping curve [curve 3 in Fig. 1.66(*b*)]. Note that in the absence of armature reaction and armature drop, the generated emf would have been curve 1.

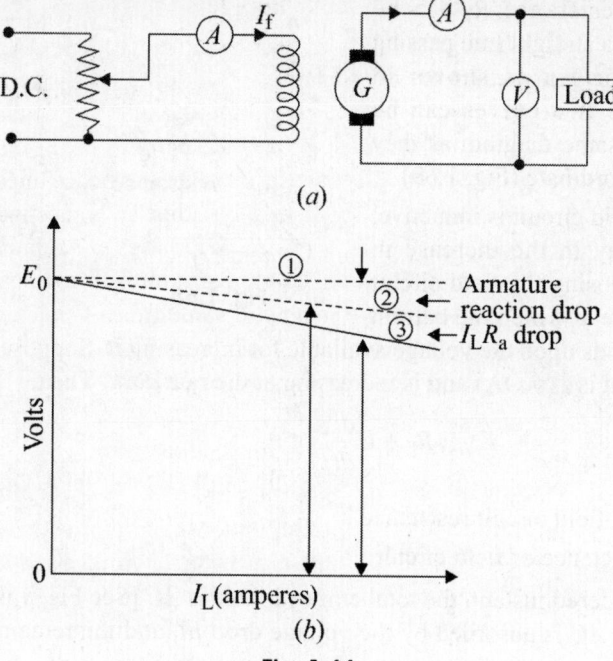

Fig. 1.66

The internal characteristic can be determined from external characteristic by adding $I_L R_a$ drop to the external characteristic. It is because armature reaction drop is included in the external characteristic. Curve 2 is the internal characteristic of the generator and should obviously lie above the external characteristic.

Power loss in the field winding,

$$P_f = V_f I_f = 120 \times 3 = 360 \text{ W} = 0.36 \text{ kW}$$

∴ Total input power, $P_f = 260 + 0.36 = 260.36$ kW

∴ Generator efficiency, $\eta = \dfrac{240}{260.36} \times 100 = 92.2\%$

(*vi*) Total resistance in the field winding is
$$R = 120/3 = 40 \text{ } \Omega$$
$$= 40 - R_f = 40 - 30 = 10 \text{ } \Omega$$

1.39 VOLTAGE BUILD-UP IN A SELF-EXCITED GENERATOR

Let us see how voltage builds up in a self-excited generator.

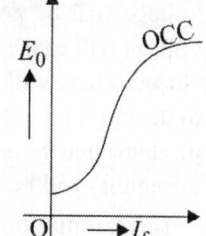

Fig. 1.67

(*i*) **Shunt generator:** Consider a shunt generator. If the generator is run at a constant speed, some emf will be generated due to residual magnetism in the main poles. This small emf circulates a field current which in turn produces additional flux to reinforce the original residual flux (provided field winding connections are correct). This process continues and the generator builds up the normal generated voltage following the OCC shown in Fig. 1.67.

The field resistance R_f can be represented by a straight line passing through the origin as shown in Fig. 1.68. The two curves can be shown on the same diagram as they have the same ordinate (Fig. 1.69).

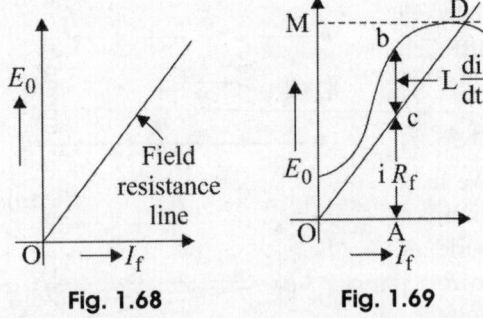

Fig. 1.68   Fig. 1.69

Since the field circuit is inductive, there is a delay in the increase in current upon closing the field circuit switch. The rate at which the current increases depends upon the voltage available for increasing it. Suppose at any instant, the field current is i (= OA) and is increasing at the rate di/dt. Then,

$$E = i R_f + L\frac{di}{dt}$$

where

R_f = total field circuit resistance

L = inductance of field circuit

At the considered instant, the total emf available is AC [See Fig. 1.69]. An amount AB of the emf. AC is absorbed by the voltage drop iR_f and the remainder part BC is available to overcome $L\frac{di}{dt}$. Since this surplus voltage is available, it is possible for the field curreitt to increase above the value OA. However, at point D, the available voltage is OM and is all absorbed by iR_f drop. Consequently, the field current cannot increase further and the generator build up stops.

We arrive at a very important conclusion that the voltage build up of the generator is given by the point of intersection of OCC and field resistance line. Thus in Fig. 1.69, D is point of intersection of the two curves. Hence, the generator will build up a voltage OM.

(*ii*) **Series generator:** During initial operation, with no current yet flowing, a residual voltage will be generated exactly as in the case of a shunt generator. The residual voltage will cause a current to flow through the whole series circuit when the circuit is closed. There will then be voltage build up to an equilibrium point exactly analogous to the build up of a shunt generator. The voltage build up graph will be similar to that of shunt generator except that now load current (instead of field current for shunt generator) will be taken along x-axis.

(*iii*) **Compound generator:** When a compound generator has its series field flux aiding its shunt field flux, the machine is said to be cumulative compound. When the series field is connected in reverse so that its field flux opposes the shunt field flux, the generator is then differential compound.

The easiest way to build up voltage in a compound generator is to start under no-load conditions. At no-load, only the shunt field is effective. When no-load voltage build up is achieved, the generator is loaded. If under load, the voltage rises, the series field connection is cumulative. If the voltage drops significiantly, the correction is differential compound.

1.40 CRITICAL FIELD RESISTANCE FOR A SHUNT GENERATOR

We have seen above that voltage build up in a shunt generator depends upon field circuit resistance. If the field circuit resistance is R_1 (line OA), then generator will build up a voltage OM as shown in Fig. 1.70. If the field circuit resistance is increased to R_2 (line OB), the generator will build up a voltage OL, slightly less than OM. As the field circuit resistance is increased, the slope of resistance line also increases. When the field resistance line becomes tangent (line OC) to OCC, the generator would just excite. If the field circuit resistance is increased beyond this point (say line OD), the generator will fail to excite. The field circuit resistance represented by line OC (tangent to OCC) is called critical field resistance R_C for the shunt generator. It may be defined as the

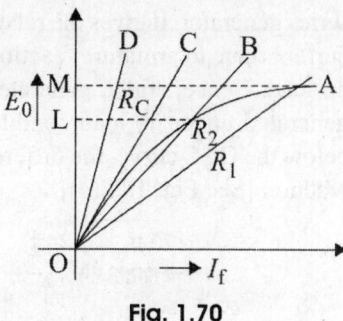

Fig. 1.70

maximum field circuit resistance (for a given speed) with which the shunt generator would just excite is known as us critical field resistance.

It should be noted that shunt generator will build up voltage only if field circuit resistance is less than critical field resistance.

1.41 CRITICAL RESISTANCE FOR A SERIES GENERATOR

Figure 1.71 shows the voltage build up in a series generator. Here R_1, R_2 etc. represent the total circuit resistance (load resistance and field winding resistance). If the total circuit resistance is R_1, then series generator will build up a voltage O_L. The line OC is tangent to OCC and represents the critical resistance R_C for a series generator. If the total resistance of the circuit is more than R_C (say line OD), the generator will fail to build up voltage. Note that Fig. 1.70 is similar to Fig. 1.71 with the following differences:

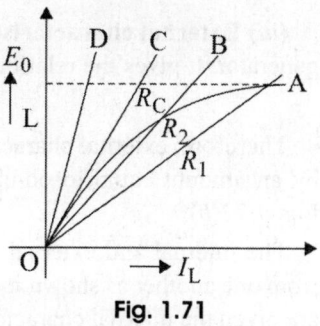

Fig. 1.71

(*i*) In Fig. 1.70, R_1, R_2, etc. represents the total field circuit resistance. However, R_1, R_2, etc. in Fig. 1.71 represents the total circuit resistance (load resistance and series field winding resistance etc.).

(*ii*) In Fig. 1.70, field current alone is represented along x-axis. However, in Fig. 1.71 load current I_L is represented along x-axis. Note that in a series generator, field current = load current I_L.

1.44 CHARACTERISTICS OF SERIES GENERATOR

Figure 1.72(*a*) shows the connections of a series wound generator. Since there is only one current (that which flows through the whole machine), the load current is the same as the exciting current.

(*i*) **OCC** curve 1 shows the open circuit characteristic (OCC) of a series generator. It can be obtained experimentally by disconnecting the field winding from the machine and exciting it from a separate DC source as discussed in Section 1.37.

(*ii*) **Internal characteristic:** Curve 2 shows the total or internal characteristic of a series generator. It gives the relation between the generated emf E on load and armature current. Due to armature reaction, the flux in the machine will be less than the flux at no load. Hence, emf E generated under load conditions will be less than the emf E_0 generated under no load conditions. Consequently, internal characteristic curve lies below the OCC curve; the difference between them representing the effect of armature reaction [See Fig. 1.72(*b*)].

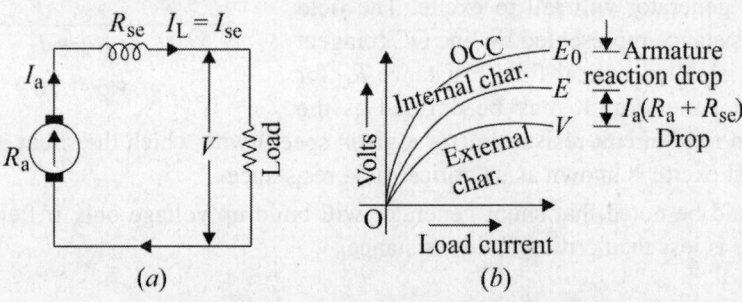

Fig. 1.72

(*iii*) **External characteristic:** Curve 3 shows the external characteristic of a series generator. It gives the relation between terminal voltage V and load current I_L.

$$V = E - I_a (R_a + R_{se})$$

Therefore, external characteristic curve will lie below internal characteristic curve by an amount equal to ohmic drop [i.e. $I_a(R_a + R_{se})$] in the machine as shown in Fig. 1.72(*b*).

The internal and external characteristics of a DC series generator can be plotted from one another as shown in Fig. 1.73. Suppose we are given the internal characteristic of the generator. Let the line OC represent the resistance of the whole machine, i.e. $R_a + R_{se}$. If the load current is OB, drop in the machine is AB i.e.

$$AB = \text{Ohmic drop in the machine}$$
$$= OB \, (R_a + R_{se})$$

Now raise a perpendicular from point B and mark a point b on this line such that $ab = AB$. Then point b

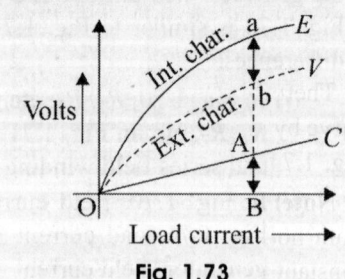

Fig. 1.73

will lie on the external characteristic of the generator. Following similar procedure, other points of external characteristic can be located. It is easy to see that we can also plot internal characteristic from the external characteristic.

1.43 CHARACTERISTIC OF A SHUNT GENERATOR

Figure 1.74(a) shows the connections of a shunt wound generator. The armature current I_a splits up into two parts; a small fraction I_{sh} flowing through shunt field winding while the major part I_L goes to the external load.

  (i) **OCC:** The OCC of a shunt generator is similar in shape to that of a series generator as shown in Fig. 1.74(b). The line OA represents the shunt field circuit resistance. When the generator is at normal speed, it will build up a voltage OM. At no-load, the terminal voltage of the generator will be constant (= OM) represented by the horizontal dotted line MC.

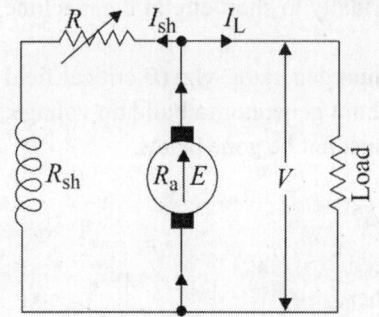
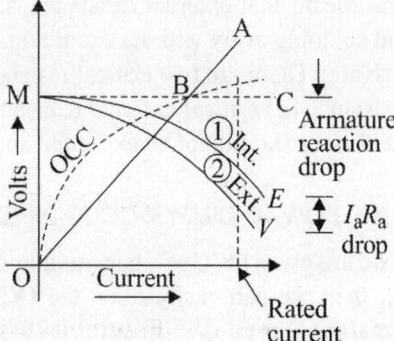

Fig. 1.74

  (ii) **Internal characteristic:** When the generator is loaded, flux per pole is reduced due to armature reaction. Therefore, emf E generated on load is less than the emf generated at no load. As a result, the internal characteristic (E/I_a) drops down slightly as shown in Fig. 1.74(b).

  (iii) **External characteristic:** Curve 2 shows the external characteristic of a shunt generator. It gives the relation between terminal voltage V and load current I_L.

$$V = E - I_a R_a$$
$$= E - (I_L + I_{sh})R_a$$

Therefore, external characteristic curve will lie below the internal characteristic curve by an amount equal to drop in the armature circuit [i.e. $(I_L + I_{sh}) R_a$) as shown in Fig. 1.74(b).

Note: It may be seen from the external characteristic that change in terminal voltage from no-load to full load is small. The terminal voltage can always be maintained constant by adjusting the field rheostat R automatically.

1.44 CRITICAL EXTERNAL RESISTANCE FOR SHUNT GENERATOR

If the load resistance across the terminals of a shunt generator is decreased, then load current increases. However, there is a limit to the increase in load current with the decrease of load resistance. Any decrease of load resistance beyond this point, instead of increasing the current, ultimately results in reduced current. Consequently, the external characteristic turns back (dotted curve) as shown in Fig. 1.75. The tangent OA to the curve represents the minimum external resistance required to excite the shunt generator on load and is called critical external resistance. If the resistance of

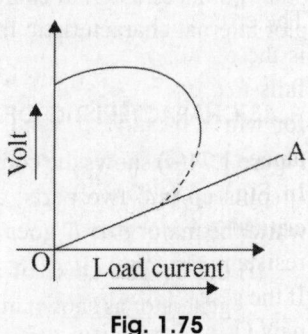

Fig. 1.75

the external circuit is less than the critical external resistance (represented by tangent OA in Fig. 1.75), the machine will refuse to excite or will de-excite if already running. This means that external resistance is so low as virtually to short circuit the machine and so doing away with its excitation.

Note: There are two critical resisiances for a shunt generator, viz. (*i*) critical field resistance (*ii*) critical external resistance. For the shunt generator to build up voltage, the former should not be exceeded and the latter must not be gone below.

1.45 HOW TO DRAW OCC AT DIFFERENT SPEEDS?

If we are given OCC of a generator at a constant speed N_1, then we can easily draw the OCC at any other constant speed N_2. Figure 1.76 illustrates the procedure. Here we are given OCC at a constant speed N_1. It is desired to find the OCC at constant speed N_2 (it is assumed that $N_2 < N_1$). For constant excitation, $E \propto N$.

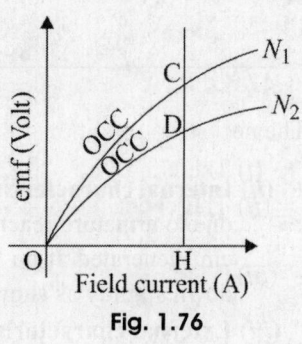

Fig. 1.76

$$\therefore \qquad \frac{E_2}{E_1} = \frac{N_2}{N_1}$$

or $$E_2 = E_1 \times \frac{N_2}{N_1}$$

As shown in Fig. 1.76, for $I_f = OH$, $E_1 = HC$. Therefore, the new value of emf (E_2) for the same I_f but at N_2 is

$$E_2 = HC \times \frac{N_2}{N_1} = HD$$

This locates the point D on the new OCC at N_2. Similarly, other points can be located taking different values of I_f. The locus of these points will be the OCC at N_2.

1.46 CRITICAL SPEED (N_C)

The critical speed of a shunt generator is the minimum speed below which it fails to excite. Clearly, it is the speed for which the given shunt field resistance represents the critical resistance. In Fig. 1.77, curve 2 corresponds to critical speed because the shunt field resistance (R_{sh}) line is tangential to it. If the generator runs at full speed N, the new OCC moves upward and the R'_{sh} line represents critical resistance for this speed.

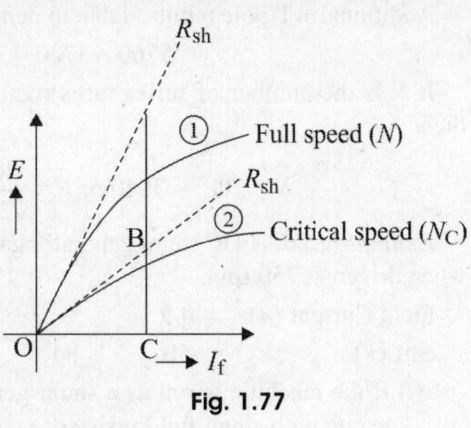

Fig. 1.77

∴   Speed ∝ critical resistance

In order to find critical speed, take any convenient point C on excitation axis and draw a perpendicular so as to cut R_{sh} and R'_{sh} lines at points B and A respectively. Then,

$$\frac{BC}{AC} = \frac{N_C}{N}$$

or

$$N_C = N \times \frac{BC}{AC}$$

1.47 CONDITIONS FOR VOLTAGE BUILD-UP OF A SHUNT GENERATOR

The necessary conditions for voltage build-up in a shunt generator are:

(*i*) There must be some residual magnetism in generator poles.

(*ii*) The connections of the field winding should be such that the field current strengthens the residual magnetism.

(*iii*) The resistance of the field circuit should be less than the critical resistance. In other words, the speed of the generator should be higher than the critical speed.

Example 1.19: A shunt generator is to be governed into a level compounded generator by the addition of a series field winding. From the test on the machine with shunt excitation only, it is found that shunt current is 3.1 A to give 400 V on no load and 4.8 A to give the same voltage when the machine is supplying its full load of 200 A. The shunt winding has 1200 turns per pole. Find the number of series turns required per pole.

Solution: Shunt field AT/pole required to produce 400 V on no-load

$$= 3.1 \times 1200 = 3720 \text{ AT}$$

Shunt field AT/pole required to produce 400 V on full-load

$$= 4.8 \times 1200 = 5760 \text{ AT}$$

Additional AT/pole required due to demagnetising effect of load current

$$= 5760 - 3720 = 2040 \text{ AT}$$

If N is the number of series turns required per pole when load current is 200 A, then,

$$N \times 200 = 2040 \text{ or } N = \frac{2040}{200} = 10.2$$

Example 1.20: A DC shunt generator gave the following open-circuit characteristic when driven at 750 rpm.

Field Current (A):	0.5	1	1.5	2	2.5
emf (V) :	50	84	105	120	131

(*i*) If the machine is run as a shunt generator at 750 rpm, to what voltage will it excite with shunt field resistance equal to 70 Ω?

(*ii*) What is the critical value of shunt field resistance?

(*iii*) What is the critical speed when the shunt field resistance is 70 Ω?

(*iv*) With shunt field resistance equal to 55 Ω, what reduction in speed be made to malce the open-circuit voltage equal to 100 V?

Solution: Plot the OCC from the given data as shown in Fig. 1.78. The line *OA* represents 70 Ω line.

(*i*) The line *OA* intersect the DCC at point B. Therefore, the voltage to which the generator excites = *OC* = 105 V.

(*ii*) Draw line *DE* tangent to OCC at the origin. Then slope of the line represents the critical shunt field resistance.

$$\therefore \text{ Critical shunt field resistance } = \frac{100 \, V}{1 \, A} = 100 \, \Omega$$

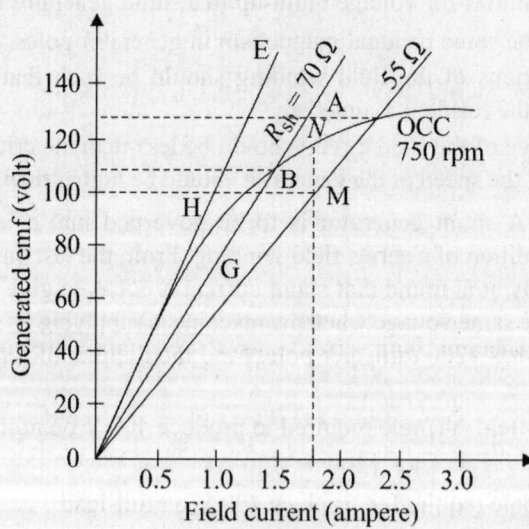

Fig. 1.78

(*iii*) To find critical speed, take any convenient point F the x-axis and draw a perpendicular which cuts the lines OA and OE at G and H respectively. Then,

$$\frac{FG}{FH} = \frac{N_C}{N}$$

or $$N_C = N \times \frac{FG}{FH} = 750 \times \frac{70}{100} = 525 \text{ rpm}$$

(*iv*) Draw 55 Ω line as shown in Fig. 1.78. With 55 Ω shunt field resistance, open-circuit voltage = OD = 128 V. To generate 100 V with field resistance of 55 Ω, the operating point R should be at M. If N is the speed, then,

$$\frac{LM}{LN} = \frac{N''}{N}$$

or $$N' = N \times \frac{LM}{LN} = 750 \times \frac{100}{115} = 652 \text{ rpm}$$

$$\text{Reduction in speed} = 750 - 652 = 98 \text{ rpm}$$

Example 1.21: The magnetisation characteristic for a 4-pole, 110 V, 1000 rpm shunt generator is as follows:

Field current (*A*) :	0	0.5	1	1.5	2	2.5	3
OC voltage (*V*) :	5	50	85	102	112	126	120

The armature is lap connected with 144 conductors residual flux per pole.

Solution: There is no need to draw the OCC. The answer can be obtained from the data given. When the field current is zero, the induced emf of 5 V is due to the residual magnetism of generator poles.

$$E_g = \frac{P \phi Z N}{60 A}$$

Here E_g = 5 V; P = 4; Z = 144; N = 1000 rpm; A = P = 4

$\therefore$ $$5 = \frac{4 \times \phi \times 144 \times 1000}{60 \times 4}$$

or $$\phi = 2.08 \times 10^{-3} \text{ Wb} = 2.08 \text{ mWb}$$

1.48 COMPOUND GENERATOR CHARACTERISTIC

In a compound generator, both series and shunt excitation are combined as shown in Fig. 1.79(*a*). The shunt winding can be connected either across the armature only (short-shunt connection S) or across armature plus series field (long-shunt connection G). The compound generator can be cumulatively compounded or differentially compounded generator. The latter is rarely used in practice. Therefore, we shall discuss the characteristics of cumulatively-compounded generator. It may be noted that external characteristics of long and short shunt compound generators are almost identical.

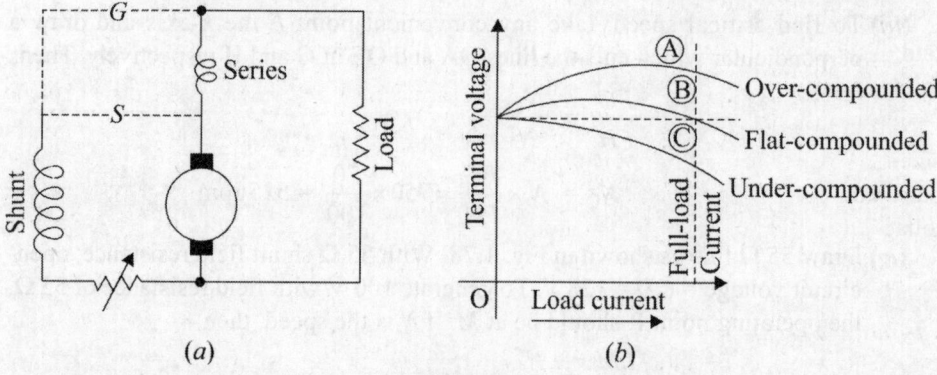

Fig. 1.79

External Characteristics: Figure 1.79(*b*) shows the external characteristics of a cumulatively compound generator. The series excitation aids the shunt excitation. The degree of compounding depends upon the increase in series excitation with the increase in load current:

(*i*) If series winding turns are so adjusted that with the increase in load current the terminal voltage increases, it is called over-compounded generator. In such a case, as the load current increases, the series field mmf increases and tends to increase the flux and hence the generated voltage. The increase in generated voltage is greater than the $I_a R_a$ drop so that instead of decreasing, the terminal voltage increases as shown by curve A in Fig. 1.79(*b*).

(*ii*) If series winding turns are so adjusted that with the increase in load current, the terminal voltage substantially remains constant, it is called flat-compounded generator. The series winding of such a machine has lesser number of turns than the one in over-compounded machine and, therefore, does not increase the flux as much for a given load current. Consequently, the full-load voltage is nearly equal to the no-load voltage as indicated by curve B in Fig. 1.79(*b*).

(*iii*) If series field winding has lesser number of turns than for a flat-compounded machine, the terminal voltage falls with increase in load current as indicated by curve C in Fig. 1.79(*b*). Such a machine is called under-compounded generator.

1.49 VOLTAGE REGULATION

The change in terminal voltage of a generator between full- and no-load (at constant speed) is called the voltage regulation, usually expressed as a percentage of the voltage at full-load.

$$\% \text{ Voltage regulation} = \frac{V_{NL} - V_{FL}}{V_{FL}} \times 100$$

where V_{NL} = Terminal voltage of generator at no load

V_{FL} = Terminal voltage of generator at full load

Note that voltage regulation of a generator is determined with field circuit and speed held constant. If the voltage regulation of a generator is 10%, it means that terminal voltage increases 10% as the load is changed from full-load to no-load.

Note: For an ideal generator (i.e. constant-voltage generator), the voltage regulation is zero. A positive voltage regulation means that terminal voltage at no load is higher than at full-load. On the other hand, a negative voltage regulation means that the terminal voltage at full load is higher than at no-load.

Example 1.22: A 60 kW DC shunt generator has 1600 turns/pole in its shunt field winding. A shunt field current of 1.25 A is required to generate 125 V at no-load and 1.75 A to generate 150 V at full-load. Calculate:

(*i*) The minimum number of series turns/pole needed to produce the required no-load and full-load voltages as a short-shunt compound generator.

(*ii*) If the generator is equipped with 3 series turns per pole having a resistance of 0.02 Ω calculate the diverter resistance required to produce the desired compounding.

(*iii*) Voltage regulation of the compound generator.

Solution:

(*i*) Extra shunt field AT required = 1600(1.75 – 1.25) = 800 AT

$$I_{se} = I_L = \frac{60 \times 10^3}{150} = 400 \text{ A}$$

(*ii*) The actual number of series turns per pole is 3. Therefore, series current required is

$$I'_{se} \times 3 = 800$$

or

$$I_{se} = 800/3 \text{ A}$$

Current in diverter, $I_d = 400 - 800/3$

$$= 400/3 \text{ A}$$

The voltages across series winding and the diverter are equal.

$$0.02 \times I'_{se} = I_d R_d$$

$$R_d = \frac{0.02 \times I'_{se}}{I_d}$$

$$= \frac{0.02 \times 800/3}{400/3} = 0.04 \text{ Ω}$$

(*iii*) Voltage regulation $= \frac{V_{NL} - V_{FL}}{V_{FL}} \times 100 = \frac{125 - 150}{150} \times 100 = -16.7\%$

Example 1.23: In a 110 V compound generator, the resistances of the armature, shunt ard series windings are 0.06 Ω, 25 Ω and 0.04 Ω respectively. The load consists of 200 lamps each rated at 55 Ω, 110 V. Find the emf generated and armature current when the machine is connected (*i*) long shunt (*ii*) short shunt (*iii*) How will the ampere-turns of series winding be changed if in (*i*) a diverter of resistance 0.1 Ω be connected in parallel with the series winding? Ignore armature reaction and brush contact drop.

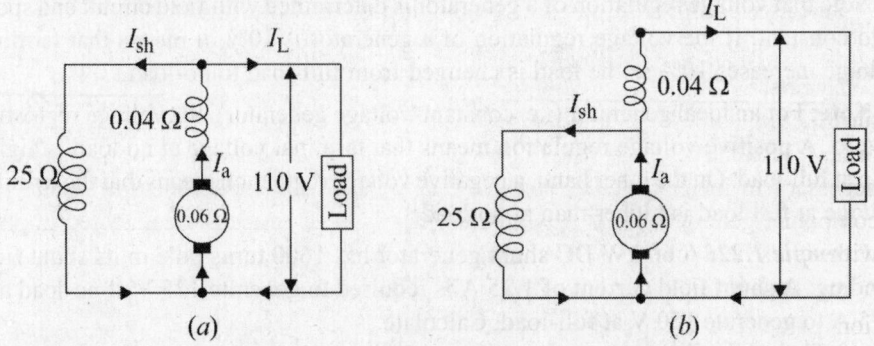

Fig. 1.80

Solution:

(*i*) Long-shunt connection [see Fig. 1.80]

$$\text{Load current, } I_L = \frac{200 \times 55}{110} = 100 \text{ A}$$

$$\text{Shunt field current, } I_{sh} = \frac{110}{25} = 4.4 \text{ A}$$

$$\text{Armature current, } I_a = I_L + I_{sh} = 100 + 4.4 = 104.4 \text{ A}$$

$$\text{Generated emf, } E_g = 110 + I_a(R_a + R_{se})$$
$$= 110 + 104.4 \,(0.06 + 0.04) = 120.4 \text{ V}$$

(*ii*) Short-shunt connection [see Fig. 1.80(*b*)]

$$I_L = I_{se} = 100 \text{ A} \qquad \text{(as calculated above)}$$

Voltage across shunt winding, $V' = 110 + I_{se}R_{se} = 110 + 100 \times 0.04 = 114$ V

$$\therefore \qquad I_{sh} = \frac{114}{25} = 4.56 A$$

$$\text{Armature current, } I_a = I_L + I_{sh} = 100 + 4.56 = 104.56 \text{ A}$$

$$\text{Generated emf, } E_g = V' + I_a R_a = 114 + 104.56 \times 0.06 = 120.3 \text{ V}$$

(*iii*) Figure 1.81 shows a diverter of resistance 0.1 Ω connected in parallel with the series winding of long-shunt connection. As calculated in part (*i*), I_a = 104.4 A. The current through series winding (current divider rule) is

$$I'_{se} = 104.4 \times \frac{0.1}{0.04 + 0.1} = 74.57 \text{ A}$$

Reduction of current in series winding

$$= \frac{104.4 - 74.57}{104.4} \times 100 = 28.57\%$$

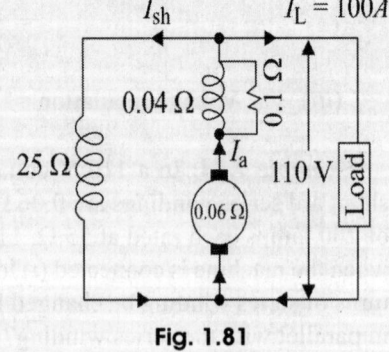

Fig. 1.81

Since series ampere-turns are proportional to series current (turns are same),

reduction in series ampere-turns = 28.57 %

Example 1.24: A long shunt compound generator has a shunt field winding of 1000 turns per pole and series field winding of 4 turns per pole and resistance of 0.05 Ω. In order to obtain the rated voltage both at no-load and full-load as shunt generator, it is necessary to increase field current by 0.2 A. The full-load armature current of the compound generator is 80 A. Calculate the diverter resistance connected in parallel with series field to obtain flat compound operation.

Solution: Figure 1.82 shows the generator connections. Additional shunt field AT required to maintain rated voltage at no-load and full-load

$$= 1000 \times 0.2 = 200 \text{ AT}$$

No. of series turns/pole = 4

Current required in series winding to produce 200 AT.

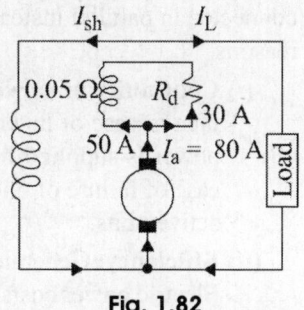

Fig. 1.82

It is given that armature current $I_a = 80$ A. Therefore, $80 - 50 = 30$ A must pass through diverter resistance.

Now $$0.5 \times 50 = 30 \times R_d$$

∴ $$R_d = \frac{0.05 \times 50}{30} = 0.0833 \Omega$$

Example 1.25: A shunt generator is to be compounded so that the no-load terminal voltage is equal to that at full-load when the output is 400 A. Number of shunt turns/pole = 640. With series field cut out and shunt field excited from an external source, it is found that this can be accomplished by increasing the field current from 6.2 A at no-load to 6.2 A, at full-load. Long-shunt connections are used.

(*i*) How many series turns are required?

(*ii*) If the generator is equipped with 5.5 turns, find the ratio of diverter to series field resistance. Neglect series AT at no-load and the drop in the series winding at full-load.

Solution:

(*i*) Additional shunt field AT required to maintain rated voltage both at no-load and full-load

$$= (6.2 - 4.2) \times 640 = 1280 \text{ AT}$$

∴ Number of series turns required to produce 1280 AT

$$= \frac{1280}{400} = 3.2$$

(*ii*) Now the actual number of series turns is 5.5. Current required to produce 1280 AT by series winning is [See Fig. 1.83]

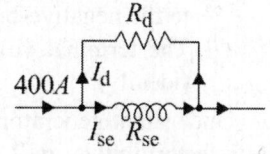

Fig. 1.83

$$5.5 \times I_{se} = 1280$$

∴ $$= 1280/5.5 = 232.7 \text{ A}$$

Current through diverter $I_d = 400 - 232.7 = 167.3$ A

Now $\qquad\qquad I_d \times I_d = I_{se} \times R_{se}$

$\therefore \qquad\qquad \dfrac{R_d}{R_{se}} = \dfrac{I_{se}}{I_d} = \dfrac{232.7}{167.3} = 1.39:1$

1.50 NEED FOR PARALLEL OPERATION OF DC GENERATORS

In a DC power plant, power is usually supplied from several generators of small ratings connected in parallel instead of from one large generator. This is due to the following reasons:

(*i*) **Continuity of service:** If a single large generator is used in the power plant, then in case of its breakdown, the whole plant will be shut down. However, if power is supplied from a number of small units operating in parallel, then in case of failure of one unit, the continuity of supply can be maintained by other active units.

(*ii*) **Efficiency:** Generators run most efficiently, when loaded to their rated capacity. Electric power costs less per kWh when the generator producing it is efficiently loaded. Therefore, when load demand on power plant decreases, one or more generators can be shut down and the remaining units can be efficiently loaded.

(*iii*) **Maintenance and repair:** Generators generally require routine maintenance and repair. Therefore, if generators are operated in parallel, the routine or emergency operations can be performed by isolating the affected generator while load is being supplied by other units. This leads to both safety and economy.

(*iv*) **Increasing plant capacity:** In the modern world of increasing population, the use of electricity is continuously increasing. When added capacity is required, the new unit can be simply paralleled with the old units.

(*v*) **Non-availability of single large unit:** In many situations, a single unit of desired large capacity may not be available. In that case a number of smaller units can be operated in parallel to meet the load requirement. Generally a single large unit is more expensive.

1.51 CONDITIONS FOR PARALLEL OPERATION OF SHUNT GENERATORS

There are two main conditions to be met when DC generators (shunt, compound or series generators) are to be paralleled viz.

(*i*) All polarities must be correct. The positive terminals of all generators must be connected to the positive bus bar and the negative terminals must be connected to the negative-bus bar.

(*ii*) The terminal voltages of the generators being paralleled must be the same (ideally).

Once shunt generators are put in parallel and adjusted so that they share the load satisfactorily, they will continue to operate properly even if the load changes. The natural tendency is for the terminal voltage of a shunt generator to fall when the load increases. This property tends to ensure stable operation.

1.54 CONNECTING SHUNT GENERATORS IN PARALLEL

The generators in a power plant are connected in parallel through bus-bars. The bus-bars are heavy thick copper bars and they act as +ve and –ve terminals. The positive terminals of the generators are connected to the +ve side of bus-bars and negative terminals to the negative side of bus-bars (Fig. 1.84).

Figure 1.84 shows shunt generator G_1 connected to the bus-bars and supplying load. When the load on the power plant increases beyond the capacity of this generator, the second shunt generator G_2 is connected in parallel with the first to meet the increased load demand. The procedure for paralleling generator G_2 with generator G_1 is as under:

   (i) The prime mover of generator G_2 is brought up to the rated speed. Now switch S_4 in the field circuit of the generator G_2 is closed.

  (ii) Next circuit breaker CB-2 is closed and the excitation of generator G_2 is adjusted till it generates voltage equal to the bus-bars voltage. This is indicated by voltmeter V_2.

 (iii) Now the generator G_2 is ready to be paralleled with generator G_1. The main switch S_3 is closed, thus putting generator G_2 in parallel with generator G_1. Note that generator G_2 is not supplying any load because its generated emf is equal to bus-bars voltage. The generator is said to be "floating" (i.e. not supplying any load) on the bus-bars.

 (iv) If generator G_2 is to deliver any current, then its generated voltage E should be greater than the bus-bars voltage V. In that case, current supplied by it is $I = (E - V)/R_a$ where R_a is the resistance of the armature circuit. By increasing

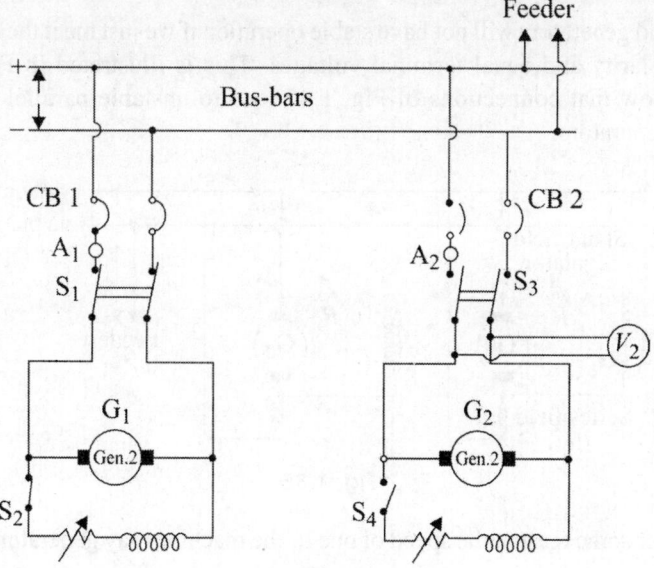

Fig. 1.84

the field current (and hence induced emf E), the generator R_a can be made to supply proper amount of load.

(v) The load may be shifted from one shunt generator to another merely by adjusting the field excitation. Thus if generator G_1 is to be shut down, the whole load can be shifted onto generator G_2 provided it has the capacity to supply that load. In that case, reduce the current supplied by generator G_1 to zero (This will be indicated by ammeter A_1), open CB-1 and then open the main switch S_1.

1.53 LOAD SHARING BY SHUNT GENERATORS IN PARALLEL

The load sharing between shunt generators in parallel can be easily regulated because of their drooping characteristics. The load may be shifted from one generator to another merely by adjusting the field excitation. Let us discuss the load sharing of two generators which have unequal no-load voltages.

$$E_1, E_2 = \text{no-load voltages of the two generators}$$
$$R_1, R_2 = \text{their armature resistances}$$
$$V = \text{common terminal voltage bus-bars voltage})$$

Then, $$I_1 = \frac{E_1 - V}{R_1} \text{ and } I_2 = \frac{E_2 - V}{R_2}$$

Thus current output of the generators depends upon the values of E_1 and E_2. These values may be changed by field rheostats. The common terminal voltage (or bus-bars voltage) will depend upon (i) the emfs of individual generators and (ii) the total load current supplied. It is generally desired to keep the bus-bars voltage constant. This can be achieved by adjusting the field excitations of the generators operating in parallel.

1.54 PARALLEL OPERATION OF COMPOUND GENERATORS

The compound generators will not have stable operation if we just meet the requirements of correct polarity and equal terminal voltages. This is illustrated in Fig. 1.85. We shall now show that connections of Fig. 1.85 lead to unstable parallel operation of compound generators.

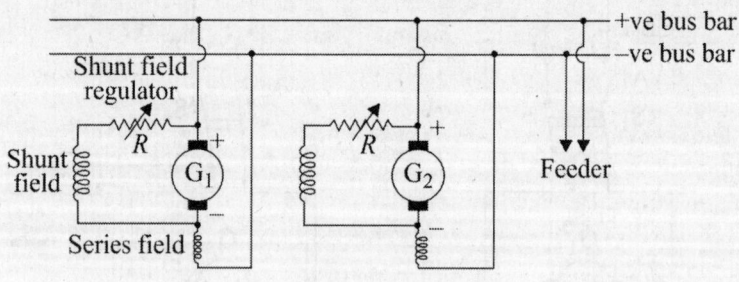

Fig. 1.85

Suppose for some reason, the speed of one of the machines say generator G_1 increases momentarily.

This will momentarily increase the generated emf of G_1 and this generator will supply more share of load current. This increase in armature current of G_1, in turn, increases the current in the series field winding of G_1. As a result, the generated emf of G_1 further increases, causing G_1 to supply more share of load current. At the same time, the armature current supplied by generator G_2 is decreased, thereby decreasing current in its series field winding and causing it to supply smaller and smaller portion of load current. This process continues until generator G_1 is supplying the entire load current and is attempting to operate machine G_2 as a motor. This causes the circuit breakers to open, disconnecting the generators from the load.

This instability of the compound generators in parallel is overcome by using an equaliser connection as shown in Fig. 1.86. The equaliser is a bus bar that connects the two series field coils in parallel. Under this condition, the voltage across the series field coils of the two machines and their currents remain the same. As a result, the series field currents of the two machines remain constant regardless of the distribution of current in the two armatures. Thus the flux produced by the two series fields remains the same for a given load current on the bus.

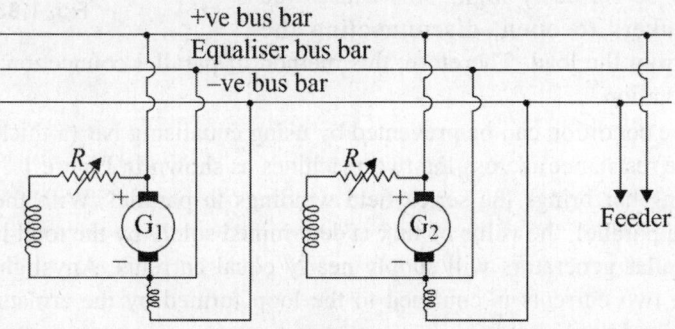

Fig. 1.86

Thus for stable parallel operation of compound generators, in addition to the requirments of correct polarity and equal terminal voltages, they require an equaliser connection before being paralleled. Compound generators are put in service and taken out of service in the same manner as shunt generators. Load is shifted from one compound generator to the other by the adjustment of shunt-field rheostats.

Note: In connecting a compound generator on to the bus bars, it is necessary that first series; field windings have equaliser connection. Then shunt field of the incoming machine is adjusted so that the voltage generated is exactly equal to the bus bar voltage. The incoming machine is now connected to the bus bars.

1.55 PARALLEL OPERATION OF SERIES GENERATORS

Although DC series generators are not usually used for supplying power, DC series motors are frequently arranged to operate as DC series generators during electric braking. For electric braking electric trains, the DC series motors are disconnected from the supply mains and are cormected in parallel across resistors. Due to kinetic energy of the train, the motors continue to run and act as setf-excited dc series generators.

Figure 1.87 shows two identical DC series generators G_1 and G_2 connected in parallel. Since the generators are identical, $E_1 = E_2 = E$ (say) and they supply equal currents i.e. $I_1 = I_1 = I$ (say). Suppose due to any reason, E_1 increases slightly so that $E_1 > E_2$. Under such conditions, I_1 becomes greater than I_2. As a result, the field of generator G_1 is strengthened thus further increasing E_1 while the field of generator G_2 is weakened thus decreasing E_2 further. The process continues until a stage is reached when generator G_1 is supplying the entire load current and is attempting to operate machine G_2 as a motor. At this stage, the two generators will form a short-circuited loop and the current will be seriously high. This causes the circuit breakers to open, disconnecting the

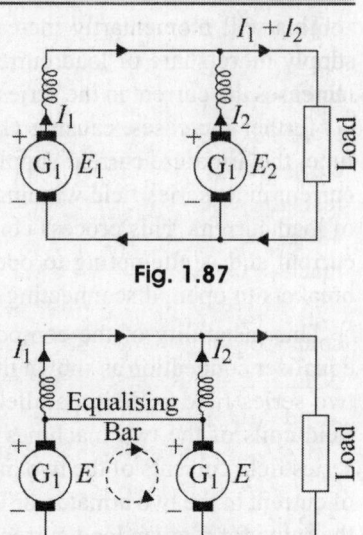

Fig. 1.87

Fig. 1.88

generators from the load. Therefore, this method of parallel connection will lead to unstable operation.

The above condition can be prevented by using equalising bar (a thick copper bar of negligible resistance) across the two machines as shown in Figure 1.88. Note that the equalising bar brings the series field windings in parallel. With the field coils connected in parallel, the value of flux is determined solely by the total load current. The two similar generators will supply nearly equal currents. Anyslight difference between the two currents is confined to the loop formed by the armatures and the equalising bar (dotted circle).

Example 1.26: Two shunt generators operating in parallel are excited so that their no-load voltages are equal. The generators have straight line drooping external characteristics. How can combined characteristic be obtained?

Solution: The way in which two, or more generators in parallel share a given load depends on the external characteristics of the generators. Figure 1.89 shows the external characteristics of two shunt generators that have the same no-load voltages. Here AB is the external characteristic of machine 1 and AD that of machine 2. For a common terminal voltage V, the generator 1 delivers I_1 amperes ($= VS$) while generator 2 delivers I_2 amperes ($= VQ$). Note that generator 1 having more drooping curve delivers less current. The total current is $I = I_1 + I_2$ ($VS + VQ = VP$). Then P is a point on the total characteristic. Other points may be

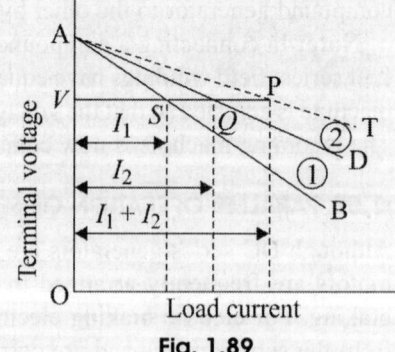

Fig. 1.89

obtained in a similar manner and so the total or combined characteristic AT may be drawn. If the characteristics can be represented With sufficient accuracy by straight lines, then results can be obtained by calculation instead of graphically.

Example 1.27: Two shunt wound generators running is parallel supply a total load current of 4000 A. Each generator has a armature resistance of 0.07 Ω and a field resistance of 40 Ω. The fields are excited so that the emf induced in one generator is 480 V and in another 490 V. Find the bus-bars voltage and kilowatt output of each machine.

Solution: Figure 1.90 shows the connections of the generators.

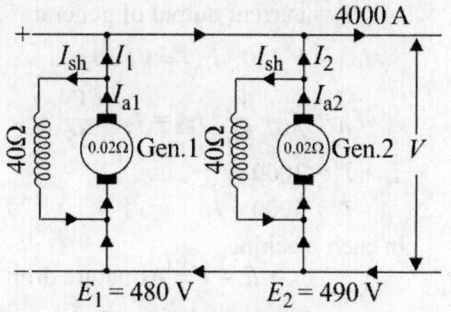

Fig. 1.90

Let V = bus-bars voltage

I_1 = current output of generator 1

I_2 = current output of generator 2

$I_1 + I_2 = 4000$

$$= \frac{V}{R_{sh}}$$

Bus-bar voltage: In each machine

bus-bars voltage + voltage drop in armature = emf generated

For gen. 1, $E_1 = V + I_{a1}R_a$; For gen. 2, $E_1 = V + I_{a2}R_a$

or

$$480 = V + \left(I_1 + \frac{V}{40}\right) \times 0.02 \qquad (i)$$

For gen. 2, $490 = V + \left(I_2 + \frac{V}{40}\right) \times 0.02 \qquad (ii)$

Substracting Eq. (i) from Eq. (ii), we have,

$$0.02I_2 - 0.02I_1 = 10$$

or

$$I_2 - I_1 = 500$$

or

$$(490 - I_1) - I_1 = 500 \quad (I_1 + I_2 = 4000)$$

$\therefore$

$$I_1 = 1750 \text{ A and } I_2 = 4000 - 1750 = 2250 \text{ A}$$

Now

$$480 = V + \left(I_1 + \frac{V}{40}\right) \times 0.02$$

or

$$480 = V + \left(1750 + \frac{V}{40}\right) \times 0.02$$

On solving, $V = 444.8$ volts

Kilowatt output: kW output of gen. 1 $= \dfrac{V I_1}{1000} = \dfrac{444.8 \times 1750}{1000} = 778.4$ kW

kW output of gen. 2 $= \dfrac{V I_2}{1000} = \dfrac{444.8 \times 2250}{1000} = 1001$ kW

Example 1.28: Two shunt generators running in parallel supply with a total load current of 3000 A. The generators have armature resistances 0.05 Ω and 0.03 Ω field resistances 30 Ω and 25 Ω and induced emfs 400 V and 380 V respectively. Calculate (*i*) current supplied by each generator (*ii*) bus-bars voltage and (*iii*) kW ouput of each generator.

Solution: Figure 1.91 shows the circuit connections.

Let V = bus-bars voltage

$\quad I_1$ = current output of generator 1

$\quad I_2$ = current output of generator 2

$I_{sh1} = V/30; \; I_{sh2} = V/25$

$I_{a1} = I_1 + \dfrac{V}{30}; \; I_{a2} = I_2 + \dfrac{V}{25}$

$I_1 + I_2 = 3000$

∴ $\qquad I_2 = 3000 - I_1$

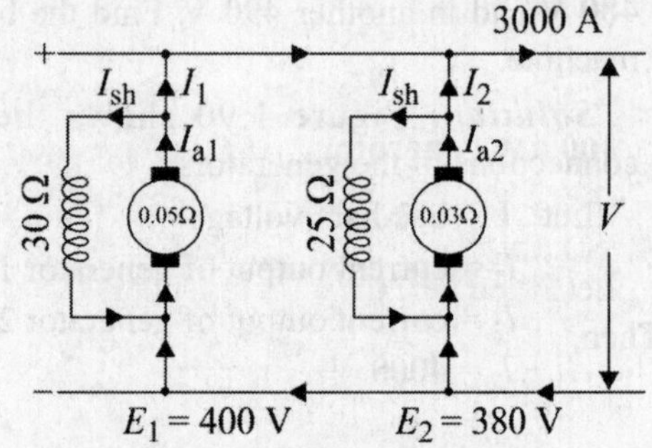

In each machine,

$\qquad E = V + \text{Armature drop}$

$400 = V + \left(1 + \dfrac{V}{30}\right) \times 0.05$

$400 = 1.0016 \, V + 0.05 \, I_1$

$0.05 \, I_1 = 400 - 1.0016 \, V$ ...(i)

Also $\qquad 380 = V + \left(I_2 + \dfrac{V}{25}\right) \times 0.03$

or $\qquad 380 = 1.0012 \, V + 0.03 \, I_2 = 1.0012 \, V + (3000 - I_1) \times 0.03$

or $\qquad 0.03 I_1 = -290 + 1 - 0012 \, V$ ...(ii)

Dividing Eq. (*i*) by Eq. (*ii*), we have,

$\dfrac{0.05 I_1}{0.03 I_1} = \dfrac{400 - 1.0016 \, V}{-290 + 1.0012 \, V}$

or $\qquad \dfrac{5}{3} = \dfrac{400 - 1.0016 \, V}{-290 + 1.0012 \, V}$

On solving, $\qquad V = 331$ volts

$\qquad 0.05 \, I_1 = 400 - 1.0016 \, V$

or $\qquad 0.05 \, I_1 = 400 - 1.016 \times 331$

∴ $\qquad I_2 = 1373.7$ A

$\qquad I_2 = 3000 - 1373.7 = 1616.3$ A

kW output of generator 1 $= \dfrac{VI_1}{1000} = \dfrac{331 \times 1373.7}{1000} = 454.7$ kW

kW output of generator 2 $= \dfrac{VI_2}{1000} = \dfrac{331 \times 1626.3}{1000} = 538.3$ kW

Fig. 1.91

$E_1 = 400$ V $\qquad E_2 = 380$ V

Example 1.29: Two shunt generators operating in parallel, deliver a total current of 250 A. One of the generator is rated 50 kW and the other 100 kW. The voltage rating of both machines is 500 V and have regulations 6% (smaller) and 4%. Assuming linear characteristics, determine (*i*) the current delivered by each machine (*ii*) terminal voltage.

Solution:

50 kW generator   FL voltage drop = 500 × 0.06 = 30 V
                       FL current = 500 × 10^3/500 = 100 A
                       Drop per ampere = 30/100 = 3/10 V/A

100 kW generator   FL voltage drop = 500 × 0.04 = 20 V
                       FL current = 500 × 10^3/500 = 200 A
                       Drop per ampere = 20/200 = 1/10 V/A

Let I_1 and I_2 be the currents supplied by two generators and V be the terminal voltage. Then,

$$V = 500 - \frac{3}{10} \times I_1 \quad \text{(first generator)}$$

$$V = 500 - \frac{1}{10} \times I_2 \quad \text{(second generator)}$$

(*i*) From Eqs (*i*) and (*ii*), we have, $I_1 = 62.5$ A ; $I_2 = 187.5$ A

(*ii*) Terminal voltage, $V = 500 - \frac{3}{10} \times I_1 = 500 - \frac{3}{10} \times 62.5 = 481.25$ V

Example 1.30: Two separately excited DC generators running in parallel each supply a current of 200 A at 200 V. The armature resistance of each machine is 0.05 Ω. Calculate (*i*) the percentage change in the generated emf of one machine necessary to reduce its load to zero (*ii*) the resulting percentage drop in terminal voltage assuming that the total load current and the emf of the other machine remain unchanged.

Solution: Initially generated emf of each machine = 200 + 0.05 × 200 = 210 V
Total load current = 200 + 200 = 400 A .

When one machine is supplying no current (*i.e.* it is floating), the other machine will supply the total load of 400 A
∴ New terminal voltage = 210 – 400 × 0.05 = 190 V
∴ Generated emf of floating machine = 190 V

(*i*) % age change (reduction) in the generated emf of floating machine

$$= \frac{210 - 190}{210} \times 100 = 9.5\%$$

(*ii*) % age reduction in the terminal voltage

$$= \frac{200 - 190}{200} \times 100 = 5\%$$

Example 1.31: A 110 V battery is connected to the same bus-bars as an AC generator. Battery resistance is 0.025 Ω and generator resistance is 0.1 Ω. The battery floats when the load current is 100 A . Find the generator output (*i*) when the battery floats (*ii*) when the load current is 50 A.

Solution: (*i*) **When the battery floats:** When the battery floats (i.e. it supplies no current), the bus-bars terminal voltage is equal to the battery emf (= 110 V).

∴   Generated emf of the generator = 110+ 100 × 0.1 = 120V

∴   Generator output = Bus-bars voltage × current supplied
$$= 110 \times 100 = 11 \times 10^3 \text{ W} = 11 \text{ kW}$$

(*ii*) **When the load current is 50 A**

Let I_1 = Current supplied by generator
I_2 = Current supplied by battery
V = New terminal or bus-bars voltage
$I_1 - I_2 = 50$

$$I_1 = \frac{120 - V}{0.1} = 1200 - 10\ V$$

$$I_2 = \frac{110 - V}{0.025} = 4400 - 40\ V$$

∴ $(1200 - 10\ V) + (4400 - 40\ V) = 50$

∴ Generator output $= VI_1 = 111 \times 90 = 9.9 \times 10^3 \text{ W} = 9.9 \text{ kW}$

Example 1.32: The terminal voltage of a generator falls from 250 V on open-circuit to 238 V when delivering 60 A. It is connected to a load in parallel with a battery having a constant emf of 245 V and a resistance of 0.1 Ω. Find the current supplied by the generator when the total load current is (*i*) 50 A (*ii*) 100 A (*iii*) zero.

Solution: Internal resistance of generator $= \dfrac{250 - 238}{60} = 0.2\ \Omega$

Let I = total load current
I_1 = current supplied by generator
I_2 = current supplied by battery
V = terminal voltage

∴ $I_1 = \dfrac{250 - V}{0.2}$ and $I_2 = \dfrac{245 - V}{0.1}$

Now $I_1 + I_2 = I$

or $\dfrac{250 - V}{0.2} + \dfrac{245 - V}{0.1} = I$

or $V = \dfrac{3700 - I}{15}$

(*i*) When total load current $I = 50$ A

$$V = \frac{3700 - 50}{15} = \frac{3650}{15} = 243.3 \text{ volts}$$

∴ $I_1 = \dfrac{250 - V}{0.2} = \dfrac{250 - 243.3}{0.2} = 33.5 \text{ A}$

(*ii*) When total load current $I = 100$ A

$$V = \frac{3700 - 100}{15} = \frac{3600}{15} = 240 \text{ volts}$$

$$\therefore \qquad I_2 = \frac{250 - V}{0.2} = \frac{250 - 240}{0.2} = 50 \text{ A}$$

(*iii*) When total load current $I = 0$

$$V = \frac{3700 - 0}{15} = 246.7 \text{ volts}$$

$$= \frac{250 - V}{0.2} = \frac{250 - 246.7}{0.2} = 16.5 \text{ A}$$

Example 1.33: two DC generaton A and B are connected to a common load. The generator *A* has a constant emf of 400 V and internal resistance of 0.25 Ω while generator *B* has a constant emf of 410 V and internal resistance of 0.4 Ω.

    (*i*) Calculate the current and power output of each generator if the terminal voltage is 390 V.

    (*ii*) What would be the current and power from each and the terminal voltage if the load is open-circuited?

Solution: Fig. 1.92 shows the conditions of the problem. The terminal voltage $V = 390$ volts.

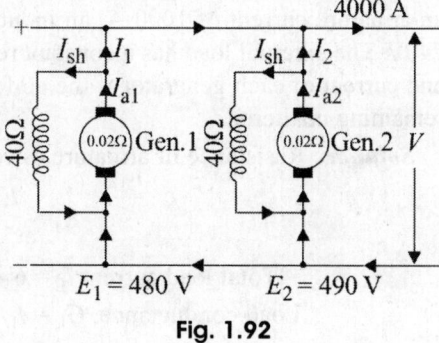

Fig. 1.92

(*i*) Current supplied by

$$A, I_A = \frac{E_A - V}{R_A} = \frac{400 - 390}{0.25} = 40 \text{ A}$$

Current supplied by

$$B, I_B = \frac{E_B - V}{R_B} = \frac{410 - 390}{0.4} = 50 \text{ A}$$

Power output from *A*,

$$P_A = VI_A = \frac{390 \times 40}{1000} \text{ kW} = 15.6 \text{ kW}$$

Power output from *B*

$$P_B = VI_B = \frac{390 \times 50}{1000} \text{ kW} = 19.5 \text{ kW}$$

(*ii*) If the load is open-circuited as shown in Fig. 1.93, the two generators are put in series and the circuit current is determined by the net circuit voltage and total circuit resistance. Note that machine B (having higher emf) acts as a generator and drives machine A as a motor.

    Net voltage in the circuit = 410 – 400 = 10 V

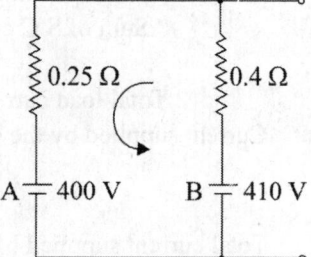

Fig. 1.93

Total circuit resistance = $R_A + R_B = 0.25 + 0.4 = 0.65 \, \Omega$

Circuit current $I = 10/0.65 = 15.4$ A

Terminal voltage $V = E_A + I R_A = 400 + 15.4 \times 0.25 = 403.8$ V

Power taken by A from B = $VI = 403.8 \times 15.4 = 6219$ W

Example 1.34: Four DC generators of open circuit voltages 220 V, 230 V, 250 V and 250 V and internal resistances 0.1 Ω, 0.1 Ω, 0.25 Ω and 0.2 Ω, respectively are joined in parallel across a 5Ω resistor. Calculate the terminal voltage and power delivered to the load.

Solution: Terminal voltage, $V = \dfrac{\text{Sum of short circuit currents}}{\text{Terminal conductances}}$

$$= \frac{(220/0.1)(230/0.1) + (240/0.25) + (250/0.2)}{(1/0.1) + (1/0.1) + (1/0.25) + (1/0.2) + (1/5)}$$

$$= \frac{6710}{29.2} = 229.8 \text{ volts}$$

Power delivered to load $= \dfrac{V^2}{R_L} = \dfrac{(229.8)^2}{5} = 10.5 \times 10^3$ W $= 10.5$ kW

Example 1.35: A station contains 6 DC generators operating in parallel, each having an armature current of 1000 A, an induced emf of 250 V, find a terminal voltage of 240V. The external load has a constant resistance. Calculate the new terminal voltage and current of each generator if the emf of one generator is raised by 5%; the others remaining unaltered.

Solution: Resistance of armature of each generator is

$$R_a = \frac{E - V}{I_a} = \frac{250 - 240}{1000} = 0.01 \, \Omega$$

Total load current $I_L = 6 \times 10000 = 6000$ A

Load conductance, $G_L = I_L/V = 6000/240 = 25$ s

emf of affected machine, $\quad E' = 250 + 5\%$ of 250

$$= 250 + 12.5 = 262.5 \text{ volts}$$

∴ New terminal voltage, $\quad V' = \dfrac{\text{Sum of short circuit currents}}{\text{Terminal conductances}}$

Terminal conductances = $(6/0.01) + 25 = 625$ s

Sum of S.C currents = $5 \times (250/0.01) + (262.5/0.01) = 151250$ A

∴ $\quad V' = 151250/625 = 242$ volts

Total load current, $I_L' = V' \times G_L = 242 \times 25 = 6050$ A

Current supplied by the generator with increased voltage

$$= \frac{E - V'}{R_a} = \frac{26.2 - 242}{0.01} = 2050 \text{ A}$$

Total current supplied by other 5 generator = $6050 - 2050 = 4000$ A

∴ Current supplied by each = $4000/5 = 800$ A

1.56 DC GENERATOR SPECIFICATION

The nominal characteristics of a DC generator are found on the nameplate that is secured to the yoke or stator of the machine. These specifications or nominal characteristics are the values guaranteed by the manufacturer. A typical 100 kW compound generator may have the following information engraved or stamped on its nameplate:

Voltage	250 V	Type	Compound
Load current	400 A	Exciting current	20 A
Output power	100 kW	Temperature rise	50°C
Speed	1200 rpm	Insulation class	A

From the nameplate data, we can infer a number of interesting conclusions:

(*i*) The DC generator can deliver continuously a power of 100 kW at a voltage of 250 V without exceeding a temperature rise of 50°C.

(*ii*) The shunt field current is 20 A.

(*iii*) The armature current is 420 A.

(*iv*) If the compound generators is connected short-shunt, the series field current is 400 A.

(*v*) The class A designation refers to the class of insulation used in the machine.

1.57 WORKING PRINCIPLE OF DC MOTOR

The working of a DC motor is based on the principle that when a current carrying conductor is placed in a magnetic field it experiences a force, whose direction is governed by Fleming's left hand rule. The magnitude of this mechanical force is given by:

$$f = BIl \text{ newton}$$

When this force acts on conductors armature starts rotating. As armature conductor rotates in magnetic field, an emf called as back emf induces in them. This back emf (E_b) opposes the applied voltage and its magnitude is same as that was for generated emf in dc generator, i.e.

$$E_b = \frac{\phi ZN}{60}\left(\frac{P}{N}\right)$$

1.58 BACK OR COUNTER EMF

When the armature of a dc motor rotates under the influence of the driving torque, the armature conductors move through the magnetic field and hence emf is induced in them as in a generator. The induced emf acts in opposite direction to the applied voltage $V(P\phi ZN/60 \text{ A})$ is always less than the applied voltage V, although this difference is small when the motor is running under normal conditions.

Consider a shunt wound motor shown in Fig. 1.94. When DC voltage V is applied across the motor terminals, the field magnets are excited the armature conductors are supplied with current. Therefore, driving torque acts on the armature which begins to rotate. As the armature rotates, back emf, E_b is induced which opposes the applied voltage V. The applied voltage V has to force current through the armature against the back emf E_b. The electric work done in overcoming and causing the current to flow against E_b is converted into mechanical energy developed in the armature. It follows, therefore, that energy conversion in a DC motor is only possible due to the production of back emf E_b.

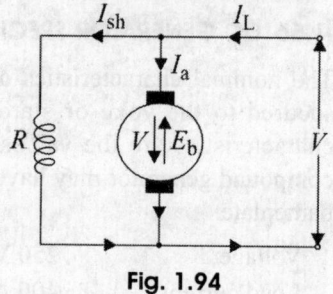

Fig. 1.94

Net voltage across armature circuit = $V - E_b$

If R_a is the armature circuit resistance, then, $I_a = \dfrac{V - E_b}{R_a}$

Since V and R_a are usually fixed, the value E_b will determine the current drawn by the motor. If the speed of the motor is high, then back emf E_b $(= P\phi ZN/60\ A)$ is large and hence the motor will draw less armature current and *vice versa*.

1.59 SIGNIFICANCE OF COUNTER EMF

The presence of back emf makes the DC motor a self-regulating machine, i.e. it makes the motor to draw as much armature current as is just sufficient to develop the torque required by the load.

$$\text{Armature current, } I_a = \frac{V - E_b}{R_a}$$

(*i*) When the motor is running on no-load, small torque is required to overcome the friction and windage losses. Therefore, the armature current I_a is small and the back emf is nearby equal to the applied voltage.

(*ii*) If the motor is suddenly loaded, the first effect is to cause the armature to slow down. Therefore, the speed at which the armature conductors move through the field is reduced and hence the back emf E_b falls. The decreased back emf allows a larger current to flow through the armature and larger current means increased driving torque. Thus, the driving torque increases as the motor slows down. The motor will stop slowing down when the armature current is just sufficient to produce the increased torque required by the load.

(*iii*) If the load on the motor is decreased, the driving torque is momentarily in excess of the requirement so that armature is accelerated. As the armature speed increases, the back emf E_b also increases and causes the armature current I_a to decrease. The motor will stop accelerating when the armature current is just sufficient to produce the reduced torque required by the load.

1.60 EQUIVALENT CIRCUIT OF A DC MOTOR

The armature of a dc motor can be represented by an equivalent circuit (Fig. 1.95).

It can be represented by three series-connected elements E, R_a and V_b.

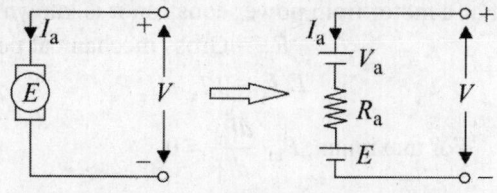

Fig. 1.95: Equivalent circuit of the armature of a DC motor

The element,

E_b = back emf
R_a = armature resistance
V_b = brush contact voltage drop

In a motor, current flows from the line into the armature against the generated voltage by KVL.

$$V = E_b + I_a R_a \qquad \qquad ...(1.2)$$

Equation (1.2) is the fundamental motor equation. It is seen that the back emf E of the motor is always less then the terminal voltage V.

If the voltage drop V_b in the brushes is also considered, then by KVL.

$$V = E_b + I_a R_a \qquad \qquad ...(1.3)$$

1.61 POWER DEVELOPED IN A DC MOTOR

Let us the voltage equation $V = E_b + I_a R_a$

Multiply both sides by I_a to get

$$VI_a = E_b I_a + I_a^2 R_a$$

The expression is known as the power equation of a DC motor. The significance of the power equation is explained as follows.

$$VI_a = E_b I_a + I_a^2 R_a$$

VI_a = electrical power supplied to the armature

$E_b I_a$ = electrical equivalent of the mechanical power produced by the dc motor (P_m)

$I_a^2 R_a$ = power loss taking place in armature winding.

In this expression VI_a represents the electrical power supplied to the armature winding and $I_a^2 R_a$ is the power loss taking place in the armature winding.

Thus $E_b I_a$ = (Input power – power loss)

Hence $E_b I_a$ represents the electrical equivalent of the gross mechanical power produced by the dc motor.

$\therefore$ Gross mechanical power $P_m = E_b I_a$

1.61.1 Condition for Maximum Power

For a motor from power equation it is known that

$$P_m = \text{Gross mechanical power developed} = E_b I_a$$

$$E_b I_a = V I_a - I_a^2 R_a$$

For maximum, P_m $\dfrac{dP_m}{dI_a} = 0$

$$dP_m = V - 2 I_a R_a$$

$$V - 2 I_a R_a = 0$$

$$\boxed{V = 2 I_a R_a \quad \therefore \ I_a = \frac{V}{2R_a}, \text{i.e. } I_a R_a = \frac{V}{2}}$$

substituting in voltage equation,

$$V = E_b + I_a R_a = E_b + \frac{V}{2}$$

then $\boxed{E_b = \dfrac{V}{2}}$ condition for maximum power

Note: This is practically impossible to achieve as for this E_b, current required is much more than its normal rated value. Large heat will be produced and efficiency of motor will be less than 50%.

1.61.2 Important Relationship between Different Types of DC Motors

Separately excited DC motor (Fig. 1.96):

Let I_a = Armature current

    R_a = Armature resistance

    I_L = Line current

    V = Input voltage

    R_f = Field resistance

    V_f = Field voltage

    I_f = Field current

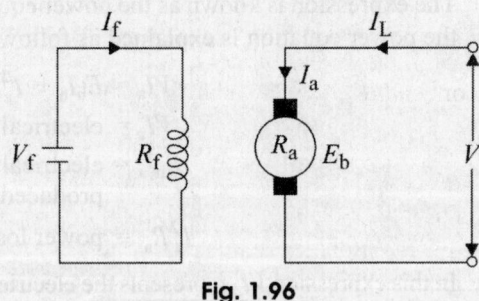

Fig. 1.96

$$I_f = \frac{V_f}{R_f}$$

$$I_a = I_L$$

$$E_b = V - I_a R_a$$

Input power $P_{in} = V . I_L$

Output power $P_{out} = E_b . I_a$

DC Shunt Motor (Fig. 1.97)

Let, I_{sh} = Shunt field current

R_{sh} = Shunt field resistance

then $I_{sh} = \dfrac{V}{R_{sh}}$

$I_L = I_a + I_{sh}$

$\boxed{E_b = V - I_a R_a}$

Input power $\boxed{P_{in} = V \cdot I_L}$

Output power $\boxed{P_{out} = E_b \cdot I_a}$

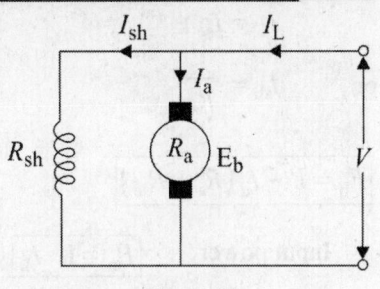

Fig. 1.97

DC Series Motor (Fig. 1.98)

Let, I_{se} = Series field current

R_{se} = Series field resistance

$I_a = I_L = I_{se}$

$E_b = V - I_a(R_a + R_{se})$

Input power $P_{in} = V \cdot I_L$

Output power $P_{out} = E_b \cdot I_a$

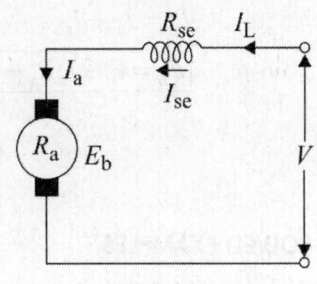

Fig. 1.98

Short Shunt DC Compound Motor (Fig. 1.99)

$I_L = I_a + I_{sh}$

$I_L = I_{se}$

$I_{sh} = \dfrac{V - I_{se} R_{se}}{R_{sh}}$

or $I_{sh} = \dfrac{E_b + I_a R_a}{R_{sh}}$

$\boxed{E_b = V - I_a R_a - I_{se} R_{se}}$

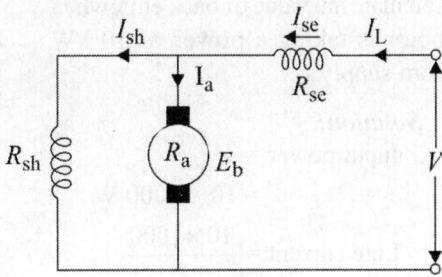

Fig. 1.99

Input power $\boxed{P_{in} = V \cdot I_L}$

Output power $\boxed{P_{out} = E_b \cdot I_a}$

Long Shunt DC Compound Motor (Fig. 1.100)

$I_L = I_{se} + I_{sh}$

$= I_a + I_{sh}$

$$I_a = I_{se}$$

or $$I_{sh} = \frac{V}{R_{sh}}$$

$$\boxed{E_b = V - I_a\left(R_a + R_{se}\right)}$$

Input power $$\boxed{P_{in} = V \cdot I_L}$$

Output power $$\boxed{P_{out} = E_b \cdot I_a}$$

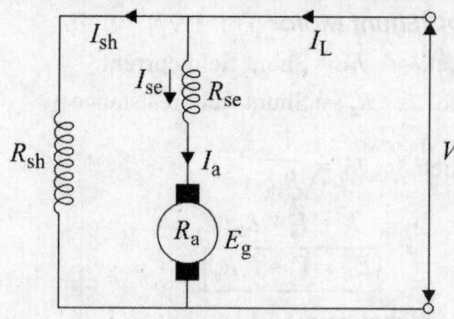

Fig. 1.100

Short Shunt DC Compound Motor (Fig. 1.101)

$$I_L = I_a + I_{sh}$$
$$\boxed{E_b = V - I_a R_a - I_L R_{se}}$$

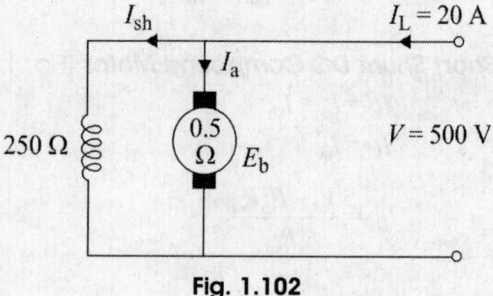

Fig. 1.101

SOLVED EXAMPLES

Example 1.36: A 500 V, DC shunt motor has an armature resistance of 0.5 Ω and field resistance of 250 Ω. Calculate the value of back emf, when motor is taking a power of 10 kW from supply.

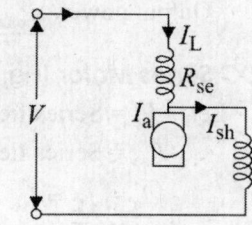

Fig. 1.102

Solution:

Input power $= V \cdot I_L$

$= 10 \times 1000$ W

Line current $= \dfrac{10 \times 1000}{V}$

$= \dfrac{10 \times 1000}{500} = 20\,\text{A}$

$I_{sh} = \dfrac{V}{R_{sh}} = \dfrac{500}{250} = 2\,\text{A}$

$I_L = I_a + I_{sh}$

or $\quad I_a = I_L I_{sh} = 200 - 2 = 18\,\text{A}$

$E_b = V - I_a R_a$

$= 500 - 18 \times 0.5 = 491\,\text{V}$ **Ans.**

Example 1.37: A 10 kW, 200 V DC shunt generator has armature and field resistance of 0.05 Ω and 200 Ω, respectively. Calculate total armature power developed when *(i)* works as a generator supplying 10 kW output and *(ii)* motor taking 10 kW input power from supply.

Solution: *(i)* When working as a generator

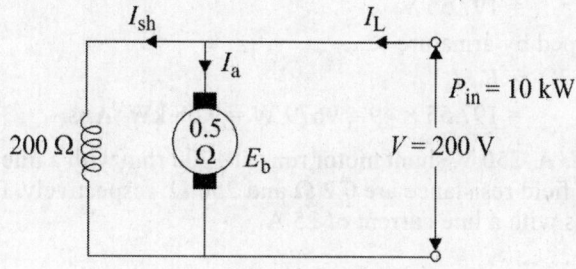

Fig. 1.103

$$I_{sh} = \frac{V}{R_{sh}} = \frac{200}{200} = 1\,A$$

Output power = $V \cdot I_L = 10 \times 1000$ W

or

$$I_L = \frac{10 \times 1000}{200} = 50A$$

$$I_a = I_L + I_{sh} = 50 + 1 = 51\,A$$

$$E_g = V + I_a R_a = 200 + 51 \times 0.05$$
$$= 202.55\,V$$

Total power developed in armature

$$= E_g I_a = 202.55 \times 51$$
$$= 10330.05\,W = 10.33\,kW \quad \textbf{Ans.}$$

(ii) When working as a motor

$$P_{in} = V \cdot I_L$$

$$I_L = \frac{P_{in}}{V} = \frac{10 \times 1000}{200} = 50\,A$$

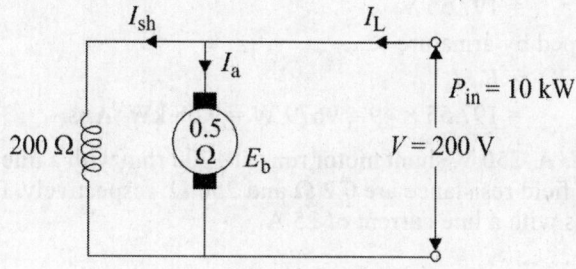

Fig. 1.104

$$I_{sh} = \frac{200}{200} = 1 \, A$$

$$I_a = I_L - I_{sh} = 50 - 1 = 49 \, A$$

$$E_b = V - I_a R_a$$

$$= 200 - 49 \times 0.05$$

$$= 197.55 \, V$$

Power developed by armature

$$= E_b \cdot I_a$$

$$= 197.55 \times 49 = 9679 \, W = 9.68 \, kW \text{ **Ans.**}$$

Example 1.38: A 250 V, shunt motor runs at 1200 rpm with a line current of 50 A. Its armature and field resistance are 0.2 Ω and 250 Ω, respectively. Find the speed at which motor runs with a line current of 25 A.

Solution:

$$\frac{N_2}{N_1} = \frac{E_{b,2}}{E_{b,1}} \times \frac{\phi_1}{\phi_2}$$

For shunt motor flux remains constant

So, $$\frac{N_2}{N_1} = \frac{E_{b,2}}{E_{b,1}}$$

$$E_{b,1} = V - I_{a,1} R_a$$

$$I_{sh} = \frac{V}{R_{sh}} = \frac{250}{250} = 1 \, A$$

$$I_{a,1} = I_{L,1} - I_{sh} = 50 - 1 = 49 \, A$$

$$E_{b,1} = 250 - 49 \times 0.2 = 240.2 \, V$$

$$I_{a,2} = I_{L,2} - I_{sh} = 25 - 1 = 24 \, A$$

$$E_{b,2} = 250 - 24 \times 0.2 = 245.2 \, V$$

$$N_2 = N_1 \times \frac{E_{b,2}}{E_{b,1}}$$

$$= 1200 \times \frac{245.2}{240.2}$$

$$= 1224.97 = 1225 \, rpm$$

1.62 TORQUE DEVELOPED IN DC MOTOR

It is seen that the turning or twisting force about an axis is called *torque*.

Consider a wheel of radius R meters acted upon by a circumferential force F newtons as shown in Fig. 1.105.

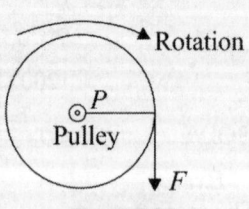

Fig. 1.105

• The wheel is rotating at a speed of N rpm. Then angular speed of the wheel is

$$\omega = \frac{2\pi N}{60} \text{ rad/s}$$

• So work done in one revolution is

$$W = F \times \text{distance travelled in one revolution}$$

$$= F \times 2\pi R \text{ Joules}$$

And $\qquad P = \text{Power developed} = \dfrac{\text{Work done}}{\text{Time}}$

$$= \frac{F \times 2\pi R}{\text{Time for one revolution}} = \frac{F \times 2\pi R}{\left(\dfrac{60}{N}\right)}$$

$$= (F \times R) \times \left(\frac{2\pi R}{60}\right)$$

$$P = T \times W \text{ watts}$$

where $\qquad T = \text{torque}$

$\qquad W = \text{angular speed in rad/sec}$

• Let T_a be the gross torque developed by the armature of the motor. It is also called armature torque. Gross mechanical power developed in the armature is $E_b I_a$ as seen from the power equation. So if speed of the motor is N rpm then

$$\text{power in armature} = \text{Armature} \times W$$

$$E_b I_a = T_a \times \frac{2\pi N}{60}$$

but E_b in a motor is given by,

$$E_b = \frac{\phi P N Z}{60\, A}$$

$$\frac{\phi P N Z}{60\, A} \times I_a = T_a \times \frac{2\pi N}{60}$$

$$T_a = \frac{1}{2\pi} \phi I_a \times \frac{PZ}{A}$$

$$\boxed{T_a = 0.159 \phi I_a \cdot \frac{PZ}{A} \text{ N} \cdot \text{m}}$$

• This is torque equation of DC motor.

SOLVED EXAMPLES

Example 1.39: A 300 V, 4 pole DC series motor has 800 wave wound conductors. The motor has total armature and series field resistance of 0.5 Ω and motor takes 50 A from supply. If flux per pole is 70 mWb, find speed and gross torque (armature torque) produced by the motor.

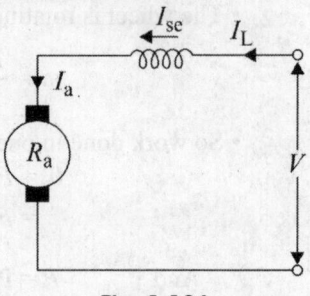

Fig. 1.106

Solution:

$$I_a = I_L = I_{se} = 50 \text{ A}$$

$$E_b = V - I_a(R_a + R_{se})$$

$$= 300 - 50 \times 0.5$$

$$= 275 \text{ V}$$

$$E_b = \frac{\phi ZN}{60}\left(\frac{P}{A}\right)$$

$$275 = \frac{70 \times 10^{-3} \times 800 \times N}{60}\left(\frac{4}{2}\right)$$

or $\quad N = 147.32$ rpm

$$T_a = 0.159\phi ZI_a\left(\frac{P}{A}\right)$$

$$= 0.159 \times 70 \times 10^{-3} \times 800 \times 50 \times \left(\frac{4}{2}\right)$$

$$= 890.4 \text{ N·m} \quad \textbf{Ans.}$$

1.63 CHARACTERISTICS OF DC MOTORS

The application of DC motor is determined by its characteristics which shows following relationships:

1. torque vs armature current (T/I_a)
2. speed vs armature current (N/I_a)
3. speed vs torque (N/T)

1.63.1 Characteristics of DC Shunt Motor

1. Torque vs Armature Current (T_a/I_a)

As $\qquad T_a \propto \phi \cdot I_a$

For shunt motors flux is assumed to be constant (although at heavy loads if reduces slightly due to armature reaction).

So $\qquad T_a \propto I_a$

This relation represents a straight line passing through the origin (Fig. 1.107).

$T_{sh} < T_a$, for same armature current.

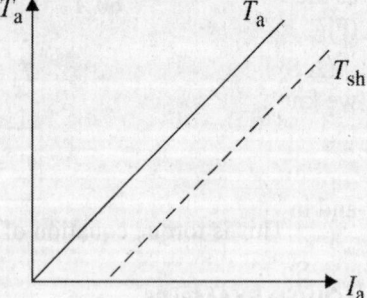

Fig. 1.107: Torque vs armature current characteristics of DC shunt motor

2. Speed vs Armature Current (N/I_a)

$$N \propto \frac{E_b}{\phi}$$

or

$$\propto \frac{V - I_a R_a}{\phi}$$

as ϕ is almost constant

$$N \propto V - I_a R_a$$

With increase in armature current, speed will drop slightly. These motors are suitable for constant speed applications (Fig. 1.108).

3. Speed vs Torque (N/T_a): In case of DC shunt motors torque is proportional to I_a. So torque follows the profile of armature current. Hence speed torque characteristic follows the similar trend as a speed-armature current characteristics as shown in Fig. 1.109.

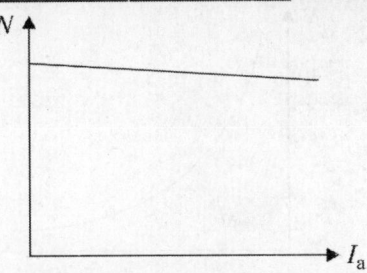

Fig. 1.108: Speed vs armature current characteristics of DC shunt motor

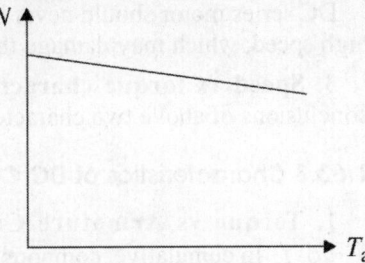

Fig. 1.109: Speed vs torque characteristics of DC shunt motor

1.63.2 Characteristics of DC Series Motor

1. Torque vs Armature Current

For series motors $\phi \propto I_a$

and $\qquad T_a \propto f$

so $\qquad T_a \propto I_a^2$

As I_a increases, T_a increases as the square of the armature current. After saturation flux becomes almost constant and characteristics becomes a straight line after magnetic saturation.

Series motors are used in trains, turns as the torque in the series motor is high (Fig. 1.110).

2. Speed vs Armature Current: As we know

$$N \propto \frac{E_b}{\phi} \propto \frac{V - I_a R_a}{\phi}$$

and in case of series motor $\phi \propto I_a$

So $N \propto \dfrac{V - I_a R_a}{I_a}$

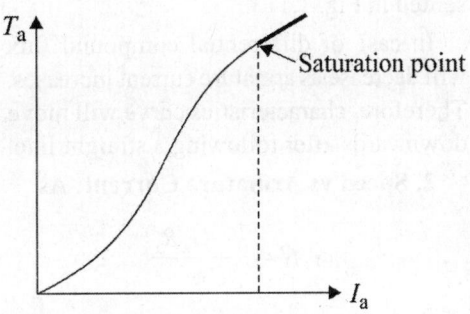

Fig. 1.110: Torque vs torque current characteristics of DC series motor

As I_a increases, $V - I_a R_a$ decreases slightly but increase in denominator (i.e. I_a), speed decreases with a much higher rate. When $I_a = 0$, speed tends towards infinity as shown in Fig. 1.111.

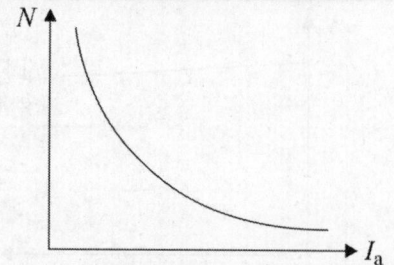

Fig. 1.111: Speed vs armature current characteristics of DC series motor

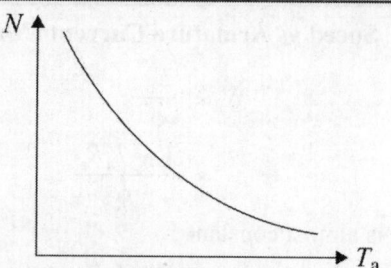

Fig. 1.112: Speed vs torque characteristics of DC series motor

DC series motor should never be run on no load otherwise it will gain dangerously high speed, which may damage the motor.

3. Speed vs torque characteristics: This characteristic curve is drawn the conclusions of above two characteristics as shown in Fig. 1.112.

1.63.3 Characteristics of DC Compound Motors

1. Torque vs Armature Current: $T \propto \phi \cdot I_a$ In cumulative: compound, series field supports shunt field. As I_a increase, net flux increases. So torque will initially follow the characteristics similar to shunt motor and as I_a increase it starts showing the domination of series field as represented in Fig. 1.113.

In case of differential compound flux will decrease as armature current increases. Therefore, characteristics curve will move downwards after following a straight line.

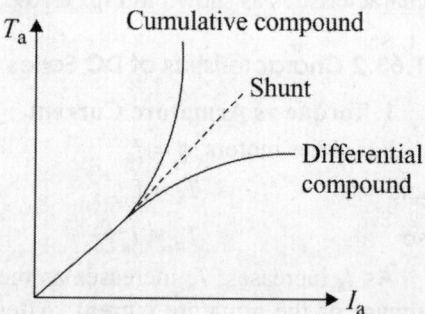

Fig. 1.113: Torque vs armature current characteristics

2. Speed vs Armature Current: As

$$N \propto \frac{V - I_a R_a}{\phi}$$

Initially motor will follow the shunt characteristics as flux increases with increase in armature current, in case of cumulative compound motor, speed will reduce with large rate. In case of differential compound, speed will increase due to reduced flux as shown in Fig. 1.114.

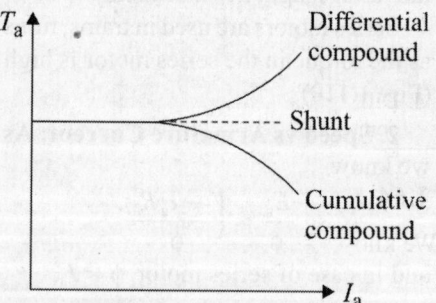

Fig. 1.114: Speed vs armature current characteristics

3. Speed vs Torque Characteristics: From above two characteristics, this curve can be drawn as shown in Fig. 1.115

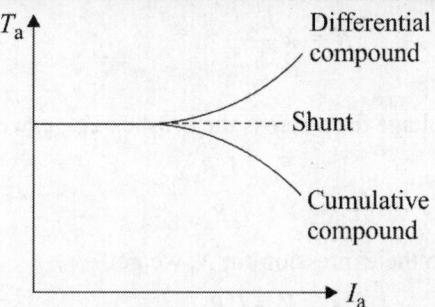

Fig. 1.115: Speed vs torque characteristics

Instead of just stating the applications, the behaviour of the various characteristics like speed, starting torque, etc. which make the motor more suitable for the applications, is shown in Table 1.1.

Table 1.1: Applications of DC motor

Type of motor	Characteristics	Applications
1. Shunt	Speed is fairly constant and medium starting torque.	1. Blowers and fans 2. Centrifugal and reciprocating pumps 3. Lathe machines 4. Machine tools 5. Milling machines 6. Drilling machines
2. Series	High starting torque. No load condition is dangerous, variable speed.	1. Cranes 2. Hoist elevators 3. Trolleys 4. Conveyors 5. Electric locomotives
3. Cumulative compound	High starting torque. No load condition is allowed.	1. Rolling mills 2. Punches 3. Shears 4. Heavy planers 5. Elevators
4. Differential compound	Speed increases as load increases.	Not suitable for any practical applications

1.64 SPEED EQUATIONS

We know that the expression for the back emf is,

$$E_b = \frac{P\phi NZ}{60A}$$

But P, Z and $60\,A$ are constants therefore we can write,

$$E_b \propto \phi_N$$

$$N \propto \frac{E_b}{\phi}$$

$$N = k\frac{E_b}{\phi}$$

But $V = E_b + I_a R_a + V_{brush}$

So neglecting the voltage drop across the brushes V_{brush}, we get

$$V = E_b + I_a R_a$$

$$\therefore \qquad E_b = V - I_a R_a$$

Substituting this into the expression for N, we get

$$N = \frac{V - I_a R_a}{\phi}$$

Since the flux ϕ is proportional to the field current, we can write that

$$N \propto \frac{(V - I_a R_a)}{I_{field}}$$

The expression for I_f will different for the various types of DC motors.

DC Shunt Motor

For a DC shunt motor the flux ϕ is constant

$$\therefore \qquad N \propto V - I_a R_a$$

DC Series Motor

For DC series motor $\qquad I_f = I_a$; therefore

$$N \propto \frac{(V - I_a R_a) - I_{se} R_{se}}{I_a}$$

where $\qquad E_b = V - I_a R_a - I_{se} R_{se}$

The expressions for N are important in understansding various characteristics of different types of DC motor.

1.65 SPEED REGULATION

The speed regulation is defined as the change in speed from no load to full load expressed as a fraction or a percentage of the full load speed. It can be written as:

$$\text{Per unit speed regulation} = \frac{N_{nl} - N_{fl}}{N_{fl}}$$

$$\text{Percentage speed regulation} = \frac{N_{nl} - N_{fl}}{N_{fl}} \times 100$$

where N_{nl} = No load speed, N_{fl} = full load speed

A motor which has a nearly constant speed is said to have a good speed regulation.

Base Speed

It is equal to rated speed of the motor at rated armature voltage and rated field current.

Constant Torque Drive

In this motor shaft torque remains constant over a given speed range and shaft power changes with speed.

Constant Power Drive

In this, motor shaft power remains, constant over a given speed range and shaft torque changes with speed.

1.66 SPEED CONTROL OF DC MOTOR

Factor controlling motor speed as follows.

It has been shown earlier that speed of a motor is given by the relation (Fig. 1.116).

$$N = \frac{V - I_a R_a}{Z \phi} \left(\frac{A}{P} \right)$$

$$N = K \frac{V - I_a R_a}{\phi} \text{ rps}$$

where R_a = armature circuit resistance

It is obvious that the speed can be controlled by varying.

(i) Flux/pole, ϕ (flux control)

(ii) Resistance R_a of armature circuit (Rheostatic control) and

(iii) Applied voltage V (voltage control)

These methods as applied to series, shunt and compound motors will be discussed below.

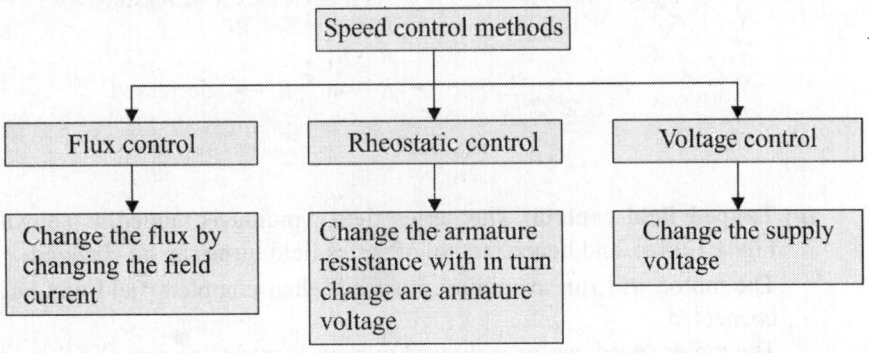

Fig. 1.116

1.66.1 Speed Control of DC Series Motor

Types of speed control of DC series motor (Fig. 1.117).

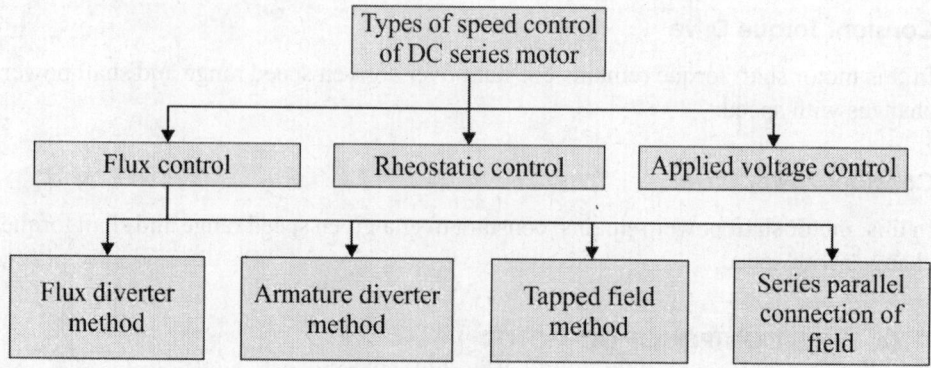

Fig. 1.117

Flux Control: Flux control method, the change in flux of series motor can be provided by any of the following ways.

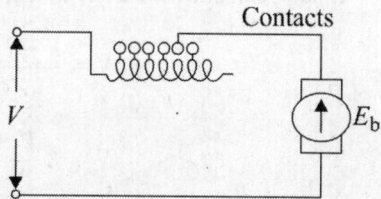

Fig. 1.118: Series motor speed control by a diverter

(a) **Diverter field control:** In this, a resistance with a diverter is connected across the series field winding as shown in Fig. 1.118.

By varying the diverter resistance, current through the series field winding is changed. Hence corresponding change in field flux and speed of the motor takes place (Fig. 1.119).

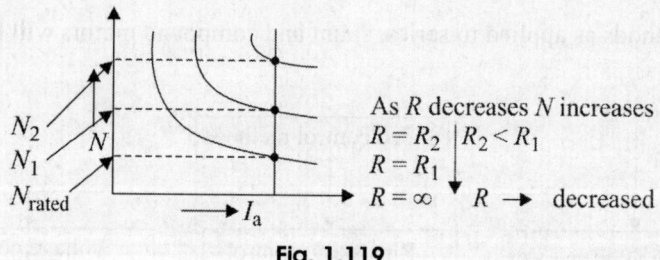

As R decreases N increases
$R = R_2 \mid R_2 < R_1$
$R = R_1$
$R = \infty \quad R \rightarrow$ decreased

Fig. 1.119

(b) **Tapped field control:** The series field winding is tapped as shown in Fig. 1.120(*a*), and hence the no. of series field turns can be changed.

The motor will run at minimum speed when complete field winding is connected.

The motor speed can be increased in steps by reducing some of the series field turns.

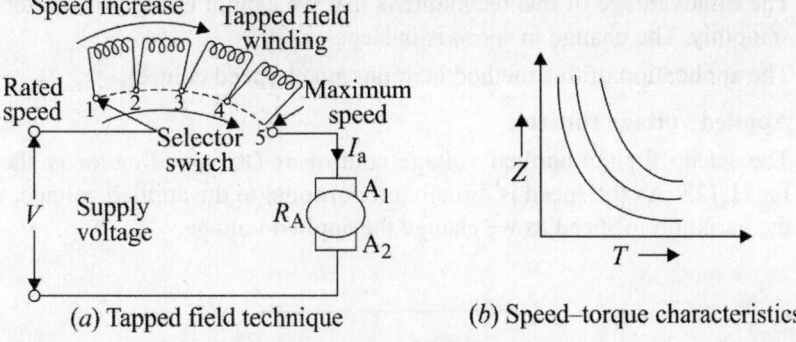

(a) Tapped field technique   (b) Speed–torque characteristics

Fig. 1.120

(c) Armature diverter control:

In this method the variable resistor R (called as diverter) is connected across the armature winding as shown in Fig. 1.121.

For series motor $T \propto I_a^2$

This method of speed control is used for the constant torque loads, and the speed control takes place as follows.

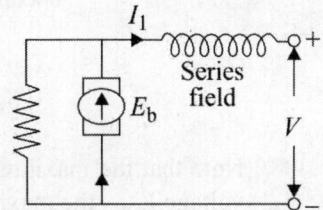

Fig. 1.121: Series motor speed control by armature diverter

As we reduce R by moving it upwards. The armature current I_a will decrease but torque produced by motor $T \propto \phi I_a$. Therefore to produce required torque motor draws more current from the source (Increases). So field current increases. Hence ϕ increases and speed decreases. Thus with reduction in R, speed decreases.

(d) Series parallel connection of field:

The field winding is divided into two or more equal parts.

These parts are then connected either in series or parallel shown in Fig. 1.122.

Due to series or parallel connections, the total mmf produced will change. This will change the flux, which will change the motor speed.

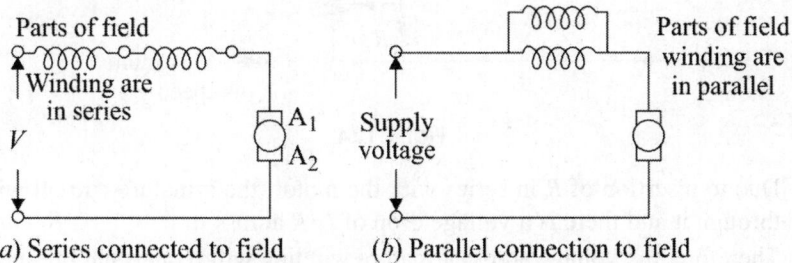

(a) Series connected to field   (b) Parallel connection to field

Fig. 1.122: Series motor speed control by armature diverter

The disadvantage of this technique is that we cannot change the motor speed smoothly. The change in speed is in steps.

The application of this method is in fan motor speed control

(e) Applied voltage control:

The set up for the applied voltage control of DC series motor is shown in Fig. 1.123. As the speed is directly proportional to the applied voltage, we get the variation in speed as we change the applied voltage.

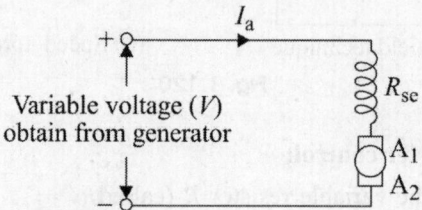

Fig. 1.123: Applied voltage

Note that the maximum voltage that can be applied to the motor is the rated voltage V_{rated} the maximum speed will be the rated speed.

Thus the speed control below the rated speed is possible to achieve.

The variable voltage is applied from generator (DC)

- As V increases N increases.

(f) Rheostatic control (armature voltage control):

- The set up for the rheostatic control of a DC series motor and is shown in Fig. 1.124.
- A variable resistance is connected in series with the armature.

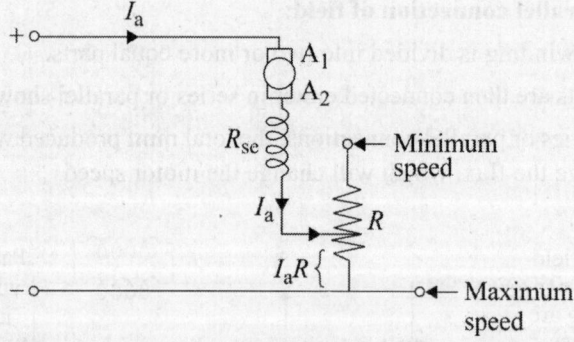

Fig. 1.124

Due to insertion of R in series with the motor, the armature current will flow through it and there is a voltage drop of $I_a \cdot R$ across it.

Therefore, the voltage across armature winding will reduce and the speed will also reduce because the speed is proportional to armature voltage.

1.66.2 Speed Control of DC Shunt Motor

Flux Control

- In this method, a variable resistance is connected in series with the shunt field winding.
- The flux of a DC motor can be changed with the help of this variable resistance.
- When shunt field resistance is increased, the field current and flux both are reduced. With the reduction of flux, back emf is also reduced.
- Hence, armature current increases. This increase in armature current is much more than the decrease in flux.
- The motor tortque is greater than the constant load torque. As a result of it, speed of the motor is increased which causes the increase in back emf and decrease in armature current.
- When, the motor torque is equal to the load torque motor start running at a constant speed which is greater than the initial speed (Fig. 1.125).

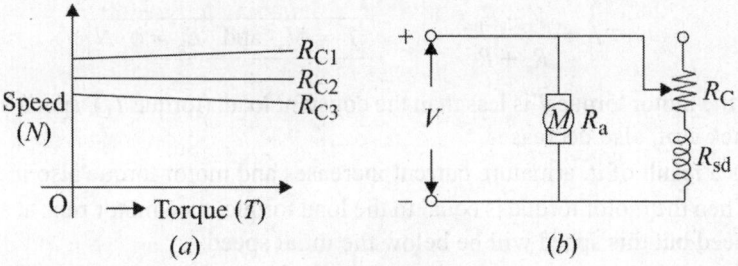

Fig. 1.125

Advantages

- It is possible to control the speed above the rated speed.
- It is possible to change the speed in a smooth manner.
- The field resistance is large so the field current is small. Hence, the power lost in the rheostat $(I_{sh}^2 R)$ is also small so flux control method is more efficient and economical as well.
- Due to small field current (I_{sh}) the size of rheostat is small.

Disadvantages

- The speed above the base speed are only possible.
- The higher speeds above rated speeds can cause problems for the commutation. This make the motor operation unstable.
- The reduction in flux will reduce the torque producing capability of the motor.

Rheostatic or Armature Control Method

- In this method, a variable resistance is connected in series with the armature circuit as shown in Fig. 1.126.

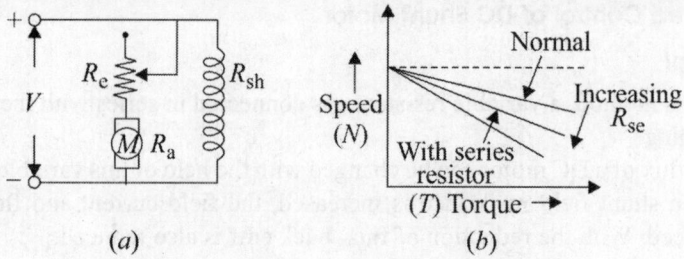

Fig. 1.126: Speed control of DC motor

- In this case, the field flux remains constant as the supply voltage is normally constant.
- By increasing the armature circuit resistance, there is reduction in armature current and torque developed in the motor.
- This can be easily explained by the following relation.

$$I_a = \frac{V - E_b}{R_a + R_g}, \qquad \boxed{T \propto \phi I_a \text{ and } E_b \propto \phi \cdot N}$$

- If the motor torque T is less than the constant load, (torque T_L), speed decreases, back emf, also decreases.
- As a result of it, armature current increases and motor torque also increases.
- When the motor torque is equal to the load torque, then motor runs at a constant speed but this speed will be below the intial speed.

Advantages

- It is possible to vary the speed smoothly.
- Speed can be controlled right from 0 to N rated.
- The potentiometer rheostat acts as the starter.

Disadvantages

- The entire armature current passes through the rheostat. Since I_a is large, there is a large power loss in the rheostat (power loss = $I_a^2 R$).
- It is a not possible to control speeds above the rated speed.
- A large size rheostat needs to be used.
- The system efficiency decreases due to the power loss taking place in the external resistance.
- The heat dissipated in the rheostat is so large that special heat dissipation arrangment needs to be done.

Armature Voltage Control

- In DC shunt motor, the flux remain constant. By changing the armature terminal voltage V, the back emf changes and hence the speed of the motor also changes.

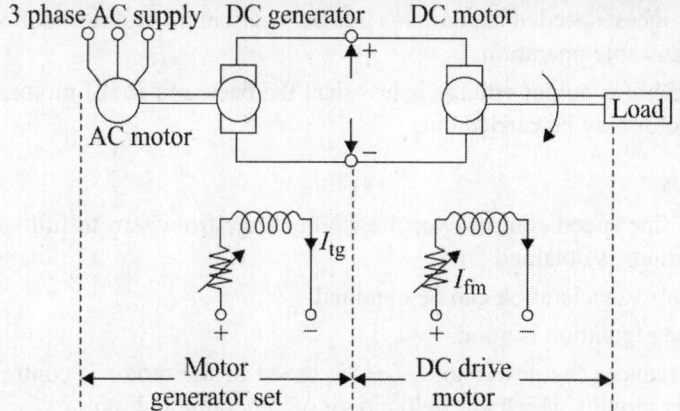

Fig. 1.127: Ward–Leonard control system

- Ward–Leonard method of speed control is based on the armature terminal voltage. The schematic diagram of this method is shown in Fig. 1.127.

- In Fig. 1.127, M is main motor whose speed control is required.

- The three phase AC motor acts as a prime mover to the generator G, whose output voltage is directly fed to the main motor M.

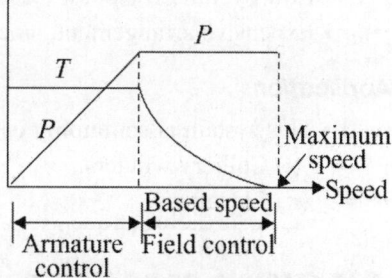

Fig. 1.128: Torque speed and power speed characteristics

- The combined AC motor and DC generator is called motor-generator set.

- The generator output voltage is changed by changing its field excitation as the motor speed is changed.

- In this method, motor armature current is maintained at rated value, by varying the generator output voltage, speed below the base speed can only be obtained.

- A constant torque is obtained up to based speed, because the speed control is carried out with rated armature current and with constant motor field flux ϕ.

- Power = (torque × speed) increases in proportion to speed.

- Thus constant torque and variable power drive is obtained up to base speed.

- The lowest speed of the motor is mainly limited by the residual magnetism of the generator.

- By decreasing field flux ϕ of the motor speeds above base speed are obtained.

- The decreasing field flux ϕ of the motor, speeds above base speed are obtained. The power (VI_a) remains constant for the speeds above base speed at rated armature current and torque (ϕI_a) decreases in flux ϕ.

- Thus constant power and variable torque drive is obtained with the above base speed.

- The highest speed of the motor is limited by armature heating, poor commutation and unstable operation.
- If generator output voltage is less than the back emf (ϕ) of motor, the braking of motor may be carried out.

Advantages

- Very fine speed control over the whole range from zero to full speed in both directions is obtained.
- Uniform acceleration can be obtained.
- Speed regulation is good.
- It is reduces the power loss, i.e. the speed of the motor is controlled by low power circuits which are field circuit of generator and motor.

Disadvantages

- Low overall efficiency.
- Expensive arrangement, since two extra machines are required.

Applications

- This system is commonly used in
  - (*i*) Colliery winders.
  - (*ii*) Elevators.

1.67 STARTING OF DC MOTOR

Need of Starter

Regarding the starter, the question arises that why starters are necessary. It can be explained with the help of an example. A DC motor has out put of 10 kW at 250V. If armature resistance and shunt field resistances are 0.1 Ω and 250 Ω respectively, then what is the armature current when motor is running at full-load at the time of starting, output power

$$P = VI$$
$$10 \times 1000 = 250 \times I$$
$$I = 40 \text{ A}$$

$$I_{sh} = \frac{V}{R_{sh}} = \frac{250}{250} = 1\text{A}$$

$$\boxed{I_a = I + I_{sh} = 40 + 1 = 41\text{A}}$$

The armature current is also given by

$$I_a = \frac{V - E_b}{R_a} \text{ and } E_b = \frac{\phi \cdot Z \cdot N}{60} \times \frac{P}{A}$$

At the time of starting $N = 0$ then back emf will be zero, i.e.

$$E_b = 0$$

and

$$I_a = \frac{V}{R_a} = \frac{250}{0.1} = 2500 \text{ A}$$

The current at the time of starting is very high and many times of rated load current. This current may damage the motor due to the excessive heat and short circuit, which results in the damage of brushes, commutator and winding, etc.

To avoid this stage, we require a starter for starting purpose, i.e. when motor runs with rated speed and back emf E_b is developed. In other words, the starter is necessary to:

• control the starting current up to safe value.

• protect the motor against the damage of brushes commutator and windings.

To protect the motor with this excessive current we have to increase the armature circuit resistance. We add a variable resistance in series with armature as shown in Fig. 1.130 of the starter.

When R_h is connected in the armature circuit, than

$$I_a = \frac{V}{R_a + R_h}$$

where R_h is called starting resistance

A DC motor starter consists of starting resistance which is properly graded. In addition to this starting resistance there are some protective device like no volt release and overload release. The starting resistances are cut-out in step till the motor get its rated speed.

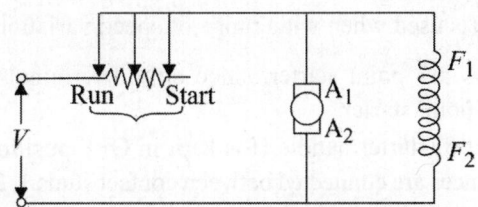

Fig. 1.129: Basic arrangement of a starter

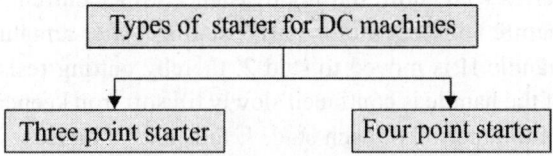

Fig. 1.130

1.67.1 Shunt and Compound Motor Starter

Primary function of a starter is to limit starting current in armature circuit. The starting resistance is arranged in steps and is gradually cut out as the motor accelerates so as to maintain a high average torque during starting. Simplest starter consist of rheostat inserted in series with armature circuit.

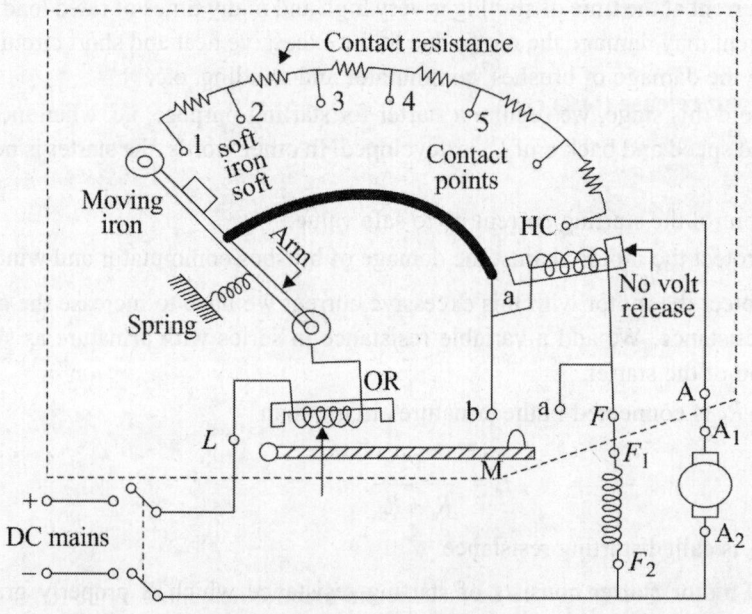

Fig. 1.131: Three point starter

There are two standard type of starter for shunt and compound motor. They are
⇒ 3 point starter
⇒ 4 point starter (is used when wide range of speed variation is required).

Figure 1.131 shows a 3 point starter, since only 3 terminals are available from starter, it is called a 3 point starter.

When motor is at rest, starter handle H is kept in OFF position by a strong spiral spring. Starting resistances are connected between contact studs 1, 2, 3, 4, 5. For starting the motor, handle is rotated to come in contact with stud 1. As soon as handle H touches stud 1, shunt field and holding coil HC get connected in series where as armature get connected in series with entire starting resistance. Since current begins to flow in both field and armature winding, motor starts rotating. After armature has picked up sufficient speed, handle H is moved to stud 2, thereby cutting resistance between 1 and 2 movement of the handle is continued slowly till soft iron keeper touches holding magnet. During waiting period of each stud, I_a falls and N increases and handle H is held in ON position and entire starting resistance is cut-out.

The holding magnet or coil HC is also called no volt release or low voltage release.

Function of holding coil or low voltage release:

⇒ In case of power failure, HC get demagnetized and spring spiral brings handle back to OFF position. If during power failure, handle fails to return to off position, motor might be damaged in case the power is restored because, then there would be no starting resistance in armature circuit.

⇒ If starting field becomes open circuited accidently, HC gets demagnetized and starter handle is required by spring pull to OFF posittion.

Overload release (OR) coil:

• It is provided in series with armature circuit OR is a small electromagnet.

• In case I_a increases preset value due to overload OR becomes more strong and attracts the movable soft iron M.

• Due to this two points ab of holding coil get short circuited.

• In this manner, HC get demagnetized and starter arm or handle H is pulled back to OFF position by sprial spring.

• Motor is automatically disconnected from the mains in case of overload.

Disadvantages of 3 point starter:

• The hold on coil and the field winding are connected in series with each other in the 3 point starter.

• So, if we reduce the field current to exercise the flux control, then the current flowing through the hold on coil will also reduce.

• If this current goes below a certain level, then the force of attraction produced by the hold on coils will be insufficient to hold the handle in the RUN position and so the handle will return back to the OFF position. The motor will be switched off.

• This disadvantage can overcome by using the four point startor.

• Figure 1.132 shows a four point starter. Four terminals (L_1, L_2, F, A) are available form this starter. It is connected to dc supply and to DC motor.

• A resistance R is used, the function of which is to prevent short circuit of supply mains in case of overload release operation.

• When HC gets short circuited by OR, current through R is limited by its own resistance and starting resistances.

• The field current and holding coil currents are not dependent in this case.

• So, speed control can be employed without affecting the current through holding coil.

Disadvantages of 4 point starter

• The only disadvantage of four point starter is that it does not provide any protection against the field failure or high speed protection to the motor.

• The field failure can be explained as follow under the running condition if the field current reduces to zero due to field winding getting open circuited or due to some other reason then the condition is called as field failure condition.

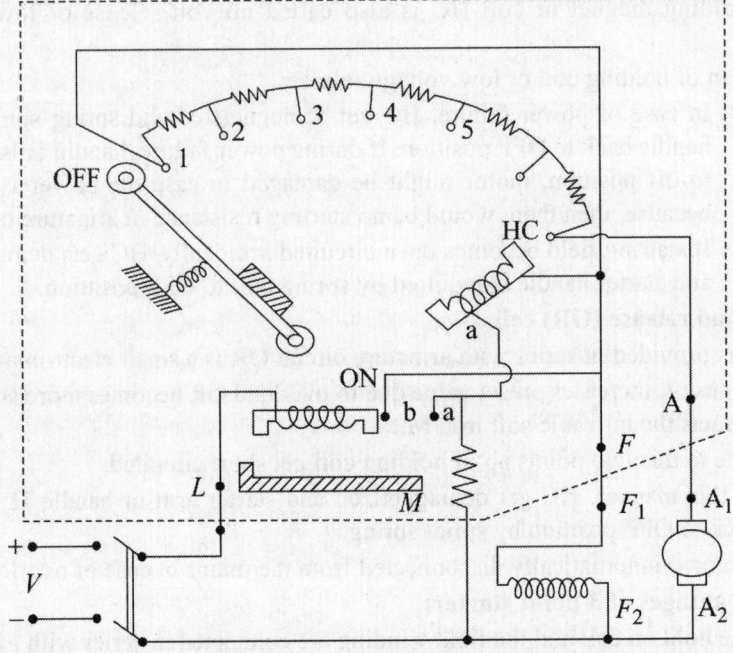

Fig. 1.132: Four-point starter

- If it reduces to zero, then the only flux present will be the leakage flux which is very small.
- Since speed $N \propto \dfrac{1}{\phi}$, the motor tries to run at a very high speed which can be very dangerous for it.
- The three point starter can protect the motor under the field failure conditions.

1.67.2 Starter Step Calculation for DC Shunt Motor

Figure 1.133 shows DC shunt motor with n resistance elements or $(n + 1)$ studs pertaining to DC shunt motor starter, studs are numbered 1,2,3,..., $n + 1$ for n resistance elements.

Note: $R_1 = r_1 + r_2 + ... + r_{n-1} + r_n + r_a, ...$ so that $R_{n+1} = r_n$

At starting when handle is in contact with stud 1. R_1 should be equal to

$$R = \frac{\text{Terminal voltage}}{\text{Maximum permissible armature current}} = \frac{V}{I_a}$$

With handle at stud 1 motor speeds up, E_n starts rising and current start decreasing as

$$I_a = \frac{V - E_n}{R_a}$$

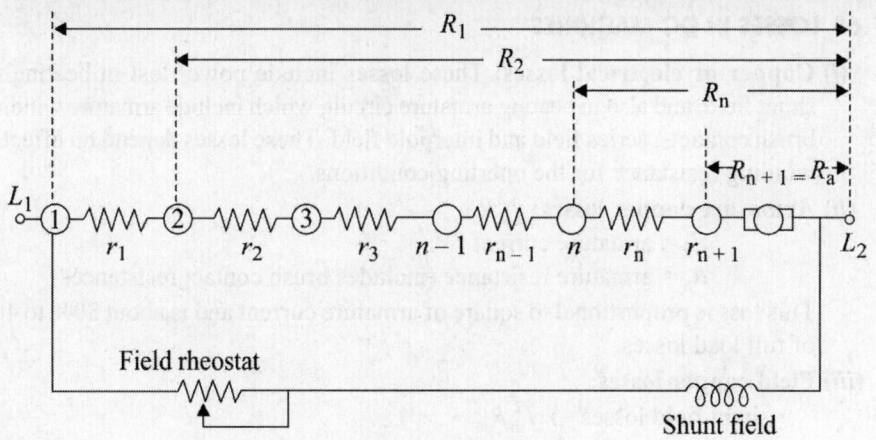

Fig. 1.133

The current is allowed to be reduced to Ia_2

Lower current limit $= \dfrac{V - E_{b1}}{R_1}$ ...(1.4)

At this instant the handle is moved to stud 2, the current increases to I_{a1}, instantaneously.

$$I_{a1} = \dfrac{V - E_{b1}}{R_2} \qquad ...(1.5)$$

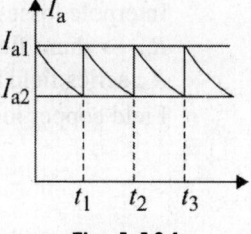

Fig. 1.134

From Eqs (1.4) and (1.5):

$$\dfrac{I_{a1}}{I_{a2}} = \dfrac{R_1}{R_2} = \dfrac{\text{Upper armature current limit } I_{a2}}{\text{Lower armature current limit}}$$

From the above relations, we get:

$$\dfrac{I_{a1}}{I_{a2}} = \dfrac{R_1}{R_2} = \dfrac{R_2}{R_3} = ... = \dfrac{R_{n-1}}{R_n} = \dfrac{R_n}{R_{n+1}} = \dfrac{R_n}{R_a}$$

$$\dfrac{R_1}{R_2} \times \dfrac{R_2}{R_3} \times ... \times \dfrac{R_{n-1}}{R_n} = \left(\dfrac{I_{a1}}{I_{a2}}\right)^{n-1} = \alpha^{n-1}$$

where $\qquad \alpha = I_{a1} / I_{a2}$

Now $\qquad \dfrac{R_1}{R_n} = \alpha^{n-1}$ or $\dfrac{R_1}{R_n} \times \dfrac{R_n}{R_a} = \alpha^n$

$$R_1 = \alpha R_2, R_3 = \alpha R_3 ...$$

Note:

1. During the waiting time at each step, the current falls and the speed rises exponentially.
2. The waiting time at each step progressively reduces.

1.68 LOSSES IN DC MACHINES

(*i*) **Copper or electrical losses:** These losses include power lost in heating the shunt field, and also in heating armature circuit, which include armature winding, brush contacts, series field and interpole field. These losses depend on effective winding resistance for the operting conditions.

(*ii*) **Armature copper losses:** $I_a^2 R_a$

I_a = armature current

R_a = armature resistance (includes brush contact resistances)

This loss is proportional to square of armature current and is about 30% to 40% of full load losses.

(*iii*) **Field copper losses:**

- shunt field losses → $I_{sh}^2 R_{sh}$
- interpole loss → $I_a^2 R_i$
- series field loss → $I_{sh}^2 R_{se}$

Interpole losses occur only in machine provide with interpoles.

R_{sh} → shunt field resistance.

R_{se} series field resistance R_i → interpole field resistance.

Field copper loss is about 20% to 30% of full load losses.

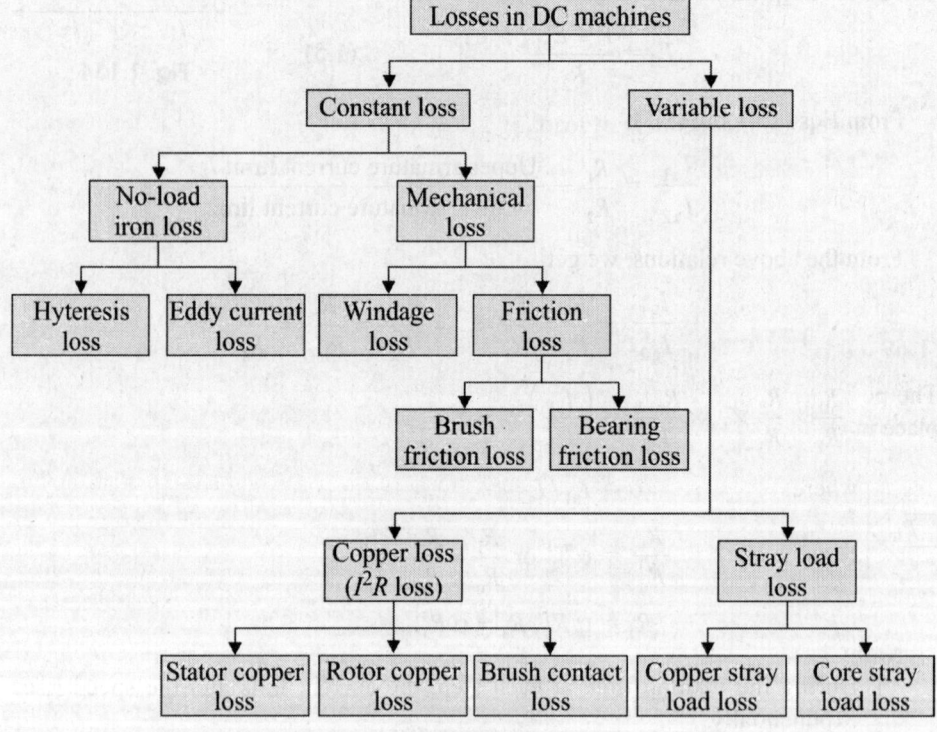

Fig. 1.135

2. Iron or magnetic losses (core loss):

These losses are present in any kind of machine that is constructed of iron and subjected to variations of magnetic flux.

⇒ **Hysteresis losses:**

Depends upon the range of variations of flux density and frequency of the variation.

With reasonable accuracy

$p_h = k_h (B_{max})^{1.6}$ fv W, where

k_h = Steimnetz hysterisis coefficient

V = volume of core (m³)

f = frequency of the magnetic cycles/second.

⇒ **Eddy current loss:**

depends upon the range of variation of flux density and the frequency of the variation of the flux, also upon the thickness of the iron laminations with reasonalable accuracy $P_e = k_e (B_{max})^2 f^2 xt^2$, where

k_e = constant (depending upon the electrical resistance of the magnetic material)

t = thickness of the magnetic material larninations

These losses are about 10% to 20% of full load losses.

3. Mechanical losses:

These losses consists of power losses due to friction of the bearings, air friction and friction between brushes and commutators or slip rings.

These losses are constant in a machine operating at approximately constant speed are independent of load.

⇒ **Friction loss:** Caused due to the friction of the bearing.

⇒ **Winding loss:** Air friction loss, as it is called caused by the noise of moving parts through the surrounding medium.

These losses are about to 10 to 20% of full load losses.

1.69 POWER FLOW IN DC MACHINES

The power flow and energy transformation diagrams at various stages, which take place in DC machines are represented diagrammatically in Figs 1.136 and 1.137.

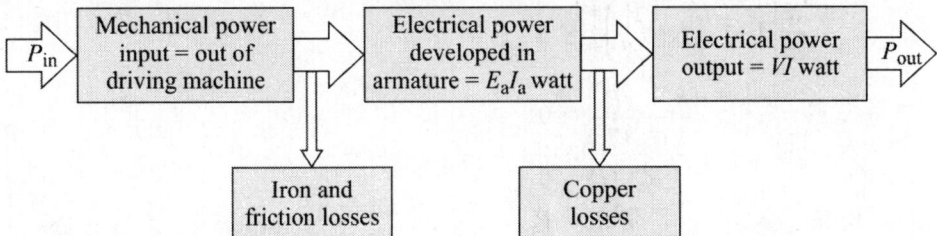

Fig. 1.136: Structure of the operational procedure of a generator

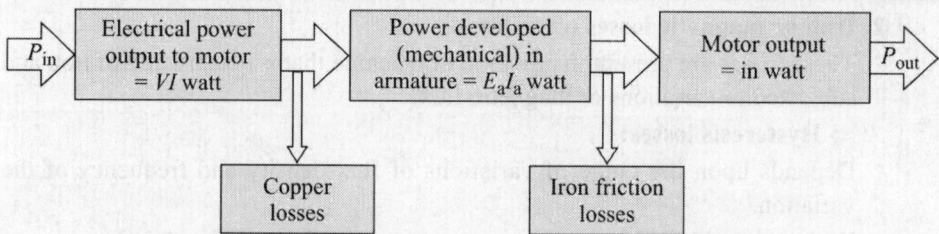

Fig. 1.137: Schematic of the operational procedure motor

1.70 EFFICIENCY OF DC MACHINES

For DC machine, its overall efficiency is given by,

$$\%\eta = \frac{\text{Total output}}{\text{Total input}} \times 100$$

Then

$$P_{in} = P_{out} + P_{Cu} + P_i$$

$$\%\eta = \frac{P_{out}}{P_{out} + P_{Cu} + P_i} \times 100$$

where P_{out} = total output of machine

P_{in} = total input of machine

P_{Cu} = variable losses

P_i = constant losses

1.70.1 Condition for Maximum Efficiency

In case of a DC generator, the output is given by

$$P_{out} = VI$$

$$P_{Cu} = \text{Variable losses} = I_a^2 R_a = I^2 R_a$$

$$I_a = I, \text{ neglecting shunt field current}$$

$$\%\eta = \frac{V \cdot I}{V \cdot I + I^2 R_a + P_i} \times 100 = \frac{1}{1 + \left(\dfrac{IR_a}{V} + \dfrac{P_i}{V \cdot I}\right)} \times 100$$

The efficiency is maximum, when the denominator is minimum. According to maxima-minima theorem.

$$\frac{d}{dI}\left[1 + \left(\frac{IR_a}{V} + \frac{P_i}{VI}\right)\right] = 0$$

$$\therefore \qquad \frac{R_a}{V} - \frac{P_i}{VI^2} = 0$$

$$\therefore \qquad I^2 R_a - P_i = 0$$

$$I^2 R_a = P_i$$

$$P_{Cu} = P_i$$

Thus for the maximum efficiency, the condition is

variable losses = constant losses

Current at maximum efficiency:

• From the condition of maximum efficiency, the current through the DC machine at the time of maximum efficiency can be obtained.

• **For shunt machines:** The I_{sh} is constant and the loss VI_{sh} is treated to be the part of constant losses. The variable losses are $I_a^2 R_a$.

• At maximum efficiency,

$$I_a^2 R_a = P_i \text{ (stray + shunt field losses)}$$

$$I_a = \sqrt{\frac{P_i}{R_a}} = \sqrt{\frac{\text{constant losses}}{\text{armature resistance}}}$$

• This is the armature current at maximum efficiency. Neglecting I_{sh}, $I_a = I_L$ is the line current of the machine.

For series machines: The current through series field is same as armature current which is same as line current.

Hence the constant losses are only mechanical losses while the variable losses are the copper losses in armature as well as series field winding due to the armature current.

At maximum efficiency,

$$I_a^2 (R_a + R_{se}) = P_i - \text{mechanical losses}$$

$$\therefore \qquad I_a = \sqrt{\frac{P_i}{R_a + R_{se}}}$$

Let us study now the various method of testing the DC motors from the losses and efficiency point of view (Fig. 1.138).

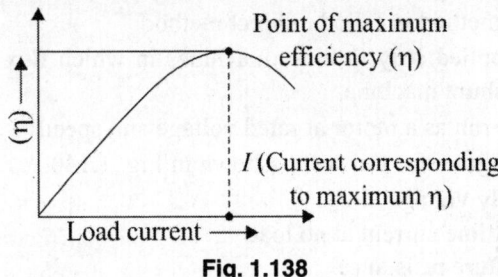

Fig. 1.138

1.71 TESTING OF DC MACHINES

DC machines can be tested by three different methods:

(a) Direct method

(b) Indirect method

(c) Regenerative method.

1.71.1 Direct Method

This method of testing is suitable for small DC machines.

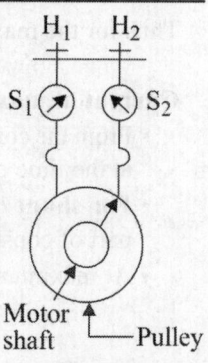

In direct method, DC machine is subjected to rated load and entire output power is wasted. Test is performed to determine efficiency.

For DC motor, brake test is carried out. A belt around the pully has its ends attached to spring balances S_1 and S_2 as shown in Fig. 1.139.

Hand wheel H_1 and H_2 are used to adjust the load on the pulley tightly by the belt.

Fig. 1.139

Motor output $= w(S_1 - S_2)r \times 9.81$ W

Here S_1 and S_2 are tension of both sides, r is effective radius of brake pulley, N is speed in rpm and $w = \dfrac{2\pi N}{60}$

$$\eta = \frac{\text{output}}{\text{input}} \times 100 = \frac{\text{output}}{VI_L} \times 100$$

1.71.2 Indirect Method

In this method, no load machine losses are first measured by suitable test and taken additional losses on load are determined by machine data to calculate machine efficiency.

The simplest method of measuring no load machine losses is by swinburne's method.

Swinburne's method:
- In this method, the no load losses of the machine are determined.
- Then the efficiency of machine is calculated with the help of machine rated data. So, this method is called indirect method.
- This test is applied only to those machines in which flux remain practically constant, i.e. shunt machine.
- The machines run as a motor at rated voltage and speed.
- The circuit diagram for this test is shown in Fig. 1.140.

   Let $V =$ supply voltage

   $I_0 =$ input line current at no load

   $R_a =$ armature resistance

   $I_{sh} =$ shunt field current

   $\therefore$ No-load armature current $= I_a = I_0 - I_{sh}$
- No load power input = iron loss + friction and winding loss + field Cu loss + armature Cu loss.
- $VI_a =$ constant losses + armature Cu loss at no-load
- Constant losses $P_i = VI_0 - I_a^2 R_a$.

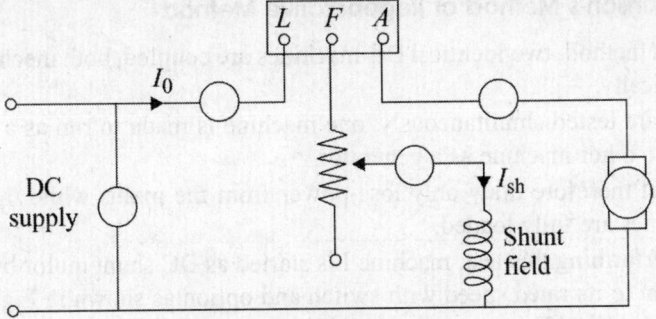

Fig. 1.140: Swinburne's test

- If the constant losses of the machine are known, then the efficiency of the machine can be determine as follows.

Motor efficiency:
- Motor armature current $I_a = (I - I_{sh})$

$$\text{Armature Cu losses} = (I - I_{sh})^2 R_a$$

$$\text{Motor input} = VI$$

$$\text{Constant losses} = P_i$$

$$\eta_m = \frac{\text{Input} - \text{losses}}{\text{Input}} = \frac{VI - (I - I_{sh})^2 R_a - P_C}{VI} \times 100$$

Generator efficiency: Armature current $I_a = I + I_{sh}$

$$\text{Generator output} = VI$$

$$\text{Armature Cu loss} = (I + I_{sh})^2 \cdot R_a$$

$$v_g = \frac{\text{output}}{\text{output} + \text{losses}} = \frac{VI}{VI + (I + I_{sh})^2 \cdot R_a + P_C} \times 100$$

Advantages
- It is a convenient and economical method of testing DC machines since power required to test a large machine is small.
- The efficiency can be predetermined at any load because constant losses are known.

Disadvantages
- No account is taken of the change in the iron loss caused due to change from no load to full-load condition.
- As the test is carried out at no-load, it does not indicate the commutation status at full-load.
- This test is applicable to machines in which flux is almost constant
- Series machines can not be tested by this method as they can not be run on light load and even the speed varies greatly.

1.71.3 Hopkinson's Method or Regenerative Method

- In this method, two identical DC machines are coupled, both mechanically and electrically.
- They are tested simultaneously, one machine is made to run as a motor drive and the other machine as a generator.
- The set therefore draw only loss-power from the mains while the individual machines are fully loaded.
- For performing this test, machine I is started as DC shunt motor by starter and brought to its rated speed with switch and option as shown in Fig. 1.141 both machines I and II run at same speed because they are mechanically coupled.

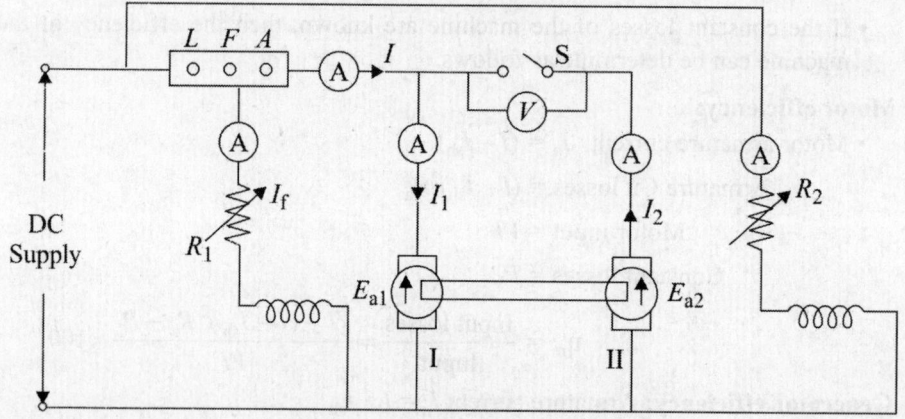

Fig. 1.141: Hopkinson's test

- The field current I_{f2} of second machine is now adjusted.
- When voltmeter across switch reads zero, switch S is closed.
- The magnitude of armature currents I_1 and I_2 can be adjusted to any value by varying I_{f1} and I_{f2}.
- The machine I running as motor drives machine II running as generator.
- Also since both machines are coupled electrically, power output of machine II is fed to the machine I.
- It is for this reason this method is also known as regenerative method.
- I_1, I_2 = armature currents of motor and generator respectively, V = terminal voltage of both machines = DC supply voltage
- Input to motor armature = VI_1
  If motor efficiency is = η_m
- Then, motor output = $\eta_m \cdot V \cdot I$
- Generator input = $\dfrac{VI_2}{n_g}$ or $\eta_m \cdot \eta_g = \dfrac{I_2}{I_1}$ if we assume $\eta_m = \eta_{g'}$

- $\eta = \sqrt{\dfrac{I_2}{I_1}} = \sqrt{\dfrac{\text{Generator armature current}}{\text{Motor armature current}}}$

- Actually the efficiencies of two machines are not equal because of following. two reasons.

- Motor armature current ($I_1 = I + I_2$) is more than generator armature current I_2. Thus armature circuit loss of motor is more than generator.

- Generator field current is greater than motor field current since both machines are running at same speed, generator iron losses are more than motor iron losses.

- So for calculating efficiencies, above two reason, can be taken into consideration.

- If R_a = resistance of armature circuit. Then $I_2^2 R_a$ = armature circuit loss is generator and $I_1^2 R_a$ = armature circuit loss in motor.

- Power drawn from supply = VI

- No load losses in two machines, $W_0 = VI - (I_1^2 + I_2^2)R_a$.
  $W_0 = VI - I_1 R_{am} - I_2^2 R_{ag}$ [if two resistances are not equal]

- No load loss of each machines = $\dfrac{W_0}{2}$

- Generator output = VI_2

$$n_g = \frac{W_0}{VI_2 + \dfrac{W_0}{2} + VI_{f2} + I_2^2 R_a}$$

[where VI_{f2} is field circuit loss]

- Total motor losses $W_m = \dfrac{W_0}{2} + VI_{f1} + I_1^2 R_a$

  motor Input = $V(I_1 + I_{f1})$

$$\eta_m = \frac{V(I_1 + I_{f1}) + \dfrac{W_0}{2} + VI_{f1} + I_1^2 R_a}{V(I_1 + I_{f1})}$$

Advantages

- The total power taken from the supply is very low. Therefore this method is very economical.

- The temperature rise and the commutation conditions can be checked under rated load conditions.

- Stray losses are considered, as both the machines are operated under rated load conditions.

- Large machines can be tested at rated load without consuming much power from the supply.

- Efficiency at different loads can be determined.

Disadvantages:

- Requirement of two indentical DC machines.
- There is no way of separating the iron-losses of the two machines which are different because of different excitations.
- Both machines are not loaded equally and this is crucial in smaller machines.
- In case of small machines, the full load speed is usually higher than the rated speed and the speed varies with load.

SOLVED UNIVERSITY PROBLEMS FOR PRACTICE

Example 1.40: A 37.3 kW, 460 V DC shunt motor takes a current of 4 A at no load and runs at 660 rpm, resistance of armature and shunt field is 0.3 Ω and 270 Ω respectively. Find the following when the motor is running at full load.

(*i*) Input current (*ii*) speed (*iii*) armature current at which the efficiency is maximum.

(RTU 2005)

Solution:

$$\text{Total no load losses} = 460 \times 4 = 1840 \text{ watts}$$

$$I_{sh} = \frac{460}{270} = 1.70A$$

or

$$I_{a0} = I - I_{sh}$$

$$= 4 - 1.70 = 2.30 \text{ A}$$

$$\text{Armature cu loss} = I_{a0}^2 R_a = (2.30)^2 \times R_a$$

$$= (2.30)^2 \times 0.3$$

$$= 1.587 \text{ watts}$$

$$\text{Field Cu loss} = (1.70)^2 \, 2 \, 270 = 0 \, 780.3 \text{ watts}$$

$$\text{Iron and friction loss} = 1840 - 780.3 - 1.587$$

$$= 1058.11 \text{ watts}$$

Iron and friction loss will assumed constant.

When delivering full load armature current.

- Armature input = $(460 \times I_a)$ watts

  Full load output = 37300 watts

- The losses in the armature are

(*i*) Armature Cu loss = $(I_a^2 \times 0.3)$ W

(*ii*) Iron friction losses = 1058.11 W

$\therefore$ $\qquad\qquad\qquad 460 \, I_a = 37300 + 1058.11 + I_a^2 \times 0.3$

or $\quad I_a^2 - 1533.33 I_a + 127860.37 = 0$

on solving $\qquad\qquad\qquad I_a = 88.5 \text{ A}$

$$\text{Input current } I_L = I_a + I_{sh} = 88.5 + 1.70 = 90.2 \text{ A}$$

Motor input = 460 × 90.2 = 41492 W

$$\text{Full load effiency} = \frac{\text{FL output}}{\text{FL input}} \times 100 = \frac{37300}{41492} \times 100 = 89.89\%$$

(*iii*) Now efficiency of the motor is maximum when armature loss is equal to constant loss, i.e.

$$I_a^2 R_a = 1840 - 1.587 = 1838.413 \text{ W}$$

or

$$I_a = \sqrt{\frac{1838.413}{0.3}} = 78.28 \text{ A}$$

Armature input = $460 I_a$ = 460 × 7828 = 36009.6 W

Armature Cu loss = $I_a^2 R_a$ = (78.28)² × 03 = 1838.33 W

Friction and windage = 1058.11 W

Armature output = 36009.6 − 1838.33 − 1058.11 = 33123.16 W

Back emf = 460 − 230 × 0.30 = 459.31 V

No. of turns = 660 rpm

Back emf at given load $E_b = V - I_a R_a$

= 460 − 88.5 × 0.30 = 433.45

Now

$$\frac{N}{660} = \frac{433.45}{439.31} \text{ or } N = 623 \text{ rpm}$$

Example 1.41: Hopkinson's test on two identical shunt machines give the following readings. [RU 2005]

Supply voltage = 240, field currents = 6A and 3A

line current = 40 A, armature current of molor = 2.40 A

Armature resistance of each machine = 0.014 A

Voltage drop/ brush = 1 V

Find the efficiency of each machines.

Solution: The connections are shown in Fig. 1.142.

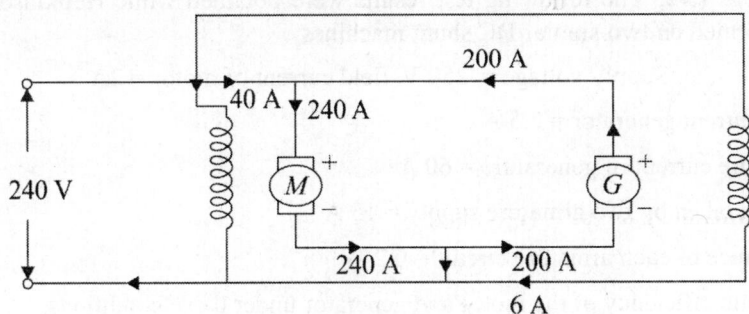

Fig. 1.142

Motor armature Cu loss = (240)² × 0.014 = 806.4 W

Generator armature Cu loss = (200)² × 0.014 = 560 W

Generator field Cu loss $= 240 \times 6 = 1440$ W

Power drawn from the supply $= 240 \times 40 = 9600$ W

Iron, friction and windage loss on for the two machines

$$= 9600 - (806.4 + 560) = 8233 \text{ W}$$

Iron, friction windage losses per machine

$$= \frac{8233.6}{2} = 4116.8 \text{ W}$$

Motor, friction and efficiency:

Motor armature Cu loss $= 806.4$ W

Motor field Cu losses $= 240 \times 5 = 1200$ W

Iron friction and windage loss $= 4116.8$ W

Total motor losses $= 806.4 + 1200 + 4116.8 = 6123.2$ W

Motor input $= 240 \times 240 + 240 \times 5 = 58800$ W

$$\text{Motor efficiency} = \frac{58800 - 6123.2}{58800} = 0.895 \text{ or } 89.5\%$$

Generator losses and efficiency:

Generator armature Cu loss $= 560$ W

Generator field Cu loss $= 240 \times 6 = 1440$ W

Iron friction and windage loss $= 4116.8$ W

Total losses $= 560 + 1440 + 4116.8 = 6116.8$ W

Generator output $= 240 \times 200 = 48000$ W

$$\text{Generature efficiency} = \frac{48000}{48000 + 6116.8} = 0.887 = 88.7\%$$

Example 1.42: The following test results were obtained while Hopkinson's test was performed on two similar DC shunt machines.

Supply voltage $= 250$ V, field current of motor $= 2$A

Field current generator $= 2.5$ A

Armature current of generature $= 60$ A

Current taken by two armature supply $= 15$ A

Resistance of each armature circuit $= 0.2$ W

Calculate efficiency of the motor and generator under these conditions.

[RU 2004]

Solution:

$$I_{am} = I_{ag} + I_a = 60 + 15 = \pm 75 \text{ A}$$

Now

$$= \frac{1}{2}\left[VI_a - I_{am}^2 \cdot R_{am} - Ia_g^2 \cdot R_{ag}\right]$$

$$= \frac{1}{2}\left[250 \times 15 - (75)^2 \times (0.2) - (60)^2 \times (0.2)\right]$$

$$= 952.5 \text{ W}$$

Input Power of motor, $P_{inm} = VI_{am} = 250 \times 75 = 1850$ W

Power accros the load of motor,

Now

$$P_{LM} = (P_{stray} + VI_{fm}) + I_{am}^2 \times R_{am}$$
$$= (952.5 + 250 \times 2) + (75)^2 \times (0.2) = 2577.5 \text{ W}$$

The motor efficiency

$$\eta_m = \frac{P_{inm} - P_{LM}}{P_{inm}} = \frac{18750 - 2577.5}{18750}$$
$$= 86.25\%$$

or

$$P_{Lg} = (P_{stray} + VI_{fg}) + Ia^2g \times P_{ag}$$

or

$$= (952.5 + 250 \times 2.5) + (60)^2 \times (0.2)\ 2297.5\text{W}$$

$$P_{outg} = V_{vag} = 250 \times 60 = 15000 \text{ W}$$

$$= \frac{P_{out}}{P_{out} + P_{Lg}} = \frac{15000}{15000 + 2297.5} = 86.7\%$$

Example 1.43: The No load test of a 44.76 kW, 220 V, DC shunt motor gave the following figures:

Input current = 13.25 A, field current = 2.55 A, resistance of armature at 75°C = 0.032 Ω and brush drop = 2 V.

Estimate the full load current and efficiency.

Solution:

$$\text{No load input} = 220 \times 133 = 2915 \text{ W}$$
$$\text{Armature current} = 13.25 - 2.25 = 10.7 \text{ A}$$
$$\text{Armature cu loss} = (10.7)^2 \times 0.03 = 3.6 \text{ W}$$
$$\text{Loss due to brush drop} = 2 \times 10.7 = 21.4 \text{ W}$$
$$\text{Variable loss} = 21.4 + 3.6 = 25$$
$$\text{Constant losses } P_i = 29.5 - 25$$
$$= 2890 \text{ W}$$

Full load condition: If I_a is the full load armature current, then full load motor input current is $(I_a + 2.55)$A.

Full load motor power input = $220 (I_a + 2.55)$

This input must be equal to the sum of

(*i*) Output = 44.76 kw

(*ii*) P_i = 2890 W

(*iii*) Brush loss = $2I_a$ watts

(*iv*) Armature Cu loss $= 0.032 I_a^2$

$$220 (I_a + 2.25) = 44.750 + 2890 + 2I_a + 0.032 I_a^2$$

or $\quad 0.03 I_a^2 - 218 I_a + 471090 = 0$

or

$$I_a = \frac{218 \pm \sqrt{218^2 - 4 \times 0.032 \times 47.090}}{2 \times 0.032} = 223.5 \text{ Amp}$$

Line input current $I = I_a + I_{sh} = 223.5 + 2.25 = 226 \text{ A}$

Full load power input $= 226 \times 220 = 49.720 \text{W}$

FL efficiency $= 44760/49720 = 0.90$ or 90%

Example 1.44: A 2000 V, shunt motor develops an output of 17.158 kw when taking 20.2 kW. The field resistance is 50Ω and armature resistance 0.06Ω. What is the efficiency and power input when the output is 7.46 kW.

Solution:

Case I:
Output $= 17158$ W; Input $= 20200$ W

Total losses $= 20200 - 17158 = 3042$ W

Input current $= 20200/200 = 101$ A

$I_{sh} = 200/50 = 4$ A; $I_a = 101 - 4 = 97$ A

Armature Cu loss $= (97)^2 \times 0.06 = 564.5$ W

Constant losses $= 3042 - 564.5 = 2477.5 = 2478$ W

Case II:

Let $\qquad I_a = $ Armature current; input current $= (I_a + 4)$ A

Now, $\qquad$ input power $=$ output $+ I_a^2 R_a +$ constant losses

$$200 (I_a + 4) = 7460 + 0.06 I_a^2 + 2478$$

or $\quad 0.06 I_a^2 - 200 I_a + 9138 = 0$

$$I_a = \frac{200 \pm \sqrt{(200)^2 - 4 \times 0.06 \times 9138}}{2 \times 0.06} = \frac{200 \pm 194}{0.12}$$

$$= 3283.3 \text{ A or } 46 \text{ A}$$

We will reject the larger value because it corresponds to unstable operation of the motor. Hence take $I_a = 46$ A.

Input current $I = I_a + I_{sh} = 46 + 4 = 50$ A

$$\text{power input} = \frac{50 \times 200}{1000} = 10 \text{ kW}$$

$$\eta = 10000$$

Example 1.45: A 500 V shunt motor takes 8 A on no load. The armature and field resistances are 0.2 ohm and 250 Ω respectively, when measured at room temperature neglect the increase in resistance due to temperature rise, find the efficiency of the machine.

(*i*) When run as a motor taking a line of 80 A at 500.

(*ii*) When run as a generator delivering a current of 90 A at 5. Assume the stray load losses to be 1.2% of the O/P.

Solution: Supply voltage $V = 500$

No load current $I_a = 8$ A

Armature resistance $R_a = 0.2\ \Omega$

Shunt field current, $R_{sh} = 250\ \Omega$

$$\text{Shunt field current } I_{sh} = \frac{V}{R_{sh}} = \frac{500}{250} = 2\,\text{Amp}$$

Shunt field losses at no-load $= VI_{sh} = 500 \times 2 = 1000$ W

Armature current at no-load $I_{a0} = I_0 - I_{sh} = 8 - 2 = 6$ A

Armature copper loss at no-load $= I_{a0}^2 R_a = 6° \times 0.2 = 7.2$ kW

Input at no-load $= VI_0 = 500 \times 8 = 4000\,\text{W}$

Machanical and iron losses = Input at no load – shunt field losses
– armature copper loss at no-load
$$= 4000 - 1000 - 7.2 = 2992.9 \text{ W}$$

Efficiency when working as motor

Input $= VI = 500 \times 80 = 40000$ W

Shunt field loss $= VI_{sh} = 500 \times 2 = 1000$ W

Armature current $= I_a = I - I_{sh} = 80 - 2 = 78$ Amp

Armature copper loss $= I_a^2 R_a = (78)^2 \times 0.2 = 1216.8$ W

Total losses = Shunt field losses + armature copper loss
+ mechanical and iron loss
$$= 1000 + 1216.8 + 2992.8 = 5209.6 \text{ W}$$

Output = Input – losses

Output $= 40000 - 5209.6 = 34790.4$ W

$$\text{Stray load loss} = \frac{1.2}{100} \times 34790.4 = 417.5 \text{ W}$$

Net output $= 34790 - 417.5 = 34373$ W

$$\text{Efficiency} = \frac{\text{Net output}}{\text{Input}} = \frac{34373}{40000} = 0.859$$

$$\eta = 85.9\%$$

Example 1.46: In an 8-pole DC machine, the flux per pole is 0.1 Wb, 400 conductor and speed 300 rpm. Calculate the generated emf when the armature is:

(*a*) lap connected

(*b*) wave connected.   **[UPTU 2005-06]**

Solution: Given, $P = 8$, $\phi = 0.1$ Wb, $Z = 400$, $N = 300$ rpm

(a)
$$E_g = \frac{\phi ZNP}{60A} \text{ for lap connected armature } A = P = 8$$

$\therefore$
$$= \frac{0.1 \times 400 \times 300 \times 8}{60 \times 8} = 200 \text{ V}$$

(b)
$$E_g = \frac{\phi ZNP}{60A} \text{ for wave connected } A = P = 2$$

$\therefore$
$$= \frac{0.1 \times 400 \times 300 \times 8}{60 \times 2} = 800 \text{ V Ans.}$$

Example 1.47: A DC shunt motor runs at 600 rpm taking 60 A from a 230 V supply armature resistance is 0.2 Ω and field resistance is 115 Ω. Find the speed when the current through the armature is 30 A. **[UPTU 2006-07, 2008]**

Solution: Given, $N_1 = 600$ rpm, $I_{L_1} = 60$ A, 230 V, $R_{sh} = 115\,\Omega$, $R_a = 0.2\,\Omega$, $N_1 = 600$ rpm, $I_a = 30$ A.

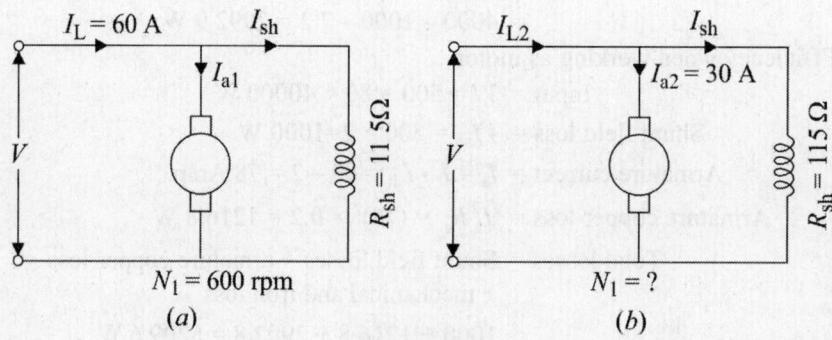

$N_1 = 600$ rpm $N_1 = ?$

(a)   (b)

Fig. 1.143

$$I_{sh} = \frac{V}{R_{sh}} = \frac{230}{115} = 2 \text{ A}$$

$$I_{a1} = I_{L1} - I_{sh}$$

$$f_{a1} = 60 - 2 = 58 \text{ A}$$

$$E_{b1} = V - I_{a1}R_a = 230 - 58 \times 0.2$$

$$E_{b1} = 21814 \text{ V}$$

$$E_{b2} = V - I_{a2}R_a$$

$$= 230 - 30 \times 0.2 = 224 \text{ V}$$

$$E_{b2} = 224 \text{ V}$$

Since $\phi = $ Constant

- We know that $\dfrac{N_2}{N_1} = \dfrac{E_{b2}}{E_{b1}}$

$$N_2 = \dfrac{E_{b2}}{E_{b1}} \times N_1 = \dfrac{224}{218.4} \times 600 = 615.38 \text{ rpm } \textbf{Ans.}$$

Example 1.48: A 6 pole lap wound shunt motor has 500 conductor in the armature. The resistance of armature path is 0.05 Ω. The resistance of shunt field is 25 Ω. Find the speed of the motor when it takes 120 A from DC mains of 100 V supply. Flux per pole is 2×10^{-2} Wb. **[UPTU 2007-08]**

Solution: Given, $P = A = 6$, $Z = 500$, $R_a = 0.05 \,\Omega$, $R_{sh} = 25\,\Omega$, $V = 100$ V, $I_L = 120$ A, $\phi = 2 \times 10^{-2}$ Wb.

- We know that

$$I_{sh} = \dfrac{V}{R_{sh}} = \dfrac{100}{25} = 4 \text{ A}$$

$$I_a = I_L - I_{sh} = 120 - 4 = 116 \text{ A}$$

$$E_b = V - I_a R_a = 100 - 116 \times 0.05 = 94.2 \text{ V}$$

$$N = \dfrac{E_b}{\phi} \times \left(\dfrac{60A}{2P}\right) = \dfrac{94.2 \times 60 \times 6}{2 \times 10^{-2} \times 6 \times 500}$$

$$= 565 \text{ rpm } \textbf{Ans.}$$

Example 1.49: A 25 kW, 250 V DC shunt generator has armature and field resistance of 0.06 Ω and 100 Ω respectively. Determine the total armature power developed when working:

(i) As a generator delivering 25 kW output and

(ii) As a motor taking 25 kW input. **[UPTU 2009-10]**

Solution:

(i) As generator $\quad P_L = 25 \times 10^3$ W

$$I_L = \dfrac{25 \times 10^3}{250} = 100 \text{ A}$$

$$I_{sh} = \dfrac{V}{R_{sh}} = \dfrac{250}{100} = 2.5 \text{ A}$$

$\therefore \qquad I_a = I_L + I_{sh} = 100 + 2.5 = 102.5 \text{ A}$

$$E_g = 250 + 102.5 \times 0.56 = 256.15 \text{ V}$$

$$P_g = E_g \times I_a$$

$$P_g = 256.15 \times 102.5 = 26.255 \text{ kW } \textbf{Ans.}$$

(ii) As motor $\qquad P_{in} = 25 \times 10^3$

$$I_L = \dfrac{25 \times 10^3}{250} = 100 \text{ A}$$

$$I_{sh} = \frac{V}{R_{sh}} = \frac{250}{100} = 2.5$$

$$I_a = I_L - I_{sh} = 100 - 2.5 = 97.5 \text{ A}$$

$$\therefore \quad E_b = V - I_a R_a$$

$$= 250 - 97.5 \times 0.06 = 244.15 \text{ V}$$

$$\therefore \quad P_m = \text{power developed} = E_b \times I_a = 244.15 \times 97.5$$

$$= 23.846 \text{ kW } \textbf{Ans.}$$

Example 1.50: The armature of a 4-pole DC machine has 100 turns and runs at 600 rpm. The emf generated in open circuit is 220 V. Find the useful flux per pole. When armature is (a) lap connected (b) wave connected. **[UPTU 2010-11]**

Solution: Given $\quad P = 4$, turns $= 100$, $N = 600$ rpm

$$E_g = 220 \text{ V}$$

Now, as we know that two conductors constitute one turn

$$Z = \text{Total conductors} = 2 \times 100$$

$$\therefore \quad = 200$$

(a) for lap connected $\quad A = P = 4$

$$E_g = \frac{\phi Z N P}{60 A}$$

$$\phi = \frac{220 \times 60 \times 4}{600 \times 4 \times 200} = 0.11 \text{Wb } \textbf{Ans.}$$

(b) for wave connected winding

$$A = 2$$

$$\phi = \frac{220 \times 60 \times 2}{600 \times 4 \times 200} = 0.055 \text{ Wb } \textbf{Ans.}$$

Example 1.51: A 6 pole lap wound DC generator has 720 conductor, a flux of 80 m Wb/pole is driven at 1000 rpm. Find the generated emf. **[GBTU 2010-11]**

Solution: Given $\quad P = 6$

$$A = P = 6 \text{ for lap}$$

$$Z = 720$$

$$\phi = 80 \times 10^{-3} \text{ Wb}$$

$$N = 1000 \text{ rpm}$$

$$E_g = \frac{\phi Z N P}{60 \text{ A}} = \frac{80 \times 10^{-3} \times 720 \times 1000 \times 6}{60 \times 6}$$

$$= 960 \text{ V } \textbf{Ans.}$$

Example 1.52: The armature of a four pole DC machine has 1000 turns and runs at 500 rpm. The emf generated in open circuit is 220 V. Find the useful flux per pole when armature is (a) lap connected (b) wave connected. **[UPTU 2011–12]**

Solution: Given $P = 4, N = 500$ rpm

$$E_g = 220 \text{ V, turns} = 1000$$

$\therefore$ $Z = 2 \times \text{turns} = 2 \times 1000 = 2000$

(a) When lap connected $A = P = 4$

$$\phi = \frac{220 \times 60 \times 4}{500 \times 4 \times 2000} = 0.012 \text{ Wb } \textbf{Ans.}$$

(b) Wave connected $A = 2$

$$\phi = \frac{220 \times 60 \times 2}{500 \times 4 \times 2000} = 0.006 \text{ Wb } \textbf{Ans.}$$

Example 1.53: A DC shunt generator delivers 50 kW at 250 V, and 400 rpm. The armature and field resistances are 0.02 Ω and 50 Ω respectively. Calculate the speed of the machine running as shunt-motor and taking 50 kW input at 250 V, allow brush drop of 1 V per brush. **[MTU 2012–13]**

Solution: The given that, $P_L = 50$ kW, $V = 250$ V, $N_g = 400$ rpm

$$R_a = 0.02 \text{ W}, R_{sh} = 50 \text{ }\Omega$$

For generator $\quad I_{sh} = \dfrac{V}{R_{sh}} = \dfrac{250}{50} = 5A$

$$I_L = \frac{P_L}{V} = \frac{50 \times 10^3}{250} = 200 \text{ A}$$

$$I_a = I_L + I_{sh} = 200 + 5 = 205 \text{ A}$$

$$E_g = V + I_a R_a + BDV$$

$$= 250 + 205 \times 0.02 + 2 \times 1$$

$$E_g = 256.1 \text{ V } \textbf{Ans.}$$

For motor $\quad I_L = 200 \text{ A}, V = 250$

$$I_{sh} = 5 \text{ A}$$

$$I_a = I_L - I_{sh} = 200 - 5 = 195 \text{ A}$$

$$E_b = V - I_a R_a - BDV$$

$$= 250 - 195 \times 0.02 - 2 \times 1 = 244.1 \text{ V}$$

$$\frac{N_m}{N_g} = \frac{E_b}{E_g} \Rightarrow N_m = \frac{E_b}{E_g} \times N_g$$

$$N_m = \frac{244.1}{256.1} \times 400 = 381.3 \text{ rpm } \textbf{Ans.}$$

Example 1.54: A DC shunt motor develops an open circuit emf of 250 V, at 1500 rpm. Find the developed torque for an armature current of 20 Amp.

[GBTU 2012–13]

Solution: Given $E_b = 250$ V, $N = 1500$ rpm, $I_a = 20$ A

$$T_e = \frac{9.55 \times E_b I_a}{N} = \frac{9.55 \times 250 \times 20}{1500}$$

$$= 31.822 \text{ N·m } \textbf{Ans.}$$

Example 1.55: The armature resistance of a 200 V DC shunt motor is 0.12 Ω. It runs at 600 rpm at constant torque load and draws a current of 21 A. Calculate its new speed if the field current is reduced to 10%. **[GBTU 2012-13]**

Solution: The given data, $V = 200$ V

$$I_{L1} = 21 \text{ A}$$

Assuming $\quad I_{a1} = I_{L1} = 21$ A

$$N_1 = 600 \text{ rpm}$$

For motor $\quad E_{b1} = V - I_{a1}R_a = 200 - 21 \times 0.12 = 197.48$ V

The given $\quad T_2 = T_1$

$$I_{a2}\phi_2 = I_{a1}\phi_1$$

$$I_{a2} = I_{a1} \times \frac{\phi_1}{\phi_2} = I_{a1} \times \frac{I_{sh1}}{I_{sh2}}$$

$$= 21 \times \frac{1}{0.9} = 23.33 \text{A} \quad \begin{bmatrix} \phi \propto I \\ I_{sh2} = 0.9 I_{sh1} \end{bmatrix}$$

$$E_{b2} = V - I_{a2}R_a$$

$$= 200 - 23.33 \times 0.12 = 197.2 \text{ V}$$

$$N_2 = \frac{E_{b2}}{E_{b1}} \times \frac{\phi_1}{\phi_2} \times N_1 = \frac{197.2}{197.4} \times \frac{1}{0.9} \times 600$$

$$= 665.7 \text{ rpm } \textbf{Ans.}$$

DESCRIPTIVE QUESTIONS

1. What is the principle of electro-mechanical energy conversion?
2. What is the principle of DC generator. Write down the emf equation of DC generator. **[UPTU 2009-10]**
3. What is back emf and its significance in D.C. motor? Write down the equation of back emf. **[UPTU 2008-09]**
4. What are the application of DC motor?
5. Explain the constructional features of the DC machine with neat sketch. **[UPTU 2003-04-05]**

EXERCISES

Short Answer Questions

1. Derive the emf equations for the DC generator. Explain with the help of neat diagram the different types of DC machine. **[UPTU 2004-05-06-08]**

2. Derive expression for the electromagnetic torque developed in a DC motor.
[UPTU 2008-09]

3. Sketch and explain the speed-current, speed-torque and torque-current characteristics of a shunt motor, series motor and compound motor.
[UPTU 2009-10, 2011-12]

4. A 10-pole DC machine having lap winding with 400 armature conductor. Calculate the emf generated when the machine is driven by a 1000 rpm and the flux per pole is 0.065 Wb.
[**Ans.** 4.33.34 V]

5. A 4-pole DC shunt generator with lap-connected armature has field and armature resistance of 80 Ω and 0.1 Ω respectively. It supplied power to 50 lamps rated for 100 volts, 60 W each. Calculate the total armature current and the generated emf by allowing a contact drop of 1 V per brush.
[**Ans.** 31.25 A, 105.125 V]   **[UPTU 2009-10]**

6. Describe methods to control the speed of DC motors.   **[2003–2004]**

7. Based on field winding armature winding connections and schematic diagrams, explain different types of DC machines.   **[2004–05]**

8. Define armature reaction and discuss its effect on the performance of DC machines.   **[2004–2005]**

9. Discuss the process of development of back emf in a DC shunt motor.
[2005–2006]

10. Explain the Ward-Leonard system for controlling the speed of DC motor.
[2005–2006]

11. Explain why a DC motor draws high current at starting? Also give construction details of a 3 point starter.   **[2006–2001, 2007–2008]**

12. Explain various losses involved in DC machine and discuss how they vary?
[2006–2007]

13. Explain the principle of operation of DC machine. Derive the expression for the back emf in a DC motor. Briefly explain the role it plays in starting and running of the motor.   **[2007–2008]**

Long Answer Questions

1. Derive the emf equation of a dc generator. What will be change in emf induced if the flux is reduced by 20% and the speed is increased by 20%?·
[UPTU 2005–06]

2. Derive the expression for torque in a dc motor. Draw the torque armature current, speed-armature current and torque speed characteristics of following dc motor.
   (a) DC shunt motor
   (b) DC series motor
   (c) DC compound motor

Single Phase Transformer

2.1 INTRODUCTION

The transformer is probably one of the most useful electrical device ever invented. It is a static device which is used to transfer electric power from one circuit to another without changing its frequency. The main function of transformer is to step-up or step-down, the voltage level within a circuit with corresponding decrease or increase in current at the same frequency. It works on the principle of Faraday's law of electromagnetic induction. Tranformer have no moving parts are rugged and durable in construction. The efficiency of transformer is very high and lie between 92% to 95%.

2.2 TRANSFORMER

- It is a static device (there is no rotating part) (Figs 2.1 and 2.2).
- It transfer the energy or power from one side to another side at same frequency.
- It has two winding; primary and secondary. There is no interconnection between primary and secondary.
- Its basic principle of operation based on mutual induction.

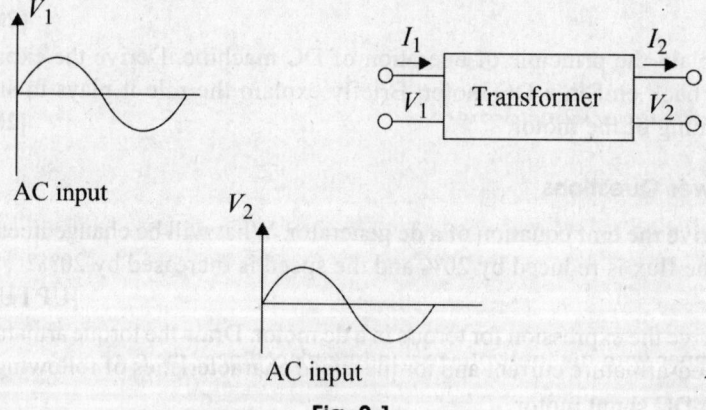

Fig. 2.1

Note: Why frequency is unchanged in transformer?
- Transformer is a static device. Frequency is related with time and time is related with speed, i.e. $N_s = \dfrac{120 f}{P}$ rpm.

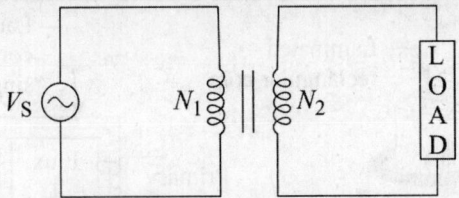

Fig. 2.2: Symbol of a transformer

2.3 CLASSIFICATION OF TRANSFORMER

  (a) On the basic of phase
     (*i*)   Single phase transformer
     (*ii*) Three phase transformer
  (b) On the basic of construction
     (*i*)   Shell-type
     (*ii*) Core-type
     (*iii*) Berry-type
  (c) On the basic of application
     (*i*)   Step-up transformer
     (*ii*) Step-down transformer

2.3.1 Types of Transformer on the Basic of Construction (Fig. 2.3)

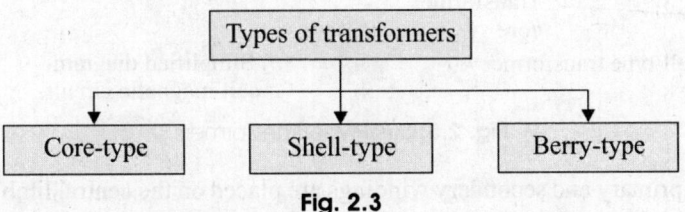

Fig. 2.3

2.3.1.1 Core-Type Transformer

- The construction of core-type transfoermer is shown in Fig. 2.4(*a*).
- The core of this transformer is in the form of a rectangular frame made from laminations. It provides a single magnetic circuit as shown in Fig. 2.4(*b*).
- The primary and secondary windings are uniformly distributed on two limbs of the core.
- Both the windings are of cylindrical shape and they are arranged in a concentric manner with the low voltage winding placed near the core.
- Here, the windings surrounded by a considerable portion of the core.

2.3.1.2 Shell-Type Transformer

Figure 2.5(*a*) shows the construction of a shell type transformer.

Figure 2.5(*b*) shows the construction of a shell-type transformer.

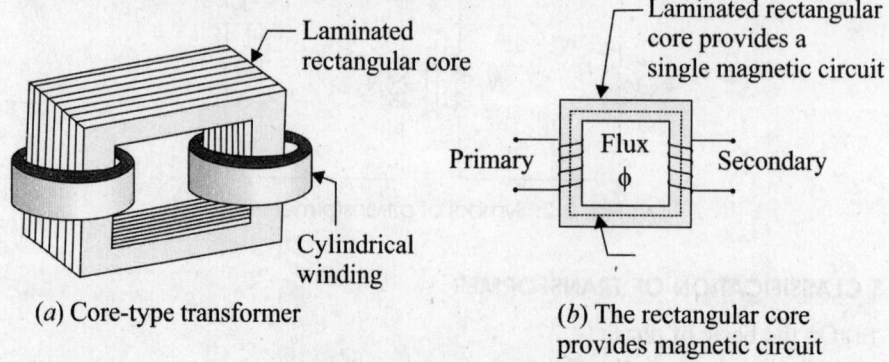

(a) Core-type transformer

(b) The rectangular core provides magnetic circuit

Fig. 2.4: Core-type transformer

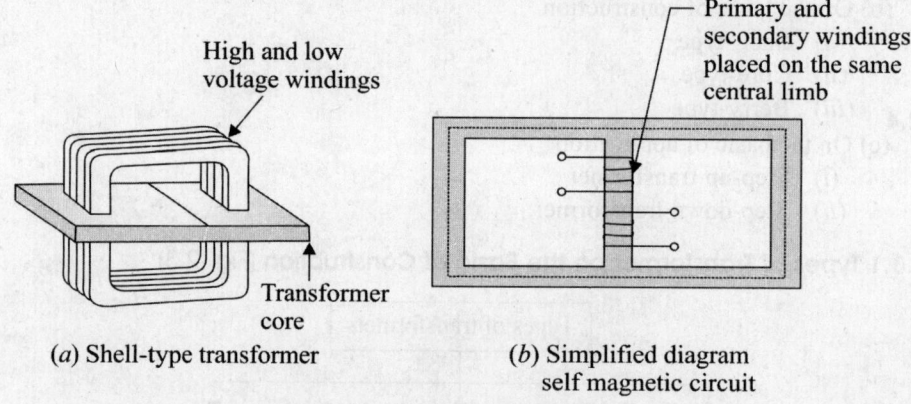

(a) Shell-type transformer

(b) Simplified diagram self magnetic circuit

Fig. 2.5: Shell-type transformers

The primary and secondary windings are placed on the central limb of the core. The high voltage and low voltage windings are of sandwich type, which are in the form of interleaved packages.

This type of core provides double magnetic circuit. This type of core provides a better mechanical support and protection for the windings.

In shell-type transformer, the core surround a considerable portion of the windings.

2.3.1.3 Berry-Type Transformers

The berry-type transformer is shown in Fig. 2.6.

The construction of the core is such that the yoke radiates out from the centre, similar to the spokes of a wheel.

Due to this construction, the berry-type transformer has a distributed magnetic circuit.

The high voltage and low voltage windings are placed as shown in Fig. 2.6 with the low voltage winding placed inside the high voltage winding placed outside.

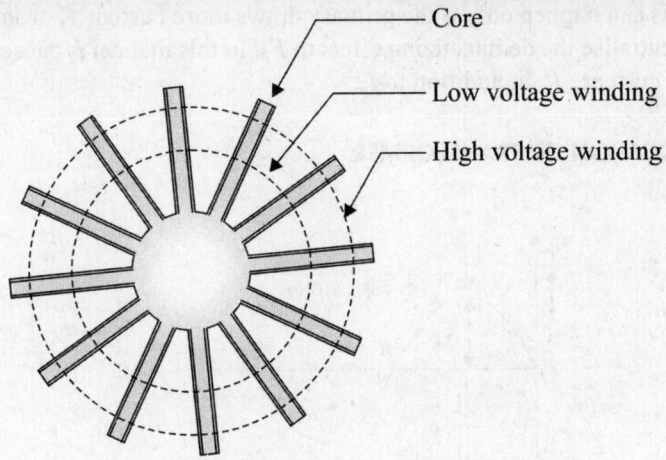

Fig. 2.6: Plane view of berry-type transformer

2.4 PRINCIPLE OF OPERATION OF TRANSFORMER

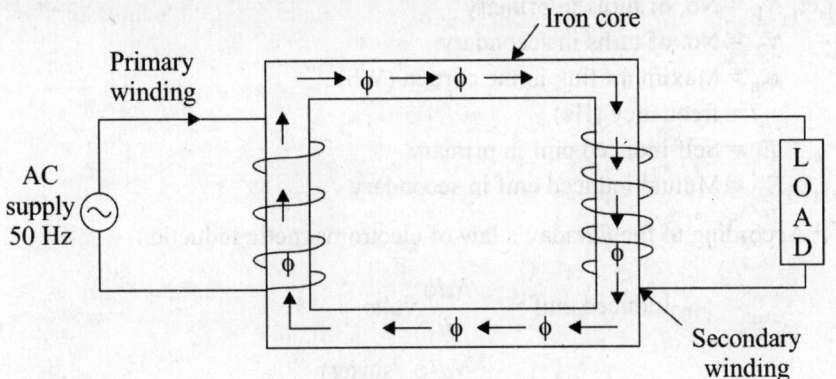

Fig. 2.7: A transformer

When alternating voltage is applied to the primary winding of a transformer, a current (termed as exciting current: I_e) flows through it. The exciting current produces an alternating flux ϕ in the core which links with both the windings. According to faraday's law of electromagnetic induction, the flux will cause self induced emf E_1, in the primary and mutually induced emf E_2 in the secondary winding. But according to Lenz's law primary induced emf will oppose the applied voltage and in magnitude this primary induced emf is equal to the applied voltage (Figs 2.7 and 2.8).

When a load is connected to the secondary side, a secondary load current (I_2) will circulate through the secondary winding. In this case the secondary mmf F_2 (I_2N_2) being opposite to f, tends to reduce E_1. For any ideal transformer, $V_1 = -E_1$. If the applied voltage V_1 is constant, E_1 and, therefore, mutual flux ϕ in the core must remain constant. This can happen only if the primary draws more current I_1' from the source, in order to neutralise the demagetizing effect to F_2. In this manner I_2 cause the primary to take more current, I_1' in addition to I_e.

2.5 EMF EQUATION OF TRANSFORMER

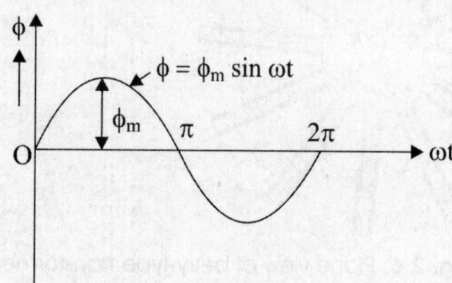

Fig. 2.8

Let, N_1 = No. of turns in primary

$\quad N_2$ = No. of turns in secondary

$\quad \phi_m$ = Maximum flux in the core in (Wb)

$\quad f$ = frequency (Hz)

$\quad E_1$ = Self induced emf in primary

$\quad E_2$ = Mutual induced emf in secondary

• According to the Faraday's law of electromagnetic induction.

$$\therefore \quad \text{Induced emf} = -\frac{N d\phi}{dt} \text{ volts}$$

$$e = -\frac{N d(\phi_m \sin \omega t)}{dt} \text{ volts}$$

$$= -N\phi_m \omega (\cos \omega t)$$

$$= N\phi_m \omega \sin\left(\omega t - \frac{\pi}{2}\right)$$

$$= E_{max} \sin\left(\omega t - \frac{\pi}{2}\right)$$

where, $\qquad E_{max} = N\phi_m 2\pi f$

$$\therefore \qquad E_{rms} = \frac{E_{max}}{\sqrt{2}} = \frac{N\phi_m 2\pi f}{\sqrt{2}}$$

$$= 4.44 \phi_m f N \text{ volts}$$

For primary side, $\quad E_1 = 4.44\phi_m f \cdot N_1$ volts

For secondary side, $\quad E_2 = 4.44\phi_m f \cdot N_2$ volts

2.6 TRANSFORMATION RATIO (K)

The transformation ratio for voltage is defined as the ratio of secondary voltage to the primary voltage of a transformer. It is denoted by

$$K = \frac{V_2}{V_1} = \frac{E_2}{E_1} = \frac{N_2}{N_1}$$

For ideal transformer, i.e. no losses

$\therefore \qquad$ Input power = output power

$$V_1 I_1 = V_2 I_2$$

$\therefore \qquad \dfrac{V_2}{V_1} = \dfrac{I_1}{I_2}$

$\therefore \qquad K = \dfrac{V_2}{V_1} = \dfrac{I_1}{I_2} = \dfrac{E_2}{E_1} = \dfrac{N_2}{N_1}$

Important note:

- $K > 1$ for step-up transformer.
- $K < 1$ for step-down transformer.

2.7 IMPEDANCE TRANSFORMATION

In Fig. 2.9, an impedance Z_2 is connected across the secondary winding at its output. The primary winding is connected to a voltage source V_1. The number of turns in the two windings are assumed to be N_1 and N_2. Induced emfs E_1 and E_2 are in phase opposition to V_1. Since V_2 is the secondary terminal voltage, it is also in the opposite phase of V_1.

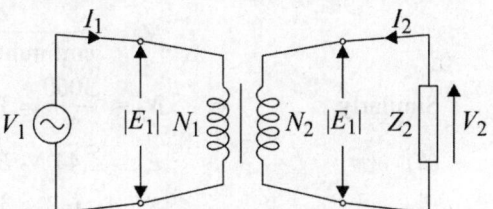

Fig. 2.9: Schematic of two-winding transformers

Assuming the transformer to be ideal

$$\frac{V_1}{V_2} = \frac{E_1}{E_2} = \frac{N_1}{N_2} = K \, (\text{turns ratio})$$

The impedance Z_2, as seen from the input side, can be obtained by voltage V_1 by I_1. Thus, we can write

$$Z_2' = \frac{V_1}{I_1} = \frac{V_1 \times (V_2 I_2)}{I_1 \times (V_2 I_2)}$$

$$= \left(\frac{V_1}{I_1}\right) \times \left(\frac{I_2}{I_1}\right) \times \left(\frac{V_2}{I_2}\right) = K \times K \times Z_2 = K^2 Z_2$$

i.e. $\qquad \dfrac{Z_2'}{Z_2} = K^2$

Therefore, impedance transformation ratio is equal to the square of turns ratio. Referring to the primary or secondary side, this transferred impedance is known as the equivalent impedance on that side. In an deal transformer thus we can note the following:

- Voltages are transformed in the direct ratio of turns ($V_1/V_2 = K$)
- Currents are transformed in the inverse ratio of turns ($I_1/I_2 = 1/K$)
- Volt–amperes of two sides are equal ($V_1 I_1 = V_2 I_2$)
- Impedances are transformed in proportion to the square of turns-ratio.

$$\left(Z_2' = K^2 Z_2; \ \ Z_1' = \frac{1}{K^2} \times Z_1 \right)$$

SOLVED EXAMPLES

Example 2.1: The maximum flux density in the core of a 250/3000 V, 50 Hz single phase transformer is 1.2 Wb/m². If the emf per tunr is 8 volt. Determine (*i*) primary and secondary turns (*ii*) area of the core.

Solution: Given, $E_1 = 250$ V, $E_2 = 3000$ V, $B_m = 1.2$ Wb/m².

(*i*) $\qquad\qquad\qquad E_1 = N_1 \times$ emf indcued/turns

$$N_1 = \frac{E_1}{\text{emf induced/turns}} = \frac{250}{8} = 32 \ \textbf{Ans.}$$

Similarly, $\qquad\qquad N_2 = \dfrac{3000}{8} = 375$ **Ans.**

(*ii*) $\qquad\qquad E_2 = 4.44 \, N_2 \, B_m \, A\phi$ [since $\phi_m = B_m \cdot A$]

$$A = \frac{3000}{4.44 \times 375 \times 50 \times 1.2}$$

$$A = 0.03 \text{ m}^2$$

Example 2.2: A single phase transformer has 400 primary and 1000 secondary turns. The net cross-sectional area of the core is 60 cm². If the primary winding be connected to a 50 Hz supply at 520 V. Calcualte (*i*) the peak value of flux density in the core (*ii*) voltage induced in the secondary winding.

Solution: Given, $N_1 = 400$, $N_2 = 1000$, $f = 50$ Hz, $E_1 = 520$ V $= V_1$.

(*i*) $\qquad\qquad\qquad E_1 = 4.44 \phi f \, N_1$

$$= 4.44 B_m A\phi N_1$$

$$520 = 4.44 \times B_m \times 60 \times 10^{-4} \times 50 \times 400$$

$\therefore \qquad\qquad\qquad B_m = 0.976$ Wb/m² **Ans.**

(ii)
$$\frac{E_2}{E_1} = K$$

∴
$$K = \frac{N_2}{N_1} = \frac{1000}{400} = 2.5$$

$$E_2 = K \cdot E_1 = 2.5 \times 520$$
$$= 1300 \text{ V Ans.}$$

2.8 IDEAL TRANSFORMER

An ideal transformer is an imaginary transformer which has the following properties:
- Its primary and secondary winding resistance are negligible.
- It has no losses thus efficiency is 100%.
- Leakage flux is zero, i.e. 100% flux produced by primary links with the secondary.

2.9 PRACTICAL TRANSFORMER

- In discussion, we considered the properties of an ideal transformer, certain assumptions were made which are not valid in a practical transformer.
- For example, in a practical transformer the windings have resistance, the core has finite permeability and there is a leakage flux.
- The efficiency of a practical transformer is not 100% due to the losses. Therefore in practical transformer we shall consider all these imperfections.

Winding Resistance

- An ideal transformer is supposed to posses no resistance, but in actual transformer there is always some resistance of primary and secondary windings as shown in the Fig. 2.10.
- The primary and secondary

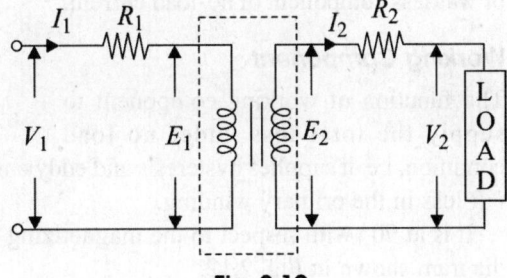

Fig. 2.10

winding resistance will causes a copper loss and voltage drop in the windings.

Leakage Reactance

- In an ideal transformer it is assumed that all the flux produced by the primary winding links both the primary and secondary winding.
- However, in an actual transformer not all of the flux remains within the magnetic core. A portion of this flux is diverted to the nonferromagnetic material surrounding the winding (generally air or oil).
- This is because the surrounding medium also has a definite permeability although it is very much less than that of the core. This small portion of the flux which travels through an external path is known as primary leakage flux.

2.10 PRACTICAL TRANSFORMER ON NO-LOAD

- A transformer is said to be at no load when the secondary winding is open circuited.
- The secondary current is thus zero. When an alternating voltage is applied to the primary, a small current I_0 is flows in the primary.
- The current I_0 is called the no-load current of the transformer. It is made up of two components as shown in the block diagram (Fig. 2.11).

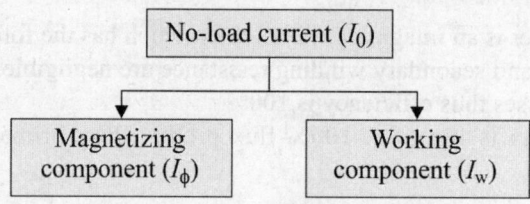

Fig. 2.11: Block diagram

Magnetizing Component

This is purely reactive component of no-load current I_0. It magnetizes the core and produces flux in the core therefore I_ϕ in phase, with flux ϕ as shown in Fig. 2.12.

The current I_ϕ is also called reactive or wattless component of no-load current.

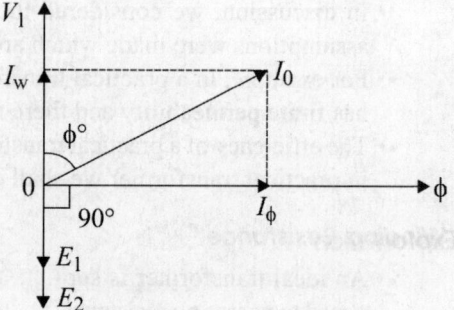

Fig. 2.12: Phasor diagram

Working Component

The function of working component to supply the total loss under no load condition, i.e. it supplies hysteresis and eddy current losses in the core and the negligible I^2R loss in the primary winding.

It is at 90° with respect to the magnetizing component (I_ϕ) as shown in the phasor diagram shown in Fig. 2.12.

It is also called wattless component or coreless component.

The total no load current I_0 is the vector addition of I_ϕ and I_w.

$$\therefore \qquad I_0 = I_\phi + I_w$$

where $I_\phi = I_0 \sin\phi_0$

$\quad\quad I_w = I_0 \cos\phi_0$

The magnitude of the no load current is given by

$$I_0 = \sqrt{I_\phi^2 + I_w^2}$$

while $\quad\quad\quad \phi_0 =$ no load primary power factor angle

The total power input on no-load is denoted by W_0

$$\therefore \qquad W_0 = V_1 I_0 \cos\phi_0$$

2.11 TRANSFORMER ON LOAD

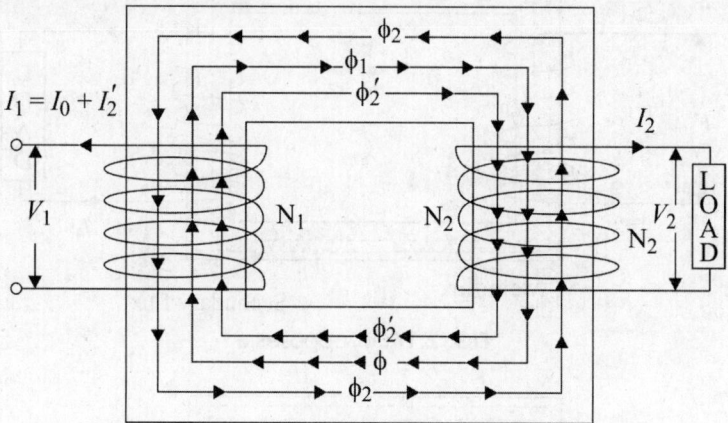

Fig. 2.13

- A transformer is said to be on load if its secondary side is connected to the load. When the secondary is loaded, the secondary current I_2 flows through it.
- The phase angle of secondary current I_2 with respect to V_2 depends upon the nature of load i.e., whether the load is resistive, inductive or capacitive.
- If load is resistive I_2 is in phase with V_2. If load is inductive, I_2 lags V_2 while for capacitive load, I_2 leads V_2.

Explanation

- The working of transformer on load can easily be explained with the help of diagram shown in Fig. 2.13.
- When the transformer is at no-load it draws no-load current I_0 from the supply mains. The no load current I_0 produces mmf $N_1 I_0$ which set up flux ϕ in the core.
- When the transformer is at load, current I_2 flows in the secondary winding. This secondary current I_2 produces an mmf $N_2 I_2$ which is set up flux in the core.
- According to Lenz's law, this flux ϕ_2 opposes the flux ϕ which is set up by the current I_0. Hence the mmf $N_2 I_2$ is called demagnetizing ampere turns. This is shown in Fig. 2.14.
- The flux ϕ_2 momentarily reduces the main flux ϕ due to which the primary induced emf E_1 also reduces. Hence the vector difference $V_1 - E_1$ increases due to which primary draws more current from the supply.
- This additional current drawn by primary is due to the load hence called load component of primary current denoted as I_2' as shown in the Fig. 2.15.
- This current I_2' is in anti-phase with I_2. The current I_2' sets up its own flux ϕ_2' which opposes the flux ϕ_2 and helps the main flux ϕ. This flux ϕ_2' neutralizes the flux ϕ_2 produced by I_2.

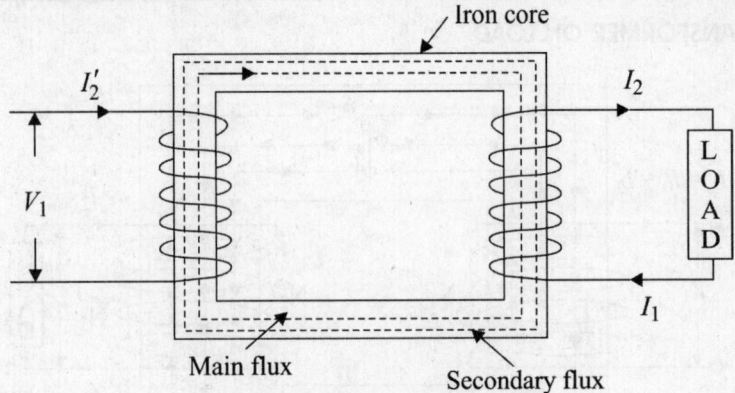

FIG. 2.14: ϕ_2 opposes ϕ

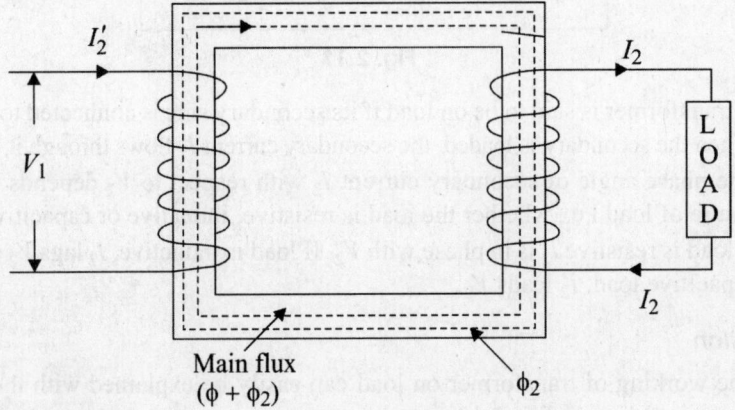

Fig. 2.15: Schematic of a transformer showing primary draws more current

- The mmf, i.e. ampere turns $N_1 I_2'$ balance the ampere turns $N_2 I_2$. Hence the net flux in the core is again maintained at constant level.
- The load components current I_2' always neutralises the changes in the load. As practically flux in core is constant, the core loss is also constant for all the loads. Hence the transformer is called *constant flux machine*.

Phasor Diagram on Load

- As the ampere-turns are balanced, we can write

$$N_2 I_2 = N_1 I_2'$$

$$\boxed{I_2' = \frac{N_2}{N_1} I_2 = K I_2}$$

- Thus when transformer is loaded, the primary current I_1 has two compenents.

- The no load current I_0 which lags V_1 by angle ϕ_0. It has two components I_m and I_c.
- The load component I_2' which is in anti-phase with I_2 and phase of I_2 is decided by the load.

  Hence primary current I_1 is vector sum of I_0 and I_2'.

  $$I_1 = I_0 + I_2'$$

- Now, let us discuss the three cases of load and phasor diagram neglecting the voltage drop in winding as shown in Fig. 2.16.

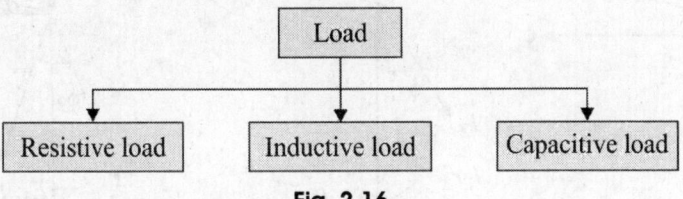

Fig. 2.16

Steps: To draw Phasor Diagram

Voltage drop in winding of transformer is neglected

$$\therefore \qquad V_1 = E_1 \text{ and } E_2 = V_2$$

- Consider flux ϕ as a reference phasor as it is linked to both primary and secondary sides. Draw I_0 as it is constant for all loads.
- E_1 lags ϕ by 90°, reverse E_1 to get $-E_1 (V_1 = -E_1)$.
- E_1 and E_2 are in phase.
- (a) For purely resistive load, I_2 in phase with V_2 or E_2.

  (b) For inductive load, I_2 lags E_2 by ϕ_2.

  (c) For capacitive load, I_2 leads E_2 by ϕ_2.
- Counter balance current $I_2' = K I_2$, let $K = 1$ then $I_2' = I_2$ is 180° out of the phase with I_2.

  (I_2' is always is anti-phase with I_2)
- The total primary current I_1 is vector sum of no load current I_0 and counter balance current I_2', i.e.

  $$I_1 = I_0 + I_2$$

2.12 EQUIVALENT CIRCUIT DIAGRAM

Importance of Equivalent Circuit Diagram

- We can draw the equivalent circuit because it makes calculation easy.
- We can analysed and investigate any electrical device with the help of equivalent circuit diagram (Figs 2.17–2.20).
- Equivalent circuit can be studied and analysed easily by the electric circuit theory.

  $$I_1 = I_0 + I_2', \ I_0 = I_w + I_\phi$$

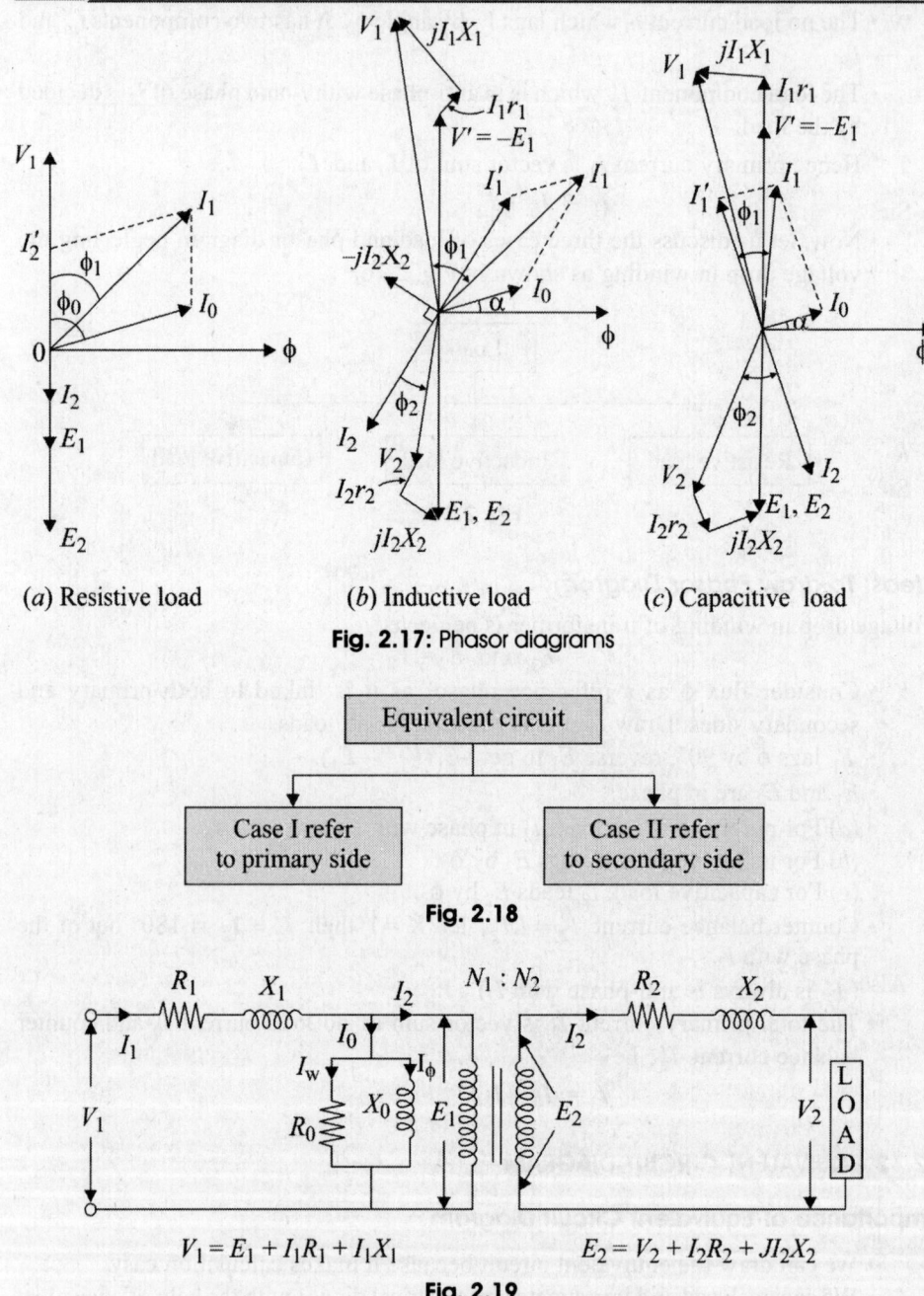

(a) Resistive load   (b) Inductive load   (c) Capacitive load

Fig. 2.17: Phasor diagrams

$$V_1 = E_1 + I_1R_1 + I_1X_1$$

$$E_2 = V_2 + I_2R_2 + JI_2X_2$$

Fig. 2.19

Case I: Equivalent Circuit Referred to Primary Side

When we are shifting all the secondary side parameters into primary side then it is known as equivalent circuit refer to primary (Fig. 2.21).

We know that, total copper loss.

$$P_{Cu} = I_1^2 R_1 + I_2^2 R_2$$

$$P_{Cu} = I_1^2 \left(R_1 + \left(\frac{I_2}{I_1} \right)^2 \right) ; R_2 = I_1^2 \left(R_1 + \frac{R_2}{K^2} \right)$$

Since

$$K = \frac{I_1}{I_2}$$

$$\therefore \qquad P_{Cu} = I_1^2 R_{01}$$

$$\therefore \qquad R_{01} = R_{1eq} = R_1 + R_2' = R_1 + \frac{R_2}{K^2}$$

$$R_2' = \frac{R_2}{K^2}$$

$R_{01} = R_{1eq} = $ Equivalent resistance of transformer transferred to primary side.
Similarly for reactance

$$X_{01} = X_1 + X_2' = X_{1eq}$$

= Equivalent reactance of transformer transferred to primary sides.

$$X_2' = \frac{X_2}{K^2}$$

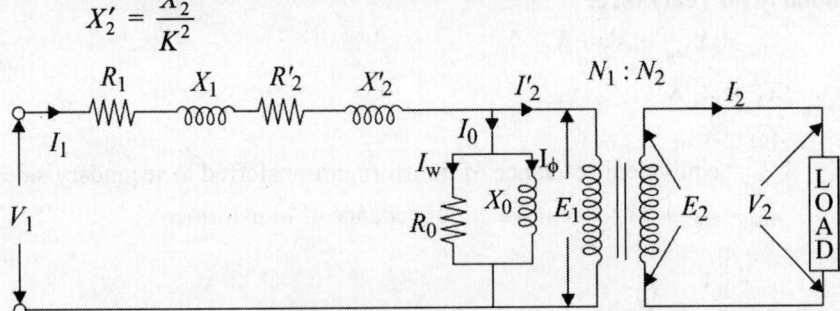

Fig. 2.20: Equivalent circuit

Also redraw the above circuit diagram.

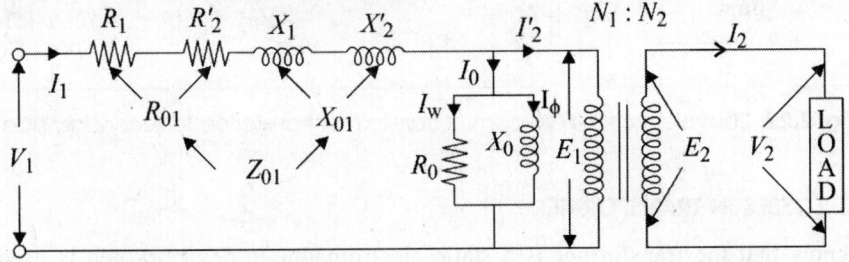

Fig. 2.21: Equivalent circuit diagram referred to primary side

$$Z_{01} = R_{01} + J_{X_{01}}$$

$$= \sqrt{(R_{01})^2 + (X_{01})^2}$$

= Equivalent impedance of transformer transferred to primary side

Case II: Equivalent Circuit Referred to Secondary Side

When we are shifting all the primary side parameters into secondary side then it is known as equivalent circuit refer to secondary side (Fig. 2.22).

We know that, total copper loss.

$$P_{Cu} = I_1^2 R_1 + I_2^2 R_2$$

$$= I_2^2 \left[\left(\frac{I_1}{I_2} \right)^2 R_1 + R_2 \right] \quad \boxed{K = \frac{I_1}{I_2}}$$

$$= I_2^2 \left(K^2 R_1 + R_2 \right)$$

$$P_{Cu} = I_2^2 (R_{02})$$

where, $R_{02} = R_{2eq} = R_2 + R_1'$

$R_1' = R_1 \cdot K^2$

$R_{02} = R_{2eq}$

= equivalent resistance of transformer transferred to secondary side

Similarly for reactance:

$$X_{02} = X_{2eq} = X_2 + X_1'$$

$$X_1' = X_1 \cdot K^2$$

$$X_{02} = X_{2eq}$$

= equivalent reactance of transformer transferred to secondary side.

$Z_{02} = R_{02} + jX_{02}$ = equivalent impedance of transformer

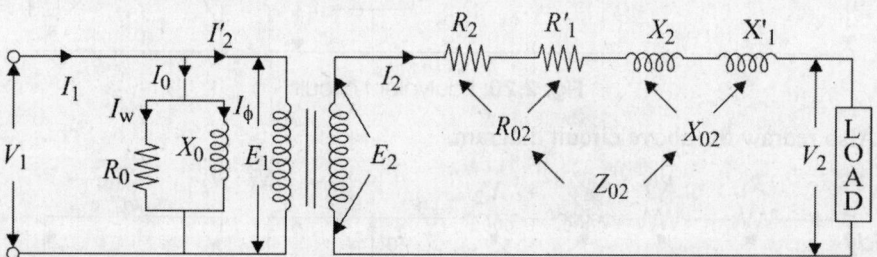

Fig. 2.22: Equivalent circuit diagram of transformer transferred to secondary side

2.13 LOSSES IN TRANSFORMER

We know that the transformer is a static electromagnetic device which is used to transfer electrical energy from one circuit to another, but whole of energy cannot be transfer into other circuit because certain amount of energy is lost in the core and winding.

There are two types of losses occur in transformer as shown in Fig. 2.23.

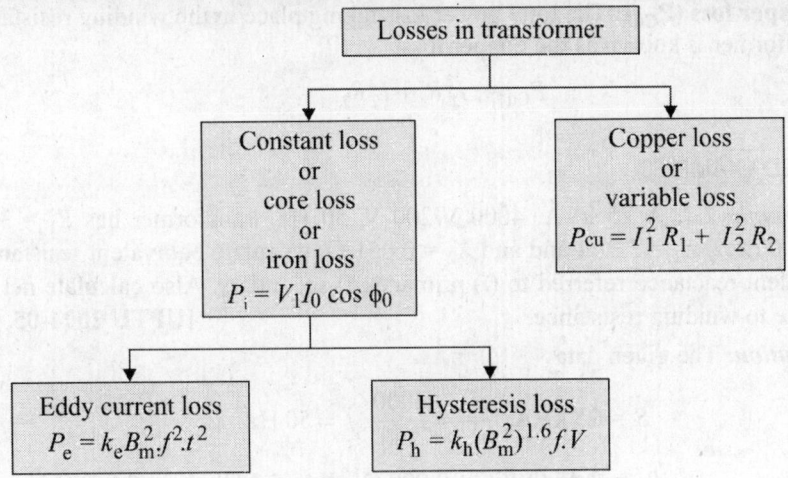

Fig. 2.23: Block diagram of losses

Iron Loss (P_i)

Iron loss P_i is the power loss taking place in the iron core of the transformer.

$\therefore$ $\qquad$ P_i = Hysteresis loss + eddy current loss

$\qquad$ $P_i = P_h + P_e$

Hysteresis Loss (P_h)

- The area enclosed by the hysteresis loop of a material represents the hysteresis loss.
- The hysteresis loss is frequency dependent. As we increase the frequency of operation, the hysterisis loss will increase proportionally.

$\therefore$ $\qquad$ $P_h = K_h (B_m)^{1.6} f \cdot V$ watts

where $\quad K_h$ = hysteresis constant depends on the material

$\qquad B_m$ = maximum flux density

$\qquad f$ = frequency

$\qquad V$ = volume of the core

Eddy Current Losses

- Due to the time varying flux, there is some induced emf in the transformer core.
- This induced emf causes some currents to flow through the core body.
- These currents are known as the eddy currents.
- The eddy current losses are minimized by using the laminated core.

$\therefore$ $\qquad$ $P_e = K_e B_m^2 \cdot f^2 \cdot t^2$ watts/unit volume

where, $\quad K_e$ = Eddy current constant

$\qquad B_m$ = maximum flux density

$\qquad f$ = frequency

$\qquad t$ = thickness of the core

Copper loss (P_{Cu}): The total power loss taking place in the winding resistances of a transformer is known as the copper loss.

$$\therefore \qquad P_{Cu} = I_1^2 R_1 + I_2^2 R_2$$

SOLVED EXAMPLES

Example 2.3: A 25 kVA, 4000 V/200 V, 50 Hz, transformer has $R_1 = 3.45\ \Omega$, $R_2 = 0.009\ \Omega$, $X_1 = 5.2\ \Omega$ and and $X_2 = 0.051\ \Omega$, calculate equivalent resistance and equivalent reactance referred to (*i*) primary (*ii*) secondary. Also calculate net power lost due to winding resistance. **[UPTU 2004-05, 09-10]**

Solution: The given data

$$S = 25\,\text{kVA}, \ \frac{E_1}{E_2} = \frac{4000}{200}, f = 50\,\text{Hz}$$

$$R_1 = 3.45\,\Omega, \ R_2 = 0.009\,\Omega, \ X_1 = 5.2\,\Omega, \ X_2 = 0.051\,\Omega$$

$$K = \frac{V_2}{V_1} = \frac{200}{4000} = 0.05$$

(*i*) $\qquad R_{01} = R_1 + \dfrac{R_2}{K^2} = 3.45 + \dfrac{0.009}{(0.05)^2} = 7.05\,\Omega$ **Ans.**

$$X_{01} = X_1 + \frac{X_2}{K^2} = 5.2 + \frac{0.051}{(0.05)^2} = 25.6\,\Omega \quad \textbf{Ans.}$$

(*ii*) $\qquad R_{02} = R_2 + R_1 \cdot K^2 = 0.009 + 3.45 \times (0.05)^2$

$$= 0.0176\,\Omega \quad \textbf{Ans.}$$

$$X_{01} = X_2 + X_1 \cdot K^2 = 0.051 + 5.2 \times (0.05)^2 = 0.064\,\Omega \quad \textbf{Ans.}$$

$$S = V_1 I_1$$

$$I_1 = \frac{25 \times 10^3}{4000} = 6.25\,\text{A} \quad \textbf{Ans.}$$

$$P_{Cu} = I_1^2 R_{01} = (6.25)^2 \times 7.05$$

$$= 275.04\ \text{watt} \quad \textbf{Ans.}$$

Example 2.4: A 30 kVA, 2000/200 V, single phase, 50 Hz transformer has a primary resistance of 2.5 Ω and reactance of 4.5 Ω. The secondary resistance and reactance are 0.015 Ω and 0.02 Ω respectively. Find: (*i*) equivalent resistance, reactance and impedance referred to the primary side (*ii*) total Cu loss in the transformer.

Solution: The given data

$$S = 30\,\text{kVA}, \ E_1 = 2000, \ E_2 = 200\,\text{V}$$

$$R_1 = 3.45\,\Omega, \ X_1 = 4.5\,\Omega, \ R_2 = 0.015\,\Omega, \ X_2 = 0.02\,\Omega$$

$$K = \frac{V_2}{V_1} = \frac{200}{2000} = 0.1$$

(i)
$$R_{01} = R_1 + \frac{R_2}{K^2} = 3.5 + \frac{0.015}{(0.1)^2} \quad \textbf{Ans.}$$

$$X_{01} = X_1 + \frac{X_2}{K^2} = 4.5 + \frac{0.02}{(0.1)^2}$$

$$R_{02} = \sqrt{(R_{01})^2 + (X_{01})^2}$$

$$S = V_1 I_1$$

$$I_1 = \frac{30 \times 10^3}{2000} = 15 \text{ A}$$

$$I_2 = \frac{30 \times 10^3}{200} = 150 \text{ A}$$

(ii)
$$P_{Cu} = I_1^2 R_1 + I_2^2 R_2$$
$$= (15)^2 \times 3.5 + (150)^2 \times 0.015$$
$$= 1125 \text{ watt } \textbf{Ans.}$$

2.14 TRANSFORMER EFFICIENCY

It is the ratio of the output power to the input power in a transformer, denoted by η.

$$\therefore \quad \eta = \frac{\text{Output power}}{\text{Input power}} = \frac{\text{Output power}}{\text{Output power} + \text{losses}}$$

Let, Power input $= V_1 I_1 \cos \phi_1$

Power output $= V_2 I_2 \cos \phi_2$

So, efficiency of transformer

$$\eta = \frac{V_2 I_2 \cos \phi_2}{V_1 I_1 \cos \phi} \times 100$$

But, we know that

Output = Input – losses

Power input = Power output + losses

$$V_1 I_1 \cos \phi_1 = V_2 I_2 \cos \phi_2 + P_i + P_{Cu}$$

$$\therefore \quad \eta = \frac{V_2 I_2 \cos \phi_2}{V_2 I_2 \cos \phi_2 + P_i + P_{Cu}}$$

Some important formula used for determining the efficiency.

(a)
$$\%\eta = \frac{s \cdot x \cdot \cos \phi_2}{s \cdot x \cdot \cos \phi_2 + P_i + x^2 P_{Cu}} \times 100$$

where,

s = rating of transformer (kVA)

$$x = \frac{\text{Actual load}}{\text{Full load}}$$

Example: $x = 1$ for full load

$$x = \frac{1}{2} \text{ for half load}$$

(b) Load at maximum efficiency $= s \times \sqrt{\dfrac{P_i}{P_{Cu}}}$

2.14.1 Condition for Maximum Power Efficiency

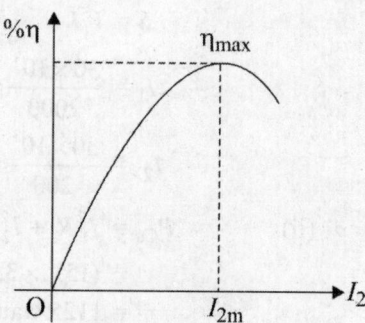

Fig. 2.24: η curve

- When a transformer works on a constant input voltage and frequency then efficiency varies with the load.
- As load increases, the efficiency increases. At a certain load current, it achieves a maximum value.
- If the transformer is loaded further the efficiency starts decreasing. The graph of efficiency against load current I_2 is shown in Fig. 2.24.

Proof:

$$\%\eta = \frac{\text{Output power}}{\text{Input power}} \times 100 = \frac{V_2 I_2 \cos\phi_2}{V_1 I_1 \cos\phi_1} \times 100$$

$$\text{Input power} = \text{Output power} + \text{losses}$$

$$V_1 I_1 \cos\phi_1 = V_2 I_2 \cos\phi_2 + P_i + P_{Cu}$$

$$V_2 I_2 \cos\phi_2 = V_1 I_1 \cos\phi_1 - (P_i + P_{Cu})$$

$$\%\eta = \frac{V_1 I_1 \cos\phi_1 - (P_i + I_1^2 R_1)}{V_1 I_1 \cos\phi_1}$$

$$\eta = 1 - \frac{P_i}{V_1 I_1 \cos\phi_1} - \frac{I_1 R_1}{V_1 \cos\phi_1}$$

For maximum

$$\frac{d\eta}{dI_1} = 0$$

$$\frac{d\eta}{dI_1} = 0 + \frac{P_i}{V_1 I_1^2 \cos\phi_1} - \frac{R_1}{V_1 \cos\phi_1}$$

$$\Rightarrow \qquad \frac{P_i}{V_1 I_1^2 \cos\phi_1} = \frac{R_1}{V_1 \cos\phi_1}$$

$$\boxed{P_i = I_1^2 R_1}$$

At maximum efficiency

$$\boxed{\text{Iron loss} = \text{copper loss}} \text{ or } P_i = I_1^2 R_1$$

and $$\boxed{P_i = x^2 P_{Cu}} \text{ at maximum efficiency}$$

2.15 VOLTAGE REGULATION OF A TRANSFORMER

- Ideally the secondary terminal voltage V_2 (or load voltage) of a transformer should remain constant independent of the load current as shown in Fig. 2.24.
- But practically the load voltage decreases with increase in load current I_1 as shown in Fig. 2.25.
- **No load voltage:** The no-load voltage is the secondary terminal voltage corresponding to zero-load current. For a transformer

$$\text{no load voltage} = E_2 \text{ volts} \qquad \qquad ...(1)$$

- **Full load voltage:** It is the secondary terminal voltage corresponding to the specified load current. Let us denote it by V_2. The percent voltage regulation is given mathematically as:

$$\% \text{ Regulation} = \frac{E_2 - V_2}{E_2} \times 100$$

- Thus with increase in load current, the value of V_2 decrease and the percent regulation increases (becomes poor). Ideal value of voltage regulation is 0%.

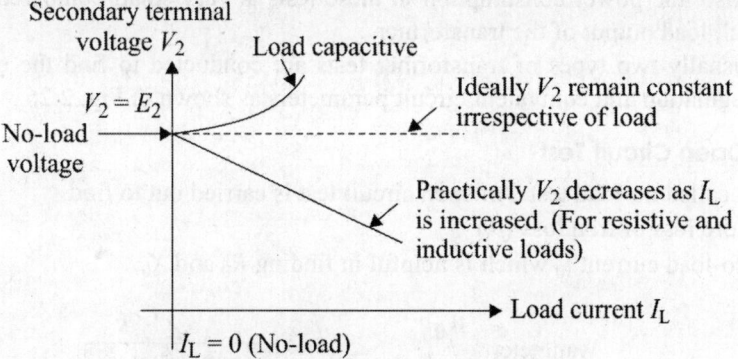

Fig. 2.25: Concept of voltage regulation

Defintion of Voltage Regulation

The voltage regulation of a transformer is defined as the change in secondary terminal voltage (V_2) from no-load to full-load with the primary source voltage (V_1) and the temperature of the transformer maintained constant.

Reason for Reduction in Voltage

- The drop in secondary terminal voltage takes place due to the voltage drop taking place across the primary and secondary impedances.
- As the load current increases, the voltage drop across these impedances will increase. This will reduce the secondary terminal voltage V_2, with increase in I_L.

Does the Secondary Terminal Voltage Depend only on the Load Current?

- The answer is NO. The secondary terminal voltage does not depend only on the magnitude of I_L.

- But it also depends on the type of the load. As shown in Fig. 2.25, the secondary terminal voltage decreases for resistive or inductive loads whereas it increases with increase in I_L if the load is capacitive.

Note: The regulation is positive for the resistive and inductive loads and it can be negative for the capacitive loads.

2.16 CALCULATION OF EFFICIENCY FROM TRANSFORMER TESTS

- Transformer tests are performed to determine efficiency, voltage regulation and equivalent circuit parameters without actually loading the transformer.

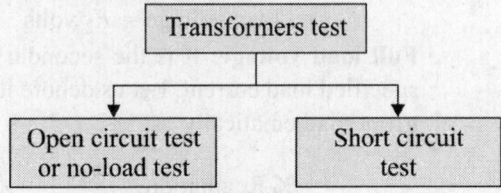

Fig. 2.26

- Transformer tests give more accurate result these obtained by taking measurements on fully loaded transformer.
- Also, the power consumption in these tests is very small compared with the full load output of the transformer.
- Usually two types of transformer tests are conducted to find the efficiency, regulation and equivalent circuit parameters as shown in Fig. 2.26.

2.16.1 Open Circuit Test

It is also called no-load test. An open circuit test is carried out to find.

$\Rightarrow$ Core loss or iron loss (p_i)

$\Rightarrow$ No-load current I_0 which is helpful in finding R_0 and X_0.

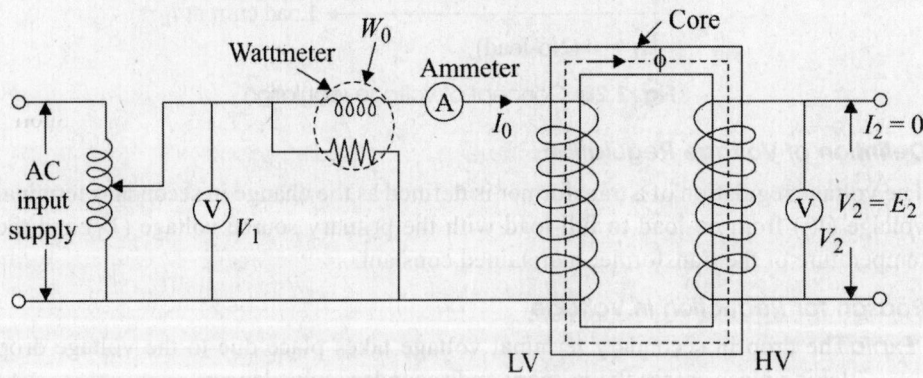

Fig. 2.27: Open circuit test

- The Fig. 2.27 shows the connection diagram for the open circuit test, the high voltage (HV) side is left open.
- In this test, the rated voltage is applied to the primary (usually low voltage side) while the secondary is left open circuited.

- The applied primary voltage V_1 is measured by the voltmeter, the no-load current I_0 by ammeter and no-load input power W_0 by wattmeter as shown in Fig. 2.27.

Calculations of Parameters

$$I_W = I_0 \sin\phi_0$$
$$I_\phi = I_0 \cos\phi_0$$
$$W_0 = V_1 I_0 \cos\phi_0$$

- No load power factor $= \cos\phi_0 = \dfrac{W_0}{V_0 I_0}$

- Magnetising current, $I_\phi = I_0 \sin\phi_0$

- Working component, $I_w = I_0 \cos\phi_0$

$$\therefore \qquad \boxed{R_0 = \frac{V_0}{I_W}\ \Omega}$$

$$\boxed{X_0 = \frac{V_0}{I_\phi}\ \Omega}$$

2.16.2 Short Circuit Test

An short circuit test is carried out to find.

⇒ Copper loss

⇒ Equivalent resistance, reactance and impedance of the transformer referred to the winding in which the measuring instrument are connected.

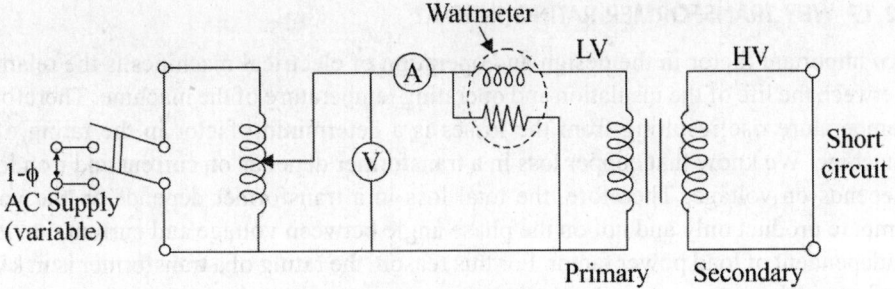

Fig. 2.28: Short circuit test

Explanation

- The high voltage winding is supplied at the reduced voltage from a variable voltage supply.
- The supply voltage is gradually increased until full load primary current flows. When the rated full load current flows in the primary winding rated full load current will flow in the secondary winding by transformer action. Readings of the ammeter, voltmeter and wattmeter are noted.

• The wattmeter reading is the power loss which is equal to full load copper loss as iron loss is very low.

$$W_{SC} = (P_{Cu}) \text{ full load} = \text{full load copper loss}$$

Calculations

$$W_{SC} = V_{SC} \cdot I_{SC} \cos\phi_{SC}$$

$$\therefore \qquad \cos\phi_{SC} = \frac{W_{SC}}{V_{SC} \cdot I_{SC}} = \text{short circuit power factor}$$

$$W_{SC} = I_{SC}^2 . R_{1e} = \text{copper loss}$$

$$\boxed{R_{1e} = \frac{W_{SC}}{I_{SC}^2} \Omega}$$

$$Z_{1e} = \frac{V_{SC}}{I_{SC}} = \sqrt{(R_{1e})^2 + (X_{1e})^2}$$

$$\therefore \qquad X_{1e} = \sqrt{(Z_{1e})^2 - (R_{1e})^2}$$

2.16.3 Calculation of Efficiency from OC and SC Test

We know that

From OC test $\quad W_O = P_i$

From SC test $\quad W_{SC} = (P_{Cu}) \text{ FL}$

$$\eta \text{ on full load} \quad = \frac{V_2 (I_2)_{FL} \cos\phi_2 \times 100}{V_2 (I_2)_{FL} \cos\phi_2 + W_O + W_{SC}}$$

2.17 WHY TRANSFORMER RATING IN KVA?

An important factor in the design and operation of electrical machines is the relation between the life of the insulation and operating temperature of the machine. Therefore, temperature rise resulting from the losses is a determining factor in the rating of a machine. We know that copper loss in a transformer depends on current and iron loss depends on voltage. Therefore, the total loss in a transformer depends on the volt-ampere product only and not on the phase angle between voltage and current, i.e. it is independent of load power factor. For this reason, the rating of a transformer is in kVA and not kW.

2.18 SUMPNER OR BACK-TO-BACK TEST

This test is conducted simultaneously on two identical transformers and provides data for finding the efficiency, regulation and temperature rise. The main advantage of this test is that the transformers are tested under full-load conditions without much expenditure of power. The power required to conduct this test is equal to the losses of the two transformers. It may be noted that two identical transformers are needed to carry out this test.

Circuit: Figure 2.29 shows the connections for back-to-back test on two identical transformers T_1 and T_2. The primaries of the two transformers are connected in parallel across the rated voltage V_1 while the two secondaries are connected in phase opposition. Therefore, there will be no circulating current in the loop formed by the secondaries because their induced emfs are equal and in opposition. There is an auxiliary low-voltage transformer T which can be adjusted to give a variable voltage and hence current is in the secondary loop circuit.

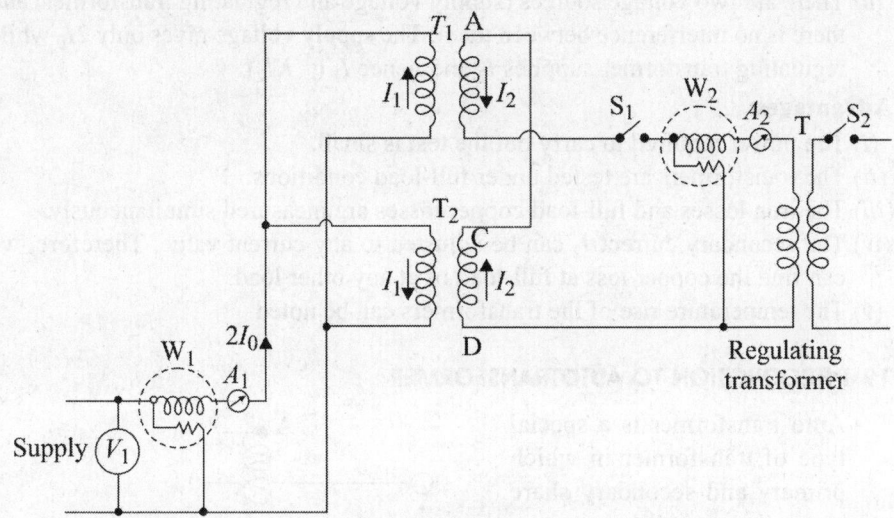

Fig. 2.29

Operation:

(i) The secondaries of the transformers are in phase opposition. With switch S_1, closed and switch S_2 open (i.e. regulating transformer not in the circuit), there will be no circulating current ($I_2 = 0$) in the secondary loop circuit. It is because the induced emfs in the secondaries are equal and in opposition. This situation is just like an open-circuit test. Therefore, the current drawn from the supply is $2I_0$ where I_0 is the no-load current of each transformer. The reading of wattmeter, will be equal to the core losses of the two transformers.

$$W = \text{Core losses of the two transformers}$$

(ii) Now switch S_2 is also closed and output voltage of the regulating transformer is adjusted till full-load current I_2 flows in the secondary loop circuit. The full-load secondary current will cause full-load current I_1 ($= KI_2$) in the primary circuit. The primary current I_1, circulates in the primary winding only and will not pass through W_1. Note that full-load currents are flowing through the primary and secondary windings. Therefore, reading of wattmeter, will be equal to the full-load copper losses of the two transformers.

$$W_2 = \text{Full-load Cu losses of two transformers}$$
$$W_1 + W_2 = \text{Total losses of two transformers at full load}$$

The following points may be noted:

(a) The wattmeter W_1 gives the core losses of the two transformers while wattmeter W_2 gives the full-load copper losses (or at any other load current I_2) of the two transformers. Therefore, power required to conduct this test is equal to the total losses of the two transformers.

(b) Although transformers are not supplying any load, yet full-load iron loss and full-load copper losses are occurring in them.

(c) There are two voltage sources (supply voltage and regulating transformer) and there is no interference between them. The supply voltage gives only $2I_0$ while regulating transformer supplies I_2 and hence I_1 $(= KI_2)$,

Advantages:

(i) The power required to carry out the test is small.

(ii) The transformers are tested under full-load conditions.

(iii) The iron losses and full-load copper losses are measured simultaneously.

(iv) The secondary current I_2 can be adjusted to any current value. Therefore, we can find the copper loss at full-load or at any other load.

(v) The temperature rise of the transformers can be noted.

2.19 INTRODUCTION TO AUTOTRANSFORMER

• Auto transformer is a special type of transformer in which primary and secondary share same single winding.

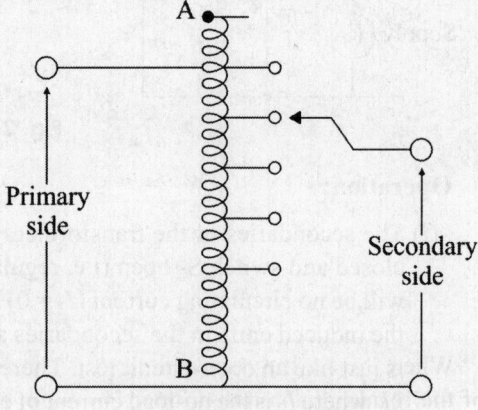

Fig. 2.30: Autotransformer

Advantages of Autotransformer

• As only one winding is used. Copper required for the transformer is very less. So size and cost is reduced compared to two winding transformer.

• The losses taking place in the winding are reduced. Hence the efficiency is higher than two winding transformer.

Disadvantages of Autotransformers

• There is no electrical isolation between the primary and secondary winding.

• Thus can prove to be dangerous for high voltage applications.

Applications of Autotransformer

• It can be used as variac, i.e. variable AC supply to vary the AC voltage applied to the load smoothly from 0 V to about 270 V.

• It is used to start to AC machines such as induction motor or synchronous motors.

• When the autotransformer is used to control the intensity of lamps in the cinema hall, etc. It is called as dimmer state.

2.20 CONVERSION OF TWO-WINDING TRANSFORMER INTO AUTO-TRANSFORMER

A two-winding transformer can be converted into an autotransformer, either step-up or step-down. Consider a 10 kVA, 2300/230 V two-winding transformer shown in Fig. 2.31(a). If we want to convert it into autotransformer, the two windings of the transformer are connected in series. If we use the additive polarity as shown in Fig. 2.31(b), we get step-up autotransformer. The voltage rating of the autotransformer is now 2300/2530. If we use subtractive polarity as shown in Fig. 2.31(c), we get a step-down autotransformer. The voltage rating of the transformer is now 2300/2070 V.

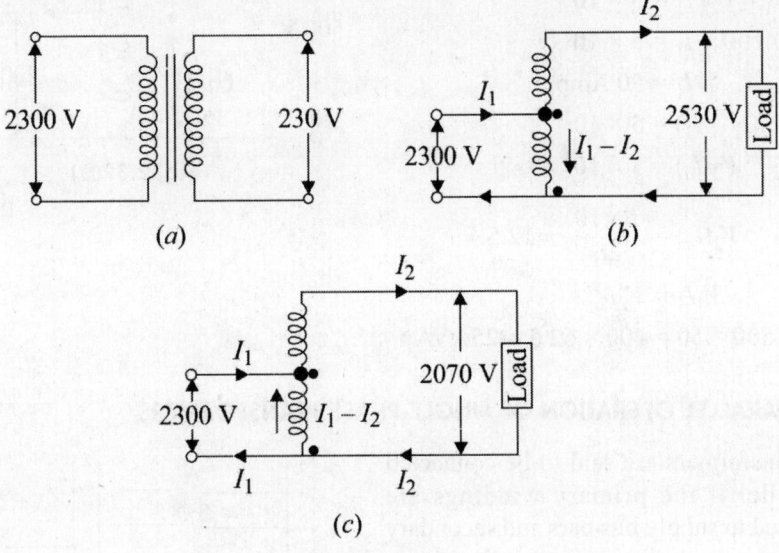

Fig. 2.31

When a two-winding transformer is converted into autotransformer, the kVA rating of the resulting autotransformer is greatly increased. This higher rating results from the conduction connection.

Example 2.5: A 400/100 V, 5 kVA, two-winding transformer is to be used as an auto transformer to supply power at 400 V from 500 V source. Draw the connection diagram and determine the kVA output of the auto transformer.

Solution: For two winding transformer
$$V_1 = 400 \text{ V}$$
$$V_2 = 100 \text{ V}$$
$$V_1 I_1 = V_2 I_2$$
$$V_1 I_1 = 5 \times 10^3$$
$$I_1 = \frac{5 \times 10^3}{400} = 12.5 \text{ A}$$

Use of two winding-transformer into autotransformer

Supply power at 400 V from a 500 V source

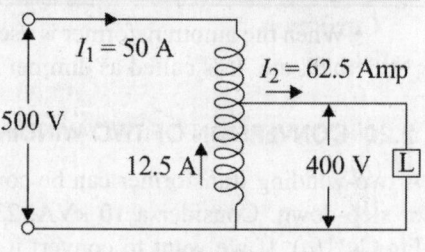

Fig. 2.32(a)

$$K = \frac{V_2}{V_1} = \frac{400}{500}$$

kVA rating of auto transformer

$$= 400 \times 62.5$$
$$= 25 \text{ kVA}$$
$$(V_1 - V_2) I_1 = V_2(I_2 - I_1)$$
$$(V_1 - V_2) I_1 = 5 \times 10^3$$
$$100 \times I_1 = 5 \times 10^3$$
$$I_1 = 50 \text{ Amp}$$
$$V_2(I_2 - I_1) = 5 \times 10^3$$
$$V_2(I_a) = 5 \times 10^3$$
$$V_2(I_a) = \frac{5 \times 10^3}{400} = 12.5 \text{ A}$$
$$V_1 I_1 = V_2 I_2$$
$$500 \times 50 = 400 \times 62.5 = 25 \text{ kVA}$$

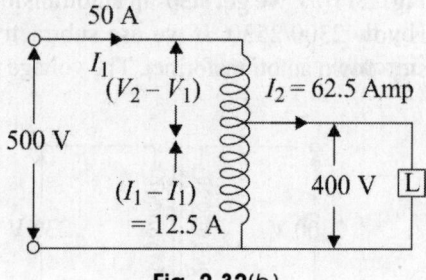

Fig. 2.32(b)

2.21 PARALLEL OPERATION OF SINGLE-PHASE TRANSFORMERS

Two transformers are said to be connected in parallel if the primary windings are connected to supply bus bars and secondary windings are connected to load busbars. Figure 2.33(a) shows two transformers A and B in parallel. While connecting two or more than two transformers in parallel, it is essential that their terminals of similar polarities are joined to the same busbars as shown in Fig. 2.33(a). The wrong connec-

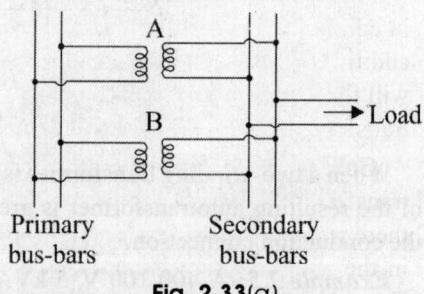

Fig. 2.33(a)

tions may result in a dead short-circuit and transformers may be damaged unless protected by fuses or circuit breakers. There are three principal reasons for connecting transformers in parallel.

Firstly, if one transformer fails, the continuity of supply can be maintained through other transformers. Secondly, when the load on the sub-station becomes more than the capacity of the existing transformers, another transformer can be added in parallel. Thirdly, any transformer can be taken out of the circuit for repair/routine maintenance without interrupting supply to the consumers.

Conditions for satisfactory parallel operation: In order that the transformers work satisfactorily in parallel, the following conditions should be satisfied:

(*i*) Transformers should be properly connected with regard to their polarities.

(*ii*) The voltage ratings and voltage ratios of the transformers should be the same.

(*iii*) The per unit or percentage impedances of the transformers should be equal.

(*iv*) The reactance/resistance ratios of the transformers should be the same.

Condition (1): This condition is absolutely essential because wrong connections may result in dead short-circuit. Figure 2.33(*b*) shows the correct method of connecting two single-phase transformer in parallel. It will be seen that around the loop formed by the secondaries, the two secondary emfs E_A and E_D oppose and there will be no circulating current.

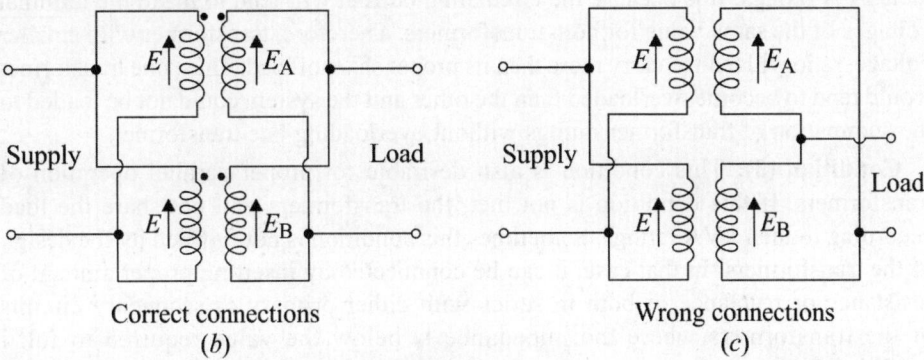

Correct connections
(*b*)

Wrong connections
(*c*)

Fig. 2.33(*b* and *c*)

Figure 2.33(*c*) shows the wrong method of connecting two single-phase transformers in parallel. Here the two secondaries are so connected that their emfs E_A, and E_B are additive. This may lead short-circuit conditions and a very large circulating current will flow in the loop formed by the two secondaries. Such a condition may damage the transformers unless they are protected by fuses and circuit breakers.

Condition (2): This condition is desirable for the satisfactory parallel operation of transformer. If this condition is not met, the secondary emfs will not be equal and there will be circulating current in the loop formed by the secondaries. This will result in the unsatisfactory parallel operation of transformers. Let us illustrate this point. Consider two single-phase transformers A and B operating in parallel as shown in Fig. 2.33(*d*). Let E_A and E_B be their no-load secondary voltages and Z_A and Z_B be their impedances referred to the secondary. Then at no-load, the circulating current in the loop formed by the secondaries is

$$\text{Circulating current, } I_C = \frac{E_A - E_B}{Z_A + Z_B}$$

Even a small difference in the induced secondary voltages can cause a large circulating current in the secondary loop because impedances of the transformers are small. This supply secondary circulating current will cause current to be drawn from the supply by the primary of each transformer. These currents will cause copper losses in both primary and secondary. This creates heating with no useful output. When load is connected to the system, this circulating current will tend to produce

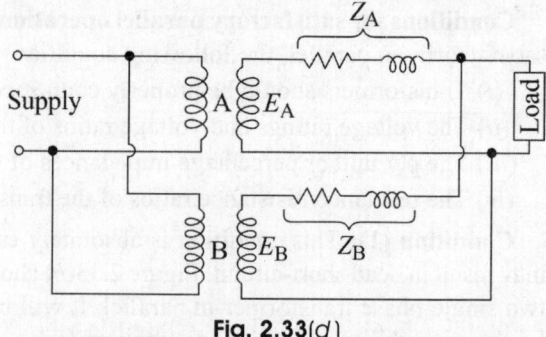

Fig. 2.33(d)

unequal loading conditions, i.e. the transformers will not share the load according to their kVA ratings. It is because the circulating current will tend to make the terminal voltages of the same value for both transformers. Therefore, transformer with smaller voltage, ratio will tend to carry more than its proper share of load. Thus, one transformer would tend to become overloaded than the other and the system could not be loaded to the summation of transformer ratings without overloading one transformer.

Condition (3): This condition is also desirable for proper parallel operation of transformers. If this condition is not met, the transformers will not share the load according to their kVA ratings. Sometimes this condition is not fulfilled by the design of the transformers. In that case, it can be connnected by inserting proper amount of resistance or reactance or both in series with either primary or secondary circuits of the transformers where the impedance is below the value required to fulfil condition 3.

Condition (4): If the reactance/resistance ratios of the two transformers are not equal, the power factor of the load supplied by the transformers will not be equal. In other words, one transformer will be operating with a higher and the other with a lower power factor than that of the load. Condition (3) is much more important than condition (4). Considerable deviation from condition (4) will result in only a small reduction in the satisfactory degree of operation. When desired, condition (4) may also be improved by inserting external impedance of proper value.

2.22 SINGLE PHASE EQUAL VOLTAGE RATIO TRANSFORMERS IN PARALLEL

Figure 2.34 shows two single phase equal voltage ratio transformers A and B in parallel. The secondary emfs of the two transformers are equal (i.e. $E_A = E_B = E$) because they have the same turns ratio and have their primaries connected to the same supply.

If the magnetising current is ignored, the two transformers can be represented by their equivalent circuits referred to secondary as shown in Fig. 2.35. It is clear that the transformers will share total load in the same way as two impedances in parallel.

Let Z_A, Z_B = Impedances of transformers referred to secondary and I_A, I_B their respective currents

V_2 = common terminal voltage

I = total load current

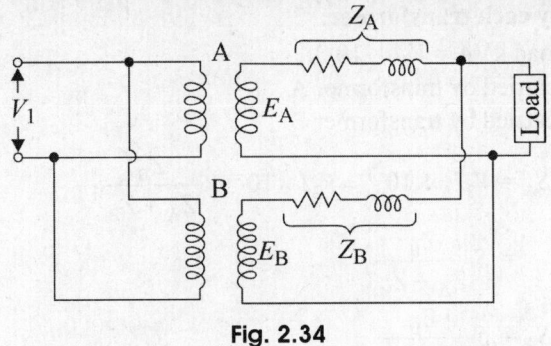

Fig. 2.34

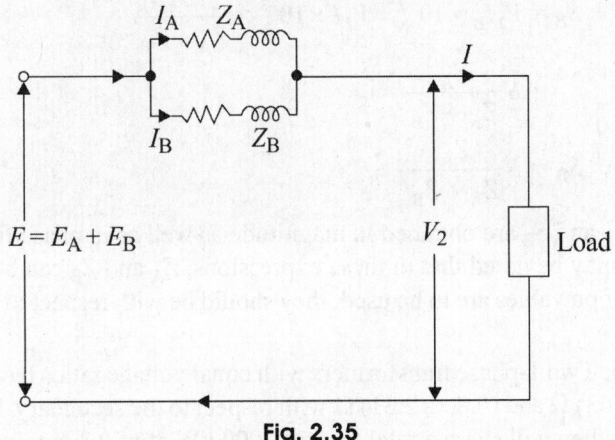

Fig. 2.35

It is clear from Fig. 2.35

$$I_A + I_B = I$$

and

$$I_A Z_A = I_B Z_B$$

$$I_A = I_B \frac{Z_B}{Z_A}$$

$$I_B \frac{Z_B}{Z_A} + I_B = I$$

$$I_B = I \frac{Z_A}{Z_A + Z_B}$$

Similarly,

$$I_A = I \frac{Z_B}{Z_A + Z_B}$$

Thus the way in which the load current I is shared by the transformers is independent of load impedance and depends only on the transformer impedances.

kVA carried by each transformer:

Let S = total load kVA = $V_2 I \times 10^{-3}$

S_A = kVA carried by transformer A

S_B = kVA carried by transformer B

$\therefore \qquad S_A = V_2 I_A \times 10^{-3} = V_2 I \times 10^{-3} \times \dfrac{Z_B}{Z_A + Z_B}$

$\qquad\qquad = S \dfrac{Z_B}{Z_A + Z_B}$

or $\qquad S_A = S \dfrac{Z_B}{Z_A + Z_B}$

Also $\qquad S_B = V_2 I_B \times 10^{-3} = V_2 I \times 10^{-3} \times \dfrac{Z_A}{Z_A + Z_B}$

$\qquad\qquad = S \dfrac{Z_A}{Z_A + Z_B}$

or $\qquad S_B = S \dfrac{Z_A}{Z_A + Z_B}$

Therefore, S_A and S_B are obtained in magnitude as well as in phase from the above expressions. It may be noted that in these expressions, Z_A and Z_B can be expressed in ohms or in pu. If pu values are to be used, they should be with respect to common base kVA.

Example 2.6: Two 1-phase transformers with equal voltage ratios have impedances of $(0.819 + j2.503)\ \Omega$ and $(0.8 + j2.31)\ \Omega$ with respect to the secondary. If they operate in parallel, how they will share a total load of 2000 kW at pf 0.8 lagging?

Solution: $\quad Z_A = (0.819 + j2.503)\ \Omega = 2.633\ \angle\ 71.88°\Omega$

$Z_B = (0.8 + j2.31)\ \Omega = 2.445\ \angle\ 70.9°$

$Z_A + Z_B = (1.619 + j4.813)\ \Omega = 5.078\ 71.4°\ \Omega$

Total load kVA = 2000/0.8 = 2500 kVA

$S = 2500\ \angle{-36.87°}$ kVA

$S_A = S \dfrac{Z_B}{Z_A + Z_B} = 2500 \angle -36.87 \times \dfrac{2.445\angle 70.9°}{5.078\angle 71.4°}$

$= 1203.7\ \angle -37.37° = 956.6\ \text{kW at } 0.79\ \text{pf lagging}$

$= \text{a load of } 1203.7°\ \cos 37.37° = 956.6\ \text{kW at } 0.79\ \text{pf}$

$= 1296.3\ \angle -36.39°\ \text{kVA}$

$= \text{a load of } 1296.3°\ \cos 36.39° = 1043.5\ \text{kW at } 0.805\ \text{pf lagging}$

Example 2.7: Two single phase transformers A and B rated at 250 kVA each are operated in parallel on both sides. Impedances for A and B are $(1 + j6)$ and $(1.2 + j4.8)\Omega$ respectively. Find the load shared by each when the total load is 500 kVA at 0.8 pf lagging.

Solution: $S = 500\angle - 36.9°$ kVA

$$\frac{Z_A}{Z_A + Z_B} = \frac{1 + j6}{2.2 + j10.8} = 0.55\angle 2.1°$$

$$= \frac{1.2 + j4.8}{2.2 + j10.8} = 0.45\angle - 2.5°$$

$\therefore \quad S_A = S\dfrac{Z_A}{Z_A + Z_B} = 500\angle - 36.9° \times 0.45\angle - 2.5° = 225\angle - 39.4°$ kVA

$\quad\quad S_B = S\dfrac{Z_A}{Z_A + Z_B} = 500\angle - 36.9° \times 0.55\angle 2.1 = 275\angle - 34.8°$ kVA

Note that transformer B is overloaded by 10% $[(275 - 250) \times 100/250 = 10\%]$.

Example 2.8: Two single-phase transformers which have the same turns ratio are connected in parallel and supply a total load of 800 kW at 0.8 power factor lagging. Their ratings are as follows

Transformer A	Rating	Pu resistance	Pu reactance
A	400 kVA	0.02	0.04
B	600 kVA	0.01	0.05

Determine the power output and the power factor of each transformer.

Solution: Total load kVA = 800/0.8 = 1000 kVA

$\therefore \quad\quad\quad\quad S = 1000 (0.8 - j\, 0.6) = (800 - j\, 600)$ kVA

The pu resistances/reactances of the two transformers should be brought to the same kVA level, say 1000 kVA.

$\therefore \quad R_A = 0.02 \times \dfrac{1000}{400} = 0.05; \quad\quad X_A = 0.04 \times \dfrac{1000}{400} = 0.1$

$\quad\quad R_B = 0.01 \times \dfrac{1000}{600} = 0.0167; \quad\quad X_B = 0.05 \times \dfrac{1000}{600} = 0.0833$

$\quad\quad S_B = S\dfrac{Z_A}{Z_A + Z_B} = (800 - j600) \times \dfrac{0.05 + j0.1}{0.05 + j0.01 + 0.0167 + j0.0833}$

$\quad\quad\quad = (800 - j600) \times \dfrac{0.05 + j0.1}{0.0667 + j0.1833} = (414 - j392) = (386 - j208)$ kVA

$\quad\quad S_A = S - S_B = (800 - j600) - (414 \cdot j392) = (386 - j208)$ kVA

Transformer A: 386 kW, 440 kVA, 0.878 pf lagging

Transformer B: 414 kW, 570 kVA, 0.726 pf lagging

Example 2.9: Two 2200/110 V single-phase transformers are connected in parallel to supply Q common load of 125 kVA at 0.8 power factor lagging. Transformers are rated as follows:

Transformer A: 100 kVA, 0.9% resistance and 10% reactance

Transformer B: 50 kVA % resistance and 5% reactance

How will the two transformers share the common load?

Solution: The percentage impedances of the two transformers should be brought to the same kVA level, say 100 kVA.

$$Z_A = (0.9 + j10)\Omega; \quad Z_A = \frac{100}{50}(1 + j5) = (2 + j10)\Omega$$

$$\cos \phi = 0.8 \therefore \phi = 37°$$

$$\therefore \quad \text{Total load, } S = 125\angle -37° \text{ kVA}$$

$$\frac{Z_A}{Z_A + Z_B} = \frac{0.9 + j10}{0.9 + j10 + 2 + j10} = \frac{0.9 + j10}{2.9 + j20} = 0.497\angle 3°$$

$$= \frac{2 + j10}{2.9 + j20} = 0.5\angle -3°$$

Load shared by transformer A is given by

$$S_A = S\frac{Z_B}{Z_A + Z_B} = 125\angle -37° \times 0.5\angle -3° = 62.5\angle -40° \text{ kVA}$$

Load shared by transformer B is given by;

$$S_B = S\frac{Z_A}{Z_A + Z_B} = 125\angle -37° \times 0.497\angle 3° = 62.12\angle -34° \text{ kVA}$$

Example 2.10: Two single phase transformers A and B rated at 600 kVA and 500 kVA respectively are operated in parallel to supply a load of 1000 kVA at 0.8 lagging power factor. The resistance and reactance of transformer A are 3% and 6.5% while that of transformer B are 1.5% and 8%. Calculate the kVA loading and the power factor at which each transformer operates.

Solution:

$$S = 1000 (0.8 - j\, 0.6) = (800 - j\, 600) \text{ kVA}$$

Choosing 1000 kVA as the base kVA, we have

$$Z_A \text{ (pu)} = \frac{1000}{600}(0.03 + j0.065) = 0.05 + j0.108$$

$$Z_B \text{ (pu)} = \frac{1000}{500}(0.015 + j0.08) = 0.03 + j0.16$$

$$\therefore \quad S = S\frac{Z_B}{Z_A + Z_B} = (800 - j600) \times \frac{0.03 + j0.16}{0.08 + j0.268} = 584.2\angle -30.9° \text{ kVA}$$

Therefore, transformer *A* will carry a load of 584.2 kVA at a pf of 0.86 lagging.

$$S_B = S\frac{Z_A}{Z_A + Z_B} = (800 - j600) \times \frac{0.05 + j0.108}{0.08 + j0.268} = 426.5\angle -45.15° \text{ kVA}$$

Therefore, transformer *B* will carry a load of 426.5 kVA at a pf of 0.7 lagging.

Example 2.11: When transformers A and B are running in parallel and supply a load of 2500 kVA at 0.8 lagging power factor.

The ratings of the transformers are as under:

	Transformer A	Transformer B
Voltage ratio	4500/16000 V	4500/16000 V
kVA rating	1500	1500
Total resistance (Ω)	0.819	0.8
Total reactance (Ω)	2.503	2.31

Find the current supplied by each transformer.

Solution:
$$Z_A = (0.08 + j\,2.503)\Omega = 2.633 \angle 71.88° \; \Omega$$
$$Z_B = (0.819 + j\,2.31) \; \Omega = 2.633 \angle 71.88° \; \Omega$$
$$Z_A + Z_B = (1.619 + j\,4.513) \; \Omega = 5.075 \angle 71.4° \Omega$$

Total current, $I = \dfrac{2500 \times 10^3}{16000} = 156.3 \, \text{A}$

Taking secondary terminal voltage as the reference phasor, we have

$$I = 156.3 \angle 36.87° \; \text{A} \quad (\cos \phi = 0.8 = \phi = 36.87°)$$

$$I = I \times \frac{Z_B}{Z_A + Z_B} = 156.3 \angle -36.87° \times \frac{2.445 \angle 70.9°}{5.078 \angle 71.4°}$$
$$= 75.26 \angle 37.37° \, \text{A}$$

$$I_a = I \times \frac{Z}{Z_A + Z_B} = 156.3 \angle -36.87° \times \frac{2.633 \angle 71.88°}{5.078 \angle 71.4°}$$
$$= 81.04 \angle -36.39° \, \text{A}$$

Note that transformer B carries about 7% more load than transformer A. Therefore working to the maximum capacity of 3000 kVA is impractical since one transformer must work above the full load.

2.23 SINGLE PHASE UNEQUAL VOLTAGE RATIO TRANSFORMERS IN PARALLEL

Figure 2.36 shows two single-phase unequal voltage ratio transformer A and B in parallel. Since the voltage ratio of the transformers are unequal, their no-load secondary voltages will also be unequal. We shall calculate how load current is shared between the two transformers.

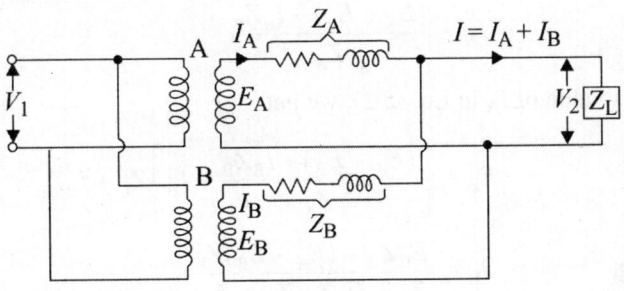

Fig. 2.36

Figure 2.37 shows the equivalent circuit of the transformers referred to secondary in a simplified way.

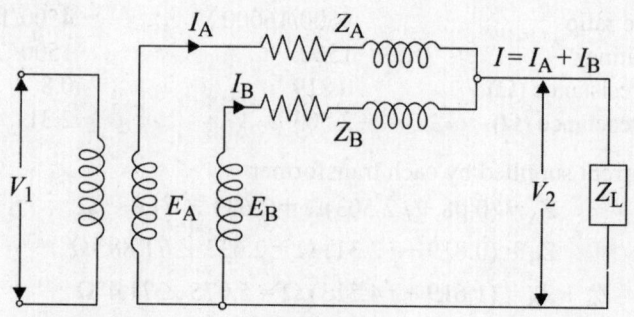

Fig. 2.37

Let E_A, E_B = no-load secondary voltages of the two transformers. It is assumed that $E_A > E_B$ and I_A, I_B = their respective currents

$\quad I$ = load current

$\quad Z_A$, Z_B = impedances of the transformers referred to secondary

$\quad Z_L$ = load impedance

$\quad V_2$ = load voltage

At no-load, the circulating current I_C is

$$I_C = \frac{E_A - E_B}{Z_A + Z_B}$$

When the system is loaded, the load current I is shared by the two transformers. By Kirchhoff's voltage law;

$$E_A = V_2 + I_A Z_A$$

and $\qquad E_B = V_2 + I_B Z_B$

But $\qquad V_2 = I Z_L = (I_A + I_B) Z_L$ ...(2.1)

∴ $\qquad E_A = (I_A + I_B)Z_L + I_A Z_A$ ...(2.2)

and $\qquad E_B = (I_A + I_B)Z_L + I_B Z_B$

Now $\qquad E_A - E_B = I_A Z_A - I_B Z_B$

or $\qquad I_A = \dfrac{(E_A - E_B) + I_B Z_B}{Z_A}$

Putting this value of I_A in Eq. (2.2), we get

$$E_B = \left[\frac{(E_A - E_B) + I_B Z_B}{Z_A} + I_B \right] Z_L + I_B Z_B$$

On solving, $\qquad I_B = \dfrac{E_B Z_A - (E_A - E_B)Z_L}{Z_A Z_B + Z_L(Z_A + Z_B)}$ ...(2.3)

From the symmetry of the expression, we get

$$I_A = \frac{E_A Z_B + (E_A - E_B)Z_L}{Z_A Z_B + Z_L(Z_A + Z_B)} \qquad \text{...(2.4)}$$

Also

$$I = I_A + I_B = \frac{E_A Z_B + E_B Z_A}{Z_A Z_B + Z_L(Z_A + Z_B)}$$

$$V_2 = IZ_L = \left[\frac{E_A Z_B + E_B Z_A}{Z_A Z_B + Z_L(Z_A + Z_B)}\right] Z_L$$

or

$$V_2 = \frac{E_A Z_B + E_B Z_A}{\dfrac{Z_A Z_B}{Z_L} + Z_A + Z_B}$$

Since the transformers have a common primary voltage, E_A and E_B will be in phase with each other.

SOLVED UNIVERSITY PROBLEMS FOR PRACTICE

Example 2.12: A transformer has copper loss of 4.5% and reactance drop of 3.5% when tested at full load calculate its full load regulation at: (*a*) 0.8 pf lagging (*b*) 0.8 pf leading **[UPTU 2003–04]**

Solution: Given: $P_{Cu} = 1.5\%$, $P_i = 3.5\%$.

(*a*) At 0.8 pf lagging

$$\% R = (V_r \cdot \cos\phi + V_x \sin\phi) + \frac{1}{200}(V_x \cos\phi - V_r \sin\phi)^2$$

$$= (1.5 \times 0.8 + 3.5 \times 0.6) + \frac{1}{200}(3.5 \times 0.8 - 1.5 \times 0.6)^2$$

$$\% R = 3.3 + 0.02 = 3.32 \text{ **Ans.**}$$

(*b*)

$$\% R = (V_r \cdot \cos\phi - V_x \sin\phi) + \frac{1}{200}(V_x \cos\phi + V_r \sin\phi)^2$$

$$= (1.5 \times 0.8 - 3.5 \times 0.6) + \frac{1}{200}(3.5 \times 0.8 + 1.5 \times 0.6)^2$$

$$\% R = -0.83 \text{ **Ans.**}$$

Example 2.13: The iron loss of a 100 kVA, 1000 V/250 V single-phase 50 Hz transformer is 1000 W. The copper loss, when primary carries a current of 50 A is 500 W. Calculate:

(*a*) Area of cross-section of limb if working flux density is 0.9 T and primary has 1000 turns.

(*b*) Primary and secondary currents.

(*c*) Efficiency at full load and 0.8 pf. **[UPTU 2005–06]**

Solution: Given: $P_i = 1000$ W, $S = 1000$ kVA, $E_1 = 1000$ V, $E_2 = 250$ V, $f = 50$ HZ, $I_1 = 50$ A, $P_{Cu} = 500$ W.

(a) $N_1 = 1000, B = 0.9$ T

We know that $E_1 = 4.44 B_m \cdot A \cdot f \cdot N_1$

$1000 = 4.44 \times 0.9 \times A \times 50 \times 1000$

$\therefore$ $A = 50$ cm^2

(b) Primary and secondary current

$S = V_1 I_1$

$\therefore$ $I_1 = \dfrac{100 \times 10^3}{1000} = 100$ A   **Ans.**

$S = V_2 I_2$

$\therefore$ $I_2 = \dfrac{100 \times 10^3}{250} = 400$ A   **Ans.**

(c) Efficiency at 0.8 pf, $x = 1$ for full load.

$\therefore$ $\eta = \dfrac{S \cdot x \cdot \cos\phi}{S \cdot x \cdot \cos\phi + P_i + x^2 P_{Cu}}$

$\therefore$ $\eta = \dfrac{100 \times 1 \times 0.8}{100 \times 1 \times 0.8 + 0.1 + (1)^2 \times .2} \times 100 = 91.38\%$

Example 2.14: A 20 kVA, 50 Hz, 2000/200 V, single phase transformer has iron loss of 120 W and full load copper loss of 300 W. Low voltage side of the transformer is loaded at 0.8 lagging pf, calculate maximum efficiency of transformer.

[UPTU 2006-07]

Solution: Given: $S = 20$ kVA, $F = 50$ Hz, $V_1 = 2000$ V, $V_2 = 200$ V, $P_i = 120$ W (FL) $= 300$ W, $\cos\phi_2 = 0.8$ lag.

kVA load for maximum efficiency $= \sqrt{\dfrac{\text{Iron loss}}{P_W \text{ loss}}} \times$ kVA

$= 20 \times \sqrt{\dfrac{120}{300}}$

$= 12.65$ kW

$\therefore$ $P_{Cu} = (0.6324)^2 \times P_{Cu} \text{(FL)}$

$= (0.6324)^2 \times 300 = 120$ W

$\%\eta_{max} = \dfrac{0.6424 \times 10^3 \times 0.8}{0.6324 \times 10^3 \times 0.8 + 120 + 120} \times 100$

$= 97.68\%$

Example 2.15: No-load measurements on a 230 V/115 V, transformer give the following readings measured on the low voltage side: 115 V, 60 Hz, 80 W and 3A.

Determine parameters and shunt branch of equivalent circuit, primary winding impedance is $(0.02 + j\,0.1)\,\Omega$.   **[UPTU 2006-07]**

Solution: Given: $V_1 = 230$ V, $V_2 = 115$ V, $f = 60$ Hz, $W_0 = 80$ W, $I_0 = 3$ A

We know that,

$$W_0 = V_2 I_0 \cos\phi_0$$
$$80 = 115 \times 3 \times \cos\phi_0$$

$\therefore$

$$\cos\phi_0 = 0.222$$
$$\phi_0 = 77.16°$$

$\Rightarrow$

$$I_W = I_0 \cos\phi_0 = 3 \times 0.222 = 0.667 \text{ A}$$
$$I_\phi = I_0 \sin\phi = 3 \times \sin(77.16) = 2.925 \text{ A}$$

$$K = \frac{V_2}{V_1} = \frac{115}{230} = 0.5$$

$$R_0 = \frac{V_2}{I_w} = \frac{115}{0.669} = 172.41\,\Omega$$

$$X_0 = \frac{V_2}{I_\phi} = \frac{115}{2.925} = 39.32\,\Omega$$

EXERCISES

1. Derive the emf equations of single-phase transformer.   [UPTU 2003–04]

2. Draw the equivalent circuit of single phase transformer. Explain the tests performed to obtain the parameters of the equivalent circuit.[UPTU 2003–04]

3. Derive voltage regulation of a transformer.   [UPTU 2004–05]

4. Define transformer. Explain its construction and working.   [UPTU 2004–05]

5. Draw no-load phasor diagram of transformer and explain significance of each phasor. Also draw corresponding equivalent circuit.   [UPTU 2004–05]

6. Discuss the effects of load and load power factor on the efficiency of transformer. Obtain expression for fraction of load at which maximum, efficiency occurs.
     [UPTU 2004–05]

7. Define voltage regulation of transformer and explain its significance. With the help of approximate equivalent circuit and phasor diagram, obtain an expression for it.   [UPTU 2004–05]

8. Give circuit diagram and explain the procedures of open circuit and short circuit tests on two winding transformer. How can parameters of equivalent circuit be determined from these tests?

9. Derive emf equation of a single-phase transformer and show that for an ideal transformer, voltage are transformed in direct ratio of turns write currents in the inverse ratio terms. Find the transformation ratio of impedances.
     [UPTU 2006–07]

10. Draw full-load diagram of a single phase transformer supplying a lagging pf load. Explain significance of each phasor.   [UPTU 2007–08]

11. Explain the operation of a single phase transformer.   [UPTU 2008–09]

Three Phase Transformer

3.1 INTRODUCTION

- In chapter 2, we have discussed about the basic of a single phase transformer.
- Generation, transmission and distribution of electrical energy is invariably done through the use of three phase system because of several advantages over single phase system.
- The electric power is generally generated in the form of three phase voltage. The distribution of the electric power also is three phase.
- The voltage generation takes place at a very high level typically at 11 kV or 33 kV. The transmission of electricity takes place at voltage levels from 11 kV to 132 kV. The three phase transformer are used for the same. Such high voltage level can not be used by the domestic or industrial users. So the voltage should be stepped down.

Three phase step-down transformer are used as distribution transformer.

3.2 CONSTRUCTION OF THREE PHASE TRANSFORMER (BANK OF SINGLE PHASE TRANSFORMERS)

- The three single phase transformers are connected such that the three phase primary windings are connected in delta or star.
- Similarly the secondary windings are connected in star or delta as shown in Fig. 3.1.
- But such a construction proves to be costly and makes the transformer bulky. Therefore, specially made three phase transformers are being used now-a-days.

3.2.1 Advantages of Bank of One Phase Transformers

- It is easy to carry spare stock of a single phase transformers.
- In the underground (mining) applications, the bank is preferred due to easy transportation.

3.2.2 Disadvantages of Bank of One Phase Transformers

- The bank of transformers cost more.
- It occupies more space.

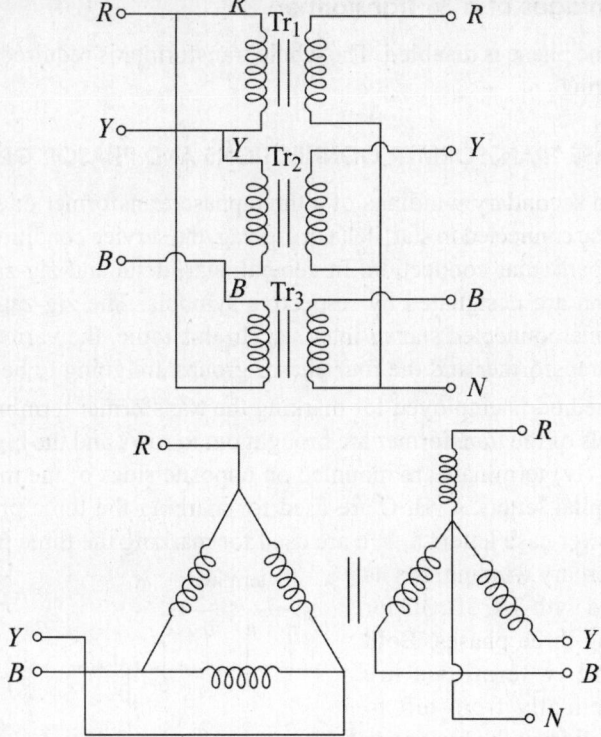

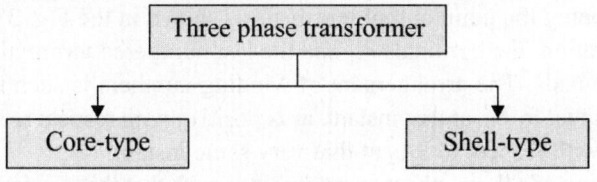

Fig. 3.1: Construction of a 3φ transformer using 1φ transformers

- The three phases are electrically connected but the three magnetic circuits are independent.

3.3 TYPES OF THREE PHASE TRANSFORMER

- The three phase transformers are of two types (Fig. 3.2).

Three phase transformer

Core-type   Shell-type

Fig. 3.2

3.3.1 Advantages of 3φ Transformer

- It occupies less space
- Less weight
- Low cost
- It acts as a single unit to handle

3.3.2 Disadvantages of a 3φ Transformer

- If even one phase is disabled. The whole transformer is required to be removed for repairing.

3.4 THREE PHASE TRANSFORMER CONNECTIONS AND PHASOR GROUPS

The primary and secondary windings of a three phase transformer or a bank of three transformer can be connected in star, delta or zig-zag, the service conditions determining the choice of a particular connection. In general, star, delta and zig-zag connections are employed and are designated by respective symbols. The zig-zag connection is also known as interconnected star or inter star. In this topic, the various connections for three phase transformer and the four phasor groups are going to be discussed.

A standard method is employed for marking the transformer terminals. As per the ISR, the terminals of the transformer are brought out in rows and the high voltage (hv) and low voltage (lv) terminals are mounted on opposite sides of the main tank of the transformer. Capital letters A, B, C are used for marking the three phases of the hv terminals and lower case letters a, b, c are used for marking the three phases of the lv terminals. The tertiary winding, has its terminals marked with 3A, 3B, 3C for its corresponding three phases. Both leads of hv and lv terminals are marked alphabetically from left to right when viewed from the hv side as depicted in the Fig. 3.3(a). The neutral, if provided, is fitted on the extreme left, see Fig. 3.3(a) and further, the letters are assigned with numbers to serve as subscript.

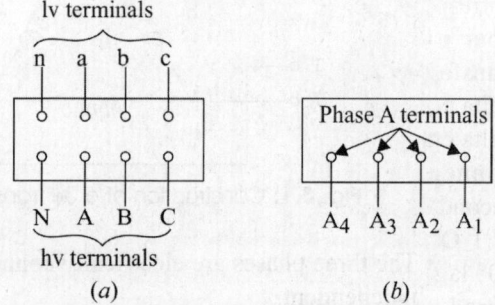

Fig. 3.3: Illustration showing the terminal markings of three phase transformers

For eaxmple, the symbol A_1 is used for the extreme right terminal, A_2 for the second right and A_4 is used for the extreme left terminal exceeding A_3. In this care, four terminals are brought out from phase A and A_1 and A_4 being the main output terminals as shown in the Fig. 3.3(b).

For generalization, the terminals A_1 and the last numbered terminal are considered as the main terminals. The arrangement of winding numbers is such that when A_1 is positive with respect to A_2, at that instant, a_1 is positive with respect to a_2 and similarly $3A_1$ is positive with respect to $3A_2$ at that very same instant.

In representation of a three–phase transformer, is such that the hv winding is indicated by a capital letter (e.g. 'Y' for star connection, 'D' for delta connection, 'Z' for zig-zag connection etc.) and lv winding by a lower case letter such as star as 'Y', delta as 'd', etc. Thus on this basis of the above assumption, the symbol "Yd" would represent a three phase transformer with its hv winding connected in star and lv winding connected in delta. In various transformer connections, its possible that there is no phase displacement between the hv line emf and lv line emf and its also possible that a particular amount of phase displacement is present. This phase displacement can be expressed in degrees, but a more convenient method is employed. This method is known as *clock method of angle designation.*

The following are the considerations of this method. The minute hand is designated as the hv line phasor and is always set at 12 O' clock position (zero hour position) and the hour hand is designated with the lv line phasor correspondingly.

As per the clock method, for a two-winding, three-phase transformer, the hv connection is represented by the first symbol, lv winding connection by the second symbol and finally, the third symbol represents the phase displacement between hv and lv line emfs which is expressed as clock hour number. For example, lets consider the symbol "Yd_{11}". This represents a polyphase transformer with hv winding connected in star (depicted by 'Y'), lv winding connected in delta (depicted by 'd') and the lv line phasor at 11 O'clock which depicts 30° ahead of the zero hour position of the hv line phasor as shown in Fig. 3.4(*a*). If a three winding, three phase transformer is considered, the first letter is concerned with the highest voltage winding connection and the next two with the other two winding connections in the decreasing order of voltage and the last two letters are considered with clock hour number of the corresponding winding, on considering the hv winding as reference along the zero-hour position. For example, symbol Dy1y1 is used to represents a three phase transformer with the hv side in delta connected and lv side star delta connected and lv side star connected with both the secondary and tertiary line emfs at 1 O'clock position (–30° or 30° lag) and hv line emf considered as reference at Yy0d1 represents a three phase transformer with three windings with its hv and lv windings connected in star and the tertiary winding connected in delta. The hv and lv line emfs are in phase with each other and the line emf of the tertiary winding lags by 30° from the reference hv line emf.

The angle of phase displacement between the hv and lv line emfs are dependent upon the method of connection of the three phases and the direction in which the windings have been wound. Thus, for a three phase

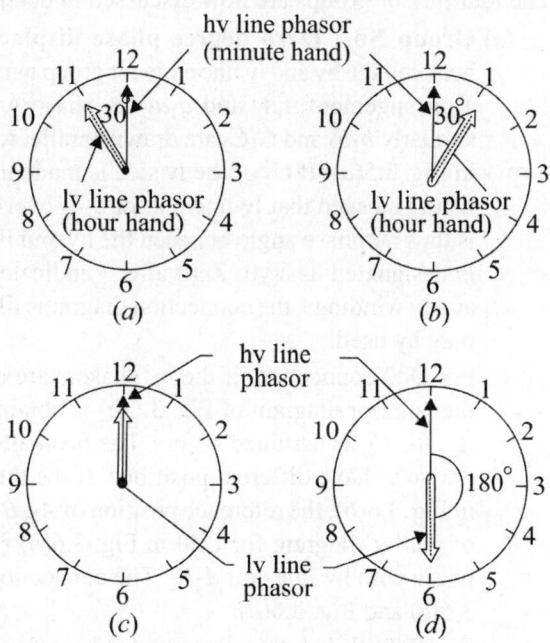

Fig. 3.4: The minute hand (long hand) represents the hv line phasor and its treated as fixed. The hour hand (short hand) represents lv line phasor, (a) Group No. 4 (b) Ground No. 3 (c) Group No. 1 (d) Group No. 2

transformer, the knowledge of connections hv and lv connections (i.e. star, delta or interconnected star) is not enough. In addition, the phase angle between the *hv* and *lv* line emfs must also be known.

The three phase transformers are classified under four phasor groups depending upon the type of connections employed and also the phase displacement angle between the hv and lv line emfs. The phase displacement for each group is different and the following points should be considered for proper understanding.

In the phasor diagram, the counter clockwise direction is the direction of rotation of the phasor. The phasors that represent the hv and lv windings induced emfs for a particular phase are drawn parallel to each other. For example, the lv phasor $a_1 a_2$ of phase A must be drawn parallel to the phasor $A_1 A_2$ of the h.v. winding, on assuming the hv winding to primary and lv winding the secondary. The position of hv terminals A_2, B_2, C_2 is assumed to be fixed, at the top, bottom right and bottom left of the equilateral triangle ($\Delta A_2 B_2 C_2$) respectively for the sake of convenience. on account of this, the phasors for star and delta connections on the hv side should be drawn in different positions.

3.4.1 Phasor Groups

The four phasor groups are now discussed in detail.

    (*a*) **Group No.1 (zero degree phase displacement):** The phase displacement between the hv and lv line emfs for group number 1 is zero as shown in Fig. 3.5 for Y_y arrangement $A_1 A_2$ and $a_1 a_2$ for phase A are drawn parallel to one another similarly $b_1 b_2$ and $C_1 C_2$ are drawn parallel to $B_1 B_2$ and C_1 respectively as shown in Fig. 3.5(*a*). If C_2 of the lv side is made to coincide C_2 of the hv side, then it would be seen that lv line phasor $c_2 b_2$ overlaps the hv line phasor $C_2 B_2$. There is thus no phase angle between the hv and lv line emfs and star-star connection is designated as Yy0. Zero after *y* indicates zero phase displacement. For hv and lv windings, the connection diagrams illustrated in Fig. 3.5(*b*) or Fig. 3.5(*c*) may by used.

    For Dd0 connection, if the hv phasors are drawn in the position of Fig. 3.5(*a*) then phasor diagram of Fig. 3.6(*a*) is obtained. Since the reference position of A_2, B_2, C_2 as assumed before, has been altered, the three hv phasors must be drawn to have different positions. If the three phasors are, drawn as indicated in Fig. 3.6(*b*), the reference position of A_2, B_2, C_2 is maintained. An examination of phasor diagram for Dd0 in Fig. 3.6(*b*) reveals that Lv. line emf $a_2 b_2$ is in phase with hv line emf $A_2 B_2$. The connection diagrams are as illustrated in Fig. 3.6(*c*) and Fig. 3.6(*d*).

    For Dz0, the phasor diagram for hv side is as shown in Fig. 3.6(*b*). The lv winding, for convenience in drawing the lv phasor diagram, is indicated in Fig. 3.7(*a*) in sectionalised form. Terminals $a_2 b_2 c_2$ are joined together to form the neutral *n*. Further, join c_1 with a_3, a_1 with b_3 and b_1 with c_3 to obtain the lv zigzag phasor diagram, which also helps in drawing the connection diagram of

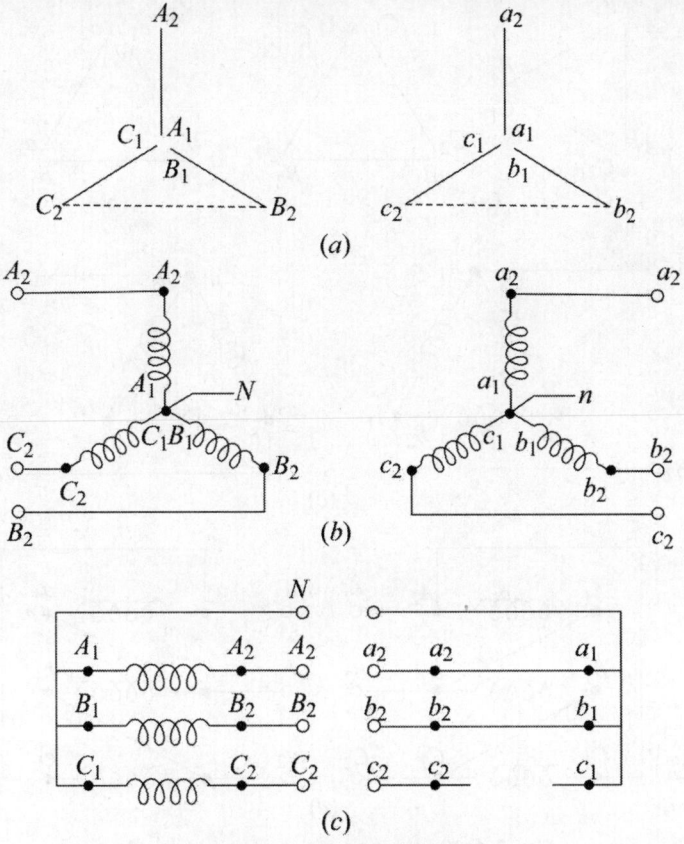

Fig. 3.5: Y_{y0} arrangement No. 1

Fig. 3.7(c). The lv line phasor C_4b_4 is parallel to the corresponding hv line phasor C_2B_2 in Fig. 3.6(b), therefore, no phase displacement between hv and lv line emfs.

Zigzag or interstar connection is primarily used to suppress the third harmonic emfs between line and neutral. The third harmonic emfs are directed in the same direction, *i.e.* from a_1, b_1, c_1 to a_4, b_4, c_4 as illustrated in Fig. 3.7(c) If the winding is traversed form neural along the path n c_2 c_1 a_3 a_4 it is seen that emf in c_2c_1 is $(-E_3)$ and in a_3 a_4, it is E_3; where E_3 is the third harmonic emf induced in each half lv section. The resultant third harmonic emf between neutral and line is, therefore, zero. If E_1 is the fundamental–frequency voltage of each half lv section, then the magnitude of line to neutral voltage is found to be $\sqrt{3}E_1$ from the phasor diagram of Fig. 3.7(b). Had the 1.v. winding been in star, the line to neutral voltage would be $2E_1$. Therefore, the phase voltage (or line voltage) in interstar is reduced to $\dfrac{\sqrt{3}E_1}{2E_1} = 0.866$ of its value in star. If the voltage

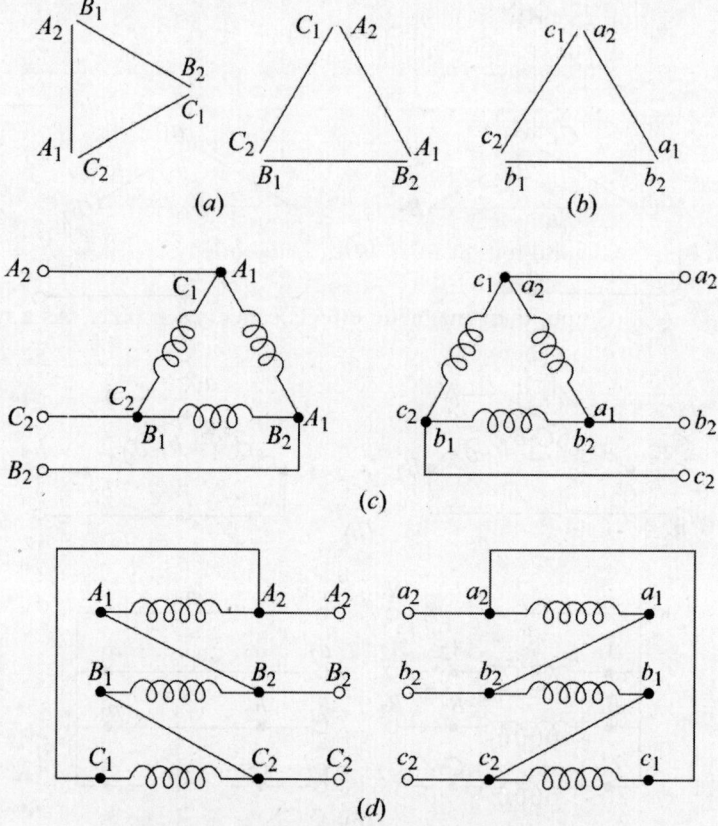

Fig. 3.6: Y_{y0} arrangement of group no. 1

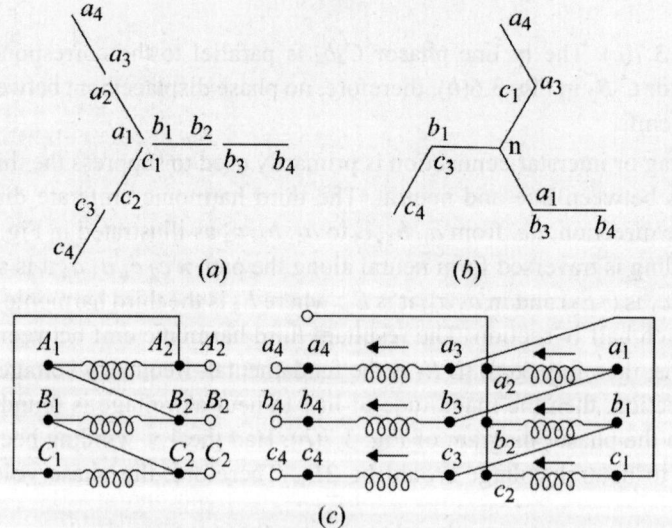

Fig. 3.7: $Dz0$ arrangement of group no. 1

$2E_1$ is to be retained in interstar, then the lv turns must be increased to $1.15 = \left(\dfrac{2}{\sqrt{3}}\right)$ times their value in star. When this is done, interstar connection becomes expensive, because it involves extra copper, additional insulation, more iron, etc. A Yy transformer cannot supply a line to neutral load, whereas Yz transformer can do so.

Interstar connections are used for rectifier circuits. The direct current, after entering the neutral n in Fig. 3.7(c), divides equally in all the three secondary phases. Since each half of the secondary winding carries direct current in opposite direction, their magnetic effects cancel each other. As a result of it, the core flux and, therefore, core loss and temperature rise remain uneffected. Had the secondary been star connected instead of zigzag, the direct current would divide equally in three phases. The flux produced by direct current would add to the alternating flux, thereby increasing the core flux, core loss and temperature rise of the transformer.

(b) **Group No. 2 (180° phase displacement):** The phase displacement between the hv and lv line emfs for group number 2 is 180° as shown in Fig. 3.4(d). For this group, the connections of all the secondary terminals are reversed with respect to the primary connections. For example, in a star–star transformer, the primary terminals $A_2B_2C_2$ are joined together to form neutral N, whereas on the secondary side. terminals a_2, b_2, C_2 must be joined together to form the neutral n as shown in Fig. 3.8.

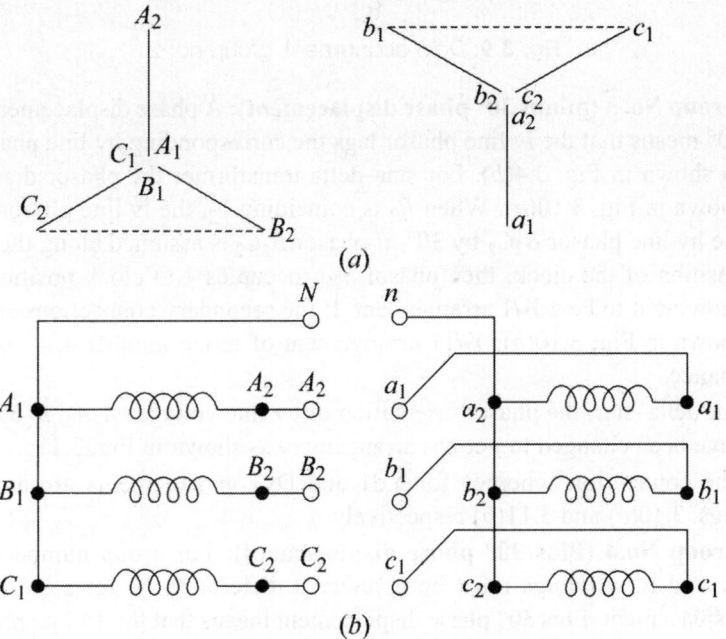

Fig. 3.8: Dd6 arrangement, group no. 2

The hv side has A_2, B_2, C_2 as the input terminals and lv side has a_1 as the output terminals. If b_1 is made to coincide B_2, then the phase angle between hv line phasor B_2C_2 and lv line phasor b_1C_1 is 180° giving Yy_6 arrangement. The connection arrangement is illustrated in Fig. 3.8(b).

For $Dd6$, the h.v. terminals A_1 and B_2 are joined together as before. For reversing the secondary connections, a_2 and b_1 should be joined together as shown in Fig. 3.9(a). The input terminals on the hv side are A_2, B_2, C_2 and the lv output terminals are a_1, b_1, c_1. The phase angle between B_2C_2 and b_1c_1 is seen to be 180°. The connection arrangement is as shown in Fig. 3.9(b).

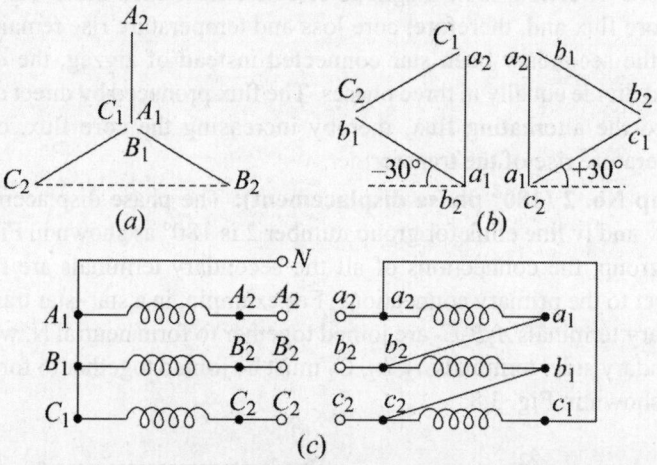

Fig. 3.9: $Dd6$ arrangment, group no. 2

(c) Group No. 3 (minus 30° phase displacement): A phase displacement of minus 30° means that the lv line phasor lags the corresponding hv line phasor by 30° as shown in Fig. 3.4(b). For star–delta transformer the phasor diagram is as shown in Fig. 3.10(a). When B_2 is coinciding b_2, the lv line phasor b_2C_2 lags the hv line phasor B_2C_2 by 30°, if phasor B_2C_2 is assumed along the zero hour position of the clock, then phasor b_2c_2 occupies 1 O'clock position, thereby showing it to be a $Yd1$ arrangement. If the secondary connections are made as shown in Fig. 3.10(b), $Yd11$ arrangement of group number 4 is obtained per chance.

For delta–star, the phasor orientation of hv line voltages A_1A_2, B_1B_2 and C_1C_2 have been changed to get Dy_1 arrangement as shown in Fig. 3.11.

The connection schemes for Yd1 and Dy1 arrangements are as shown in Figs. 3.10(c) and 3.11(b) respectively.

(c) Group No.4 (Plus 30° phase displacement): For group numbers 3 and 4, hv and lv windings must be connected differently to get a suitable phase displacement. Plus 30° phase displacement means that the lv line phasor leads the corresponding hv line phasor by 30° as shown in Fig. 3.4(a). For $Dy1$

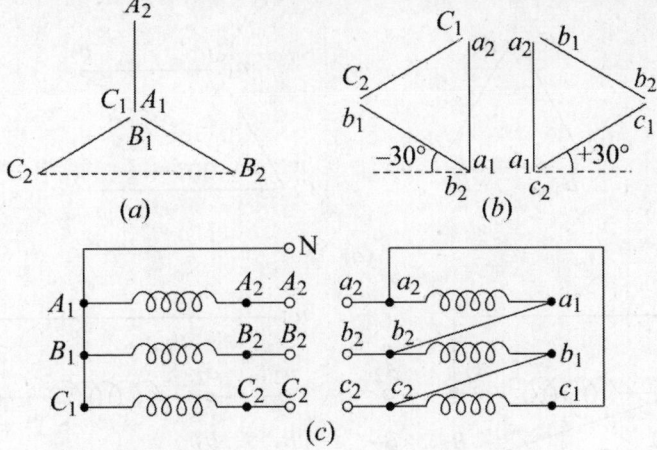

Fig. 3.10: Yd1 arrangement group no. 3

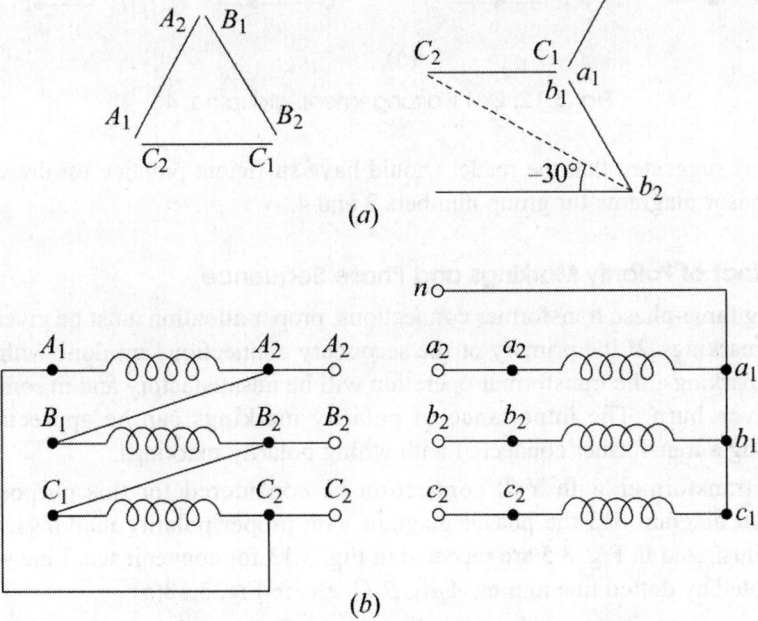

Fig. 3.11: Dy1 arrangement, group no. 3

arrangement, the phasor diagram is drawn in Fig. 3.12(*a*) When C_2 coincides c_2, lv phasor c_2b_2 is seen to lead lv phasor C_2B_2 by 30° and this proves it to be *Dy*11 arrangement. The connection scheme for *Dy*11 is drawn in Fig. 3.12(*b*). The phasor diagram for the lv side of *Yd*11 is already shown in Fig. 3.10(*b*) for which the connection diagram can be drawn accordingly.

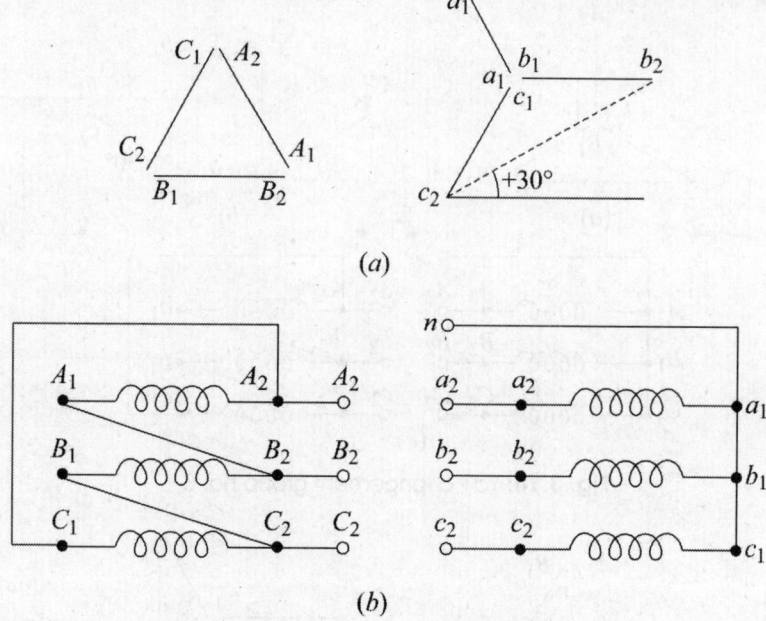

(a)

(b)

Fig. 3.12: Dy11 arrangement, group no. 4.

It is suggested that the reader should have sufficient practice for drawing the phasor diagrams for group numbers 3 and 4.

3.4.2 Effect of Polarity Markings and Phase Sequence

In making three-phase transformer connections, proper attention must be given to the polarity markings. If the primary or the secondary connections are done with wrong polarity markings, the transformer operation will be unsatisfactory and in some cases it may even burn. The importance of polarity markings can be appreciated by considering a transformer connected with wrong polarity markings.

Let a transformer with Yy0 connection be considered for this purpose. The connection diagram and the phasor diagram with proper polarity markings, though already illustrated in Fig. 3.5 are repeated in Fig. 3.13 for convenience. Line voltages are indicated by dotted line joining A_2B_2, B_2C_2 etc. in Fig. 3.13(b).

Now suppose the lv (secondary) winding connection a_1 and a_2 are reversed without altering lv winding connections A_1 and A_2 as shown in Fig. 3.14(a). With this change, lv phasor diagram of Fig. 3.13(b), gets modified to that shown in Fig. 3.14(b). Note that phasor a_2a_1 is reversed from the position it occupied in Fig. 3.13(b). The secondary line voltages are indicated by dotted lines in the phasor diagram of Fig. 3.14(b). In order to show the line voltages clearly, the phasor diagram of Fig. 3.14(c) is drawn. The three line voltages on the lv sides are seen to be unbalanced. The line voltages Va_1b_2 and Vc_2a_1 have their magnitudes equal to the phase voltage and they are time

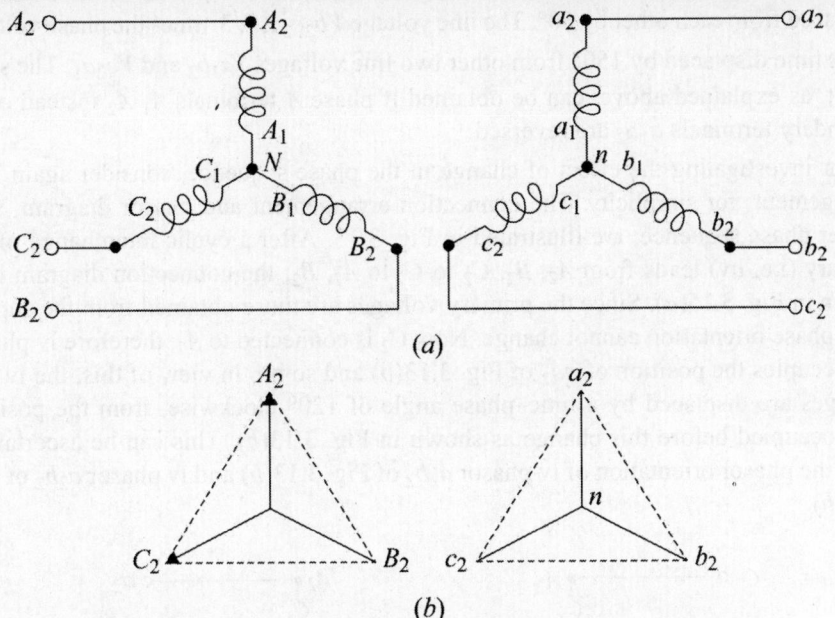

Fig. 3.13: Yy0 arrangement with correct polarity markings

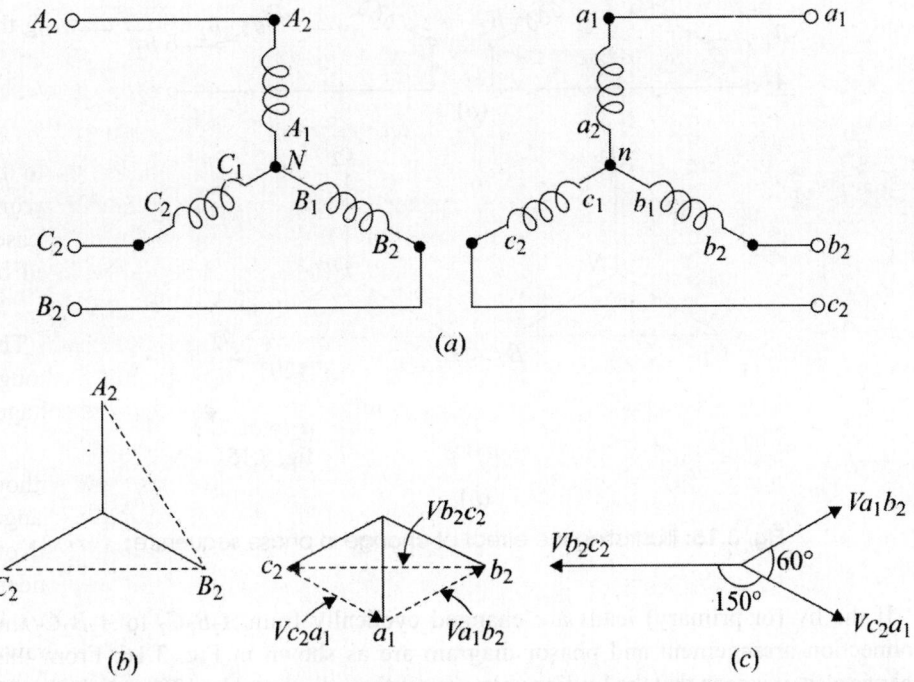

Fig. 3.14: Illustration pertaining to the effect of connecting three secondary windings inaccurately

displaced from each other by 60°. The line voltage Vb_2c_2 is $\sqrt{3}$ times the phase voltage, but is time displaced by 150° from other two line voltages Va_1b_2 and Vc_2a_1. The same effect, as explained above, can be obtained if phase A terminals A_1 A_2 instead of its secondary terminals a_1a_2 are reversed.

For investigating the effect of change in the phase sequence, consider again Yy0 arrangement, for simplicity. The connection arrangement and phasor diagram, with proper phase sequence, are illustrated in Fig. 3.13. After a cyclic interchange of the primary (i.e. hv) leads from A_2, B_2, C_2 to C_2 to A_2, B_2; the connection diagram is as shown in Fig. 3.15(a). Since the primary voltages are those obtained from the supply, their phase orientation cannot change. Now C_2 is connected to A_2, therefore lv phasor nc_2 occupies the position of na_2 of Fig. 3.13(b) and so on. In view of this, the lv line voltages are displaced by a time–phase angle of 120° clockwise, from the position they occupied before this change as shown in Fig. 3.13(b). This can be ascertained from the phasor orientation of lv phasor a_2b_2 of Fig. 3.13(b) and lv phasor a_2b_2 of Fig. 3.15(b).

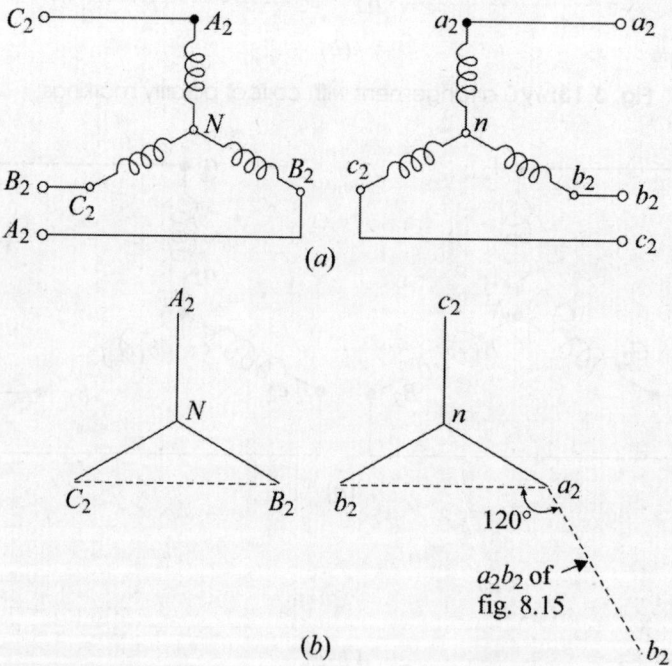

Fig. 3.15: Illustrating the effect of change in phase sequence

If the hv (or primary) leads are changed cyclically from $A_2B_2C_2$ to $A_2B_2C_2$ the connection arrangement and phasor diagram are as shown in Fig. 3.16. From this change also, it is seen that the l.v. line voltages are time-displaced by 120° anticlockwise from the position they occupied before this change. Hence, it may be concluded from

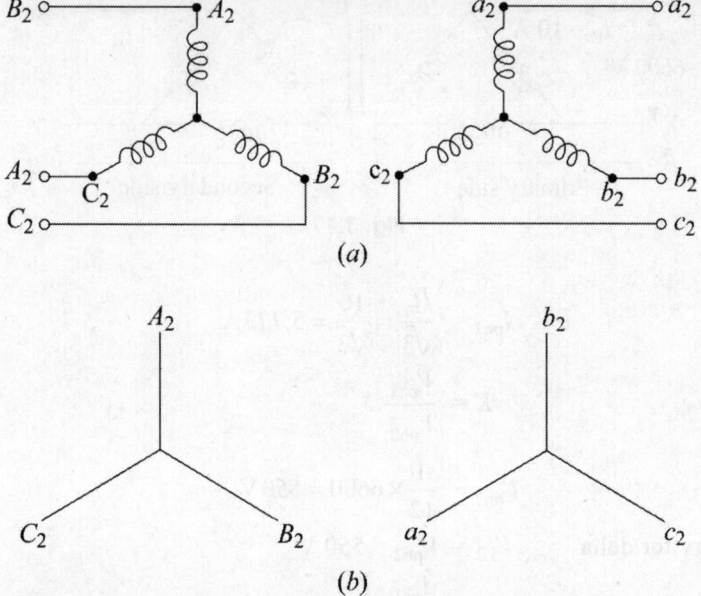

(a)

(b)

Fig. 3.16: Illustrating the effect of change in phase sequence

above that if the primary leads are interchanged cyclically once, the secondary line phasors are rotated through a time-phase angle of 120°.

In case the primary leads are interchanged cyclically twice, for example from $A_2B_2C_2$ to $C_2A_2B_2$ and then from $C_2A_2B_2$ to $B_2C_2A_2$, in Fig. 3.15 the secondary line phasors are rotated through same phase angle of 240°.

SOLVED EXAMPLES

Example 3.1: A 3-phase step down tranformer is connected to 6600 V mains and it takes 10 A. Calculate the secondary line voltage, line current, and output for following:

(a) Delta–Delta   (c) Star–Delta

(b) Star–Star   (d) Delta–Star

The turns ratio at 1 phase is 12. Draw connection diagrams.

[UPTU 2006–07; 10 marks]

Solution: Given,

$$V_{L1} = 6600 \text{ V},$$

$$K = \frac{1}{12}$$

$$I_L = 10 \text{ A}$$

(a) Delta–Delta

Primary for delta

∴

$$V_{L1} = V_{ph1} = 6600 \text{ V}$$

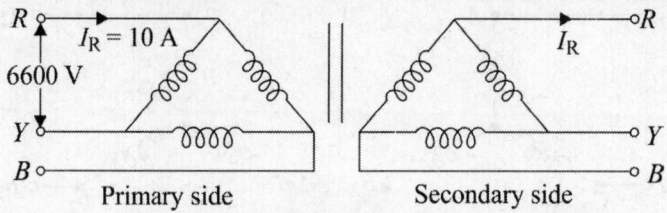

Fig. 3.17

$$I_{ph1} = \frac{I_{L_1}}{\sqrt{3}} = \frac{10}{\sqrt{3}} = 5.773\,A$$

$$K = \frac{V_{ph1}}{V_{ph2}}$$

$$V_{ph2} = \frac{1}{12} \times 6600 = 550\,V$$

Secondary for delta $\quad V_{L2} = V_{ph2} = 550$ V

$$K = \frac{I_{ph1}}{I_{ph2}}$$

$$\therefore \quad I_{ph2} = \frac{I_{ph1}}{K} = \frac{5.773}{\dfrac{1}{12}} = 69.28\,A$$

$$\therefore \quad I_{L2} = \sqrt{3} \times 550 \times 120 = 114.315\,kVA$$

(b) **Star–star connection** $V_{L1} = 6600$ V

$$I_{ph1} = \frac{V_{L1}}{\sqrt{3}} = \frac{6600}{\sqrt{3}} = 3810.5\,V$$

$$I_{L1} = I_{ph1} = 10\,A$$

$$K = \frac{I_{ph1}}{I_{ph2}}$$

• Secondary is connected in star

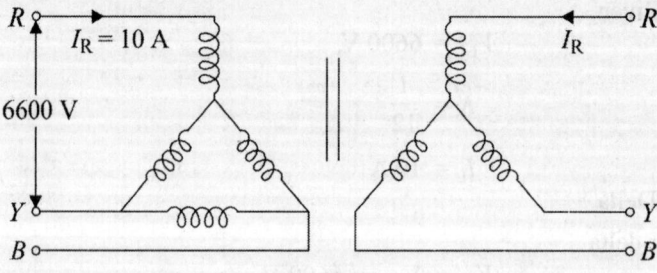

Fig. 3.18

$$I_{ph2} = \frac{I_{ph1}}{K} = \frac{10}{1/12} = 120\,\text{A}$$

$$\therefore \qquad I_{L2} = I_{ph2} = 120\,\text{V}$$

$$V_{ph2} = KV_{ph1}$$

$$V_{ph2} = \frac{1}{12} \times 3810.5 = 317.5\,\text{kV}$$

$$V_{L2} = \sqrt{3}\,V_{ph2}$$

$$V_{L2} = \sqrt{3} \times 317.54 = 550\,\text{V}$$

• Secondary output $= \sqrt{3}V_{L2}I_{L1} = \sqrt{3} \times 550 \times 120 = 114.315\,\text{kVA}$

(c) Star–delta connection

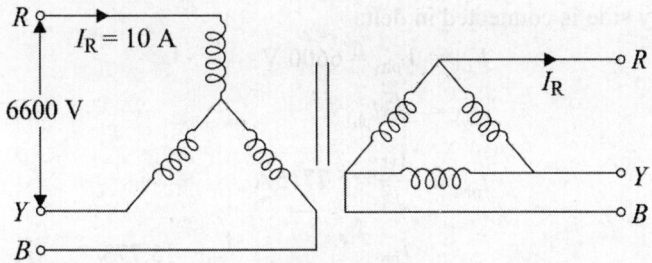

$I_R = 10\,\text{A}$

6600 V

Fig. 3.19

• Primary is connected in star

$$V_{L1} = 6600\,\text{V}$$

$$\therefore \qquad V_{ph1} = \frac{V_{L_1}}{\sqrt{3}} = \frac{6600}{\sqrt{3}} = 3810.51\,\text{V}$$

$$\therefore \qquad V_{ph2} = KV_{ph_1} = \frac{1}{12} \times 3810.51 = 317.54\,\text{V}$$

• Secondary is connected in delta

$$V_{L2} = V_{ph2} = 317.54\,\text{V}$$

$$I_{L1} = 10\,\text{A} = I_{ph1}$$

$$I_{ph2} = \frac{I_{ph_1}}{K} = \frac{10}{\dfrac{1}{12}} = 120\,\text{A}$$

$$I_{L2} = \sqrt{3}\,I_{ph_2} = \sqrt{3} \times 120 = 207.84\,\text{A}$$

• Secondary output $= \sqrt{3}V_{L2}I_{L2} = \sqrt{3} \times 317.54 \times 207.84$

$$= 114.315 \text{ kVA}$$

(*d*) Delta–star connection

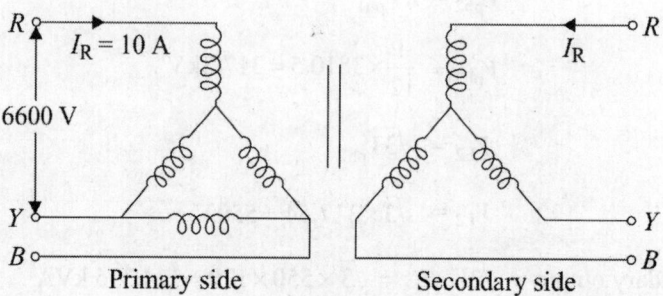

R $\quad I_R = 10 \text{ A}$

6600 V

Y

B

Primary side

I_R

R

Y

B

Secondary side

Fig. 3.20

• Primary side is connected in delta

∴ $\qquad V_{L1} = V_{ph1} = 6600 \text{ V}$

$\qquad I_{L1} = \sqrt{3}I_{ph1}$

∴ $\qquad I_{ph1} = \dfrac{10}{\sqrt{3}} = 5.7715 \text{ A}$

$\qquad K = \dfrac{I_{ph1}}{I_{ph2}} \qquad I_{ph2}l = \dfrac{I_{ph1}}{K} = \dfrac{5.773}{\left(\dfrac{1}{12}\right)} = 69.28 \text{ A}$

• Secondary is connected in star

∴ $\qquad I_{L2} = I_{ph2} = 69.282 \text{ A}$

$\qquad V_{ph2} = KV_{ph1} = V_{ph2} = \dfrac{1}{12} \times 6600 = 550 \text{ V}$

$\qquad V_{L2} = \sqrt{3}I_{ph2} = \sqrt{3} \times 550 = 952.62 \text{ V}$

• Secondary output $= \sqrt{3}\,V_{L2}I_{L2} = \sqrt{3} \times 952.62 \times 69.282 = 114.315 \text{ kVA}$

Example 3.2: A three phase transformer has delta connected primary and it star connected secondary working on 50 Hz three phase supply. The line voltages or primary and secondary are 3300V and 400V respectively. The line current on primary side is 12A and secondary has a balanced load at 0.8 lagging pf. Determine the secondary phase voltage, line current and the output. **[UPTU 2008–09]**

Solution: Given: $V_{L1} = 3300$ $I_{L1} = 12$A, $V_{L2} = 400$ V, pf $= 0.8$ (lagging)

$$V_{ph_2} = \dfrac{V_{L2}}{\sqrt{3}} = \dfrac{400}{\sqrt{3}} = 230.94 \text{ V}$$

$$I_{L1} = 12 \text{ A}$$

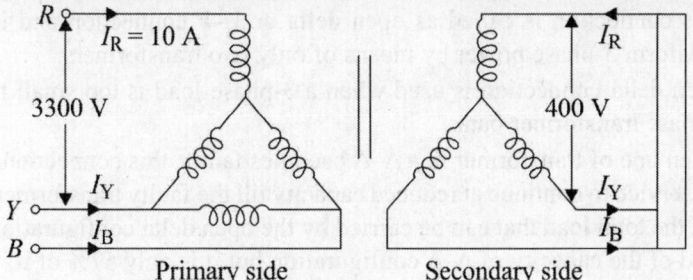

Fig. 3.21

$$I_{ph1} = \frac{I_{L_1}}{\sqrt{3}} = \frac{12}{\sqrt{3}} = 6.928\,\text{A}$$

$$K = \frac{I_{ph1}}{I_{ph2}} = \frac{V_{ph2}}{V_{ph1}}$$

∴ $$K = \frac{230.94}{3300} = 0.0699$$

∴ $$I_{ph2} = \frac{I_{ph1}}{K} = \frac{6.928}{0.069} = 99.11\,\text{A}$$

$$\text{output} = \sqrt{3}V_{L_2}I_{L_2}\cos\phi = \sqrt{3} \times 400 \times 99.11 \times 0.8 = 54.93\,\text{kW}$$

3.5 THREE PHASE TRANSFORMATION WITH TWO TRANSFORMERS

It is possible to transform three phase power by means of only two single phase transformers. There are two methods of doing so, namely (*i*) open delta connection method, and (*ii*) T-connection method, Both of these methods result in slightly unbalanced output voltage under load because of the unsymmetrical relations. This is not serious in commerical transformers, as their regulation is seldom poorer than 2 or 3%.

3.5.1 Open Delta or V–V Connection

- If we remove anyone transformer of a Δ–Δ configuration and 3-phase is connected to the primary windings as shown in Fig. 3.22, then equal 3-phase voltages are available at the secondary terminal on no-load.

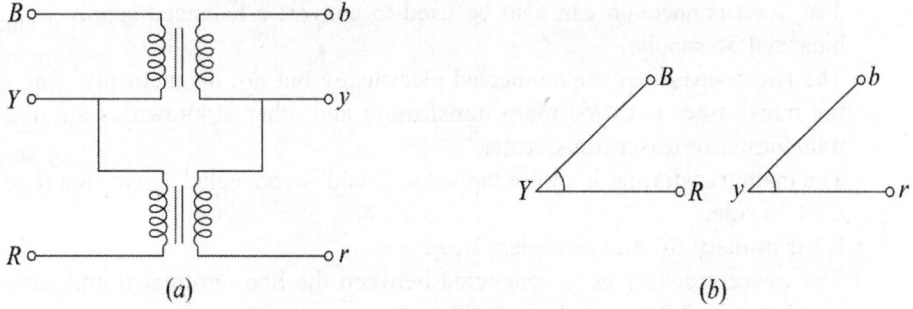

Fig. 3.22: Open-delta or V-V connection

- This connection is called as open delta or *V–V* connection and it is used to transform 3-phase power by means of only two transformers.
- Open delta connection is used when a 3-phase load is too small to use a full 3-phase transformer bank.
- When one of transformer in a Δ–Δ becomes faulty, this connections will allow the service to continue at reduced capacity till the faulty transformer is required and the total load that can be carried by the open delta configuration is not two third of the capacity of Δ–Δ configuration but it is only 57% of it.

Disadvantages

- The actual power factor at which the open delta configuration operates, is less than that of the load. It is 86.6% of the balanced load power factor.
- With increased load, the secondary terminal voltages tend to become unbalanced even with the balanced load.

3.5.2 Scott Connection/T–T Connection

- The scott connection is the most common method of connecting two single-phase transformer to perform the 3-phase to two-phase conversion and *vice versa*.
- This connection is used to convert a balanced 3ϕ supply into a balanced 2ϕ supply (Fig. 3.23).

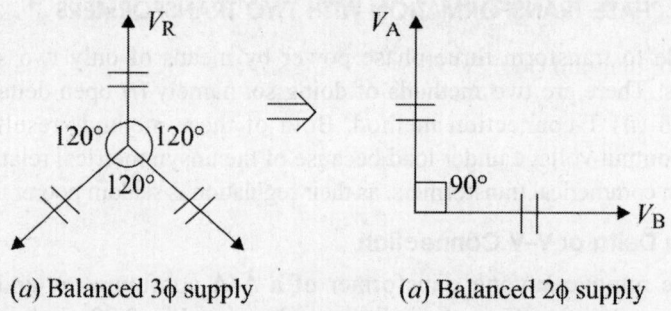

(*a*) Balanced 3ϕ supply   (*a*) Balanced 2ϕ supply

Fig. 3.23

- The Scott connection can also be used to convert a balanced supply into a balanced 3ϕ supply.
- The two transformer are connected electrically but not magnetically. One of the transformer is called main transformer and other is known as auxillary transformer or teaser transformer.
- The main transformer is centre tapped at *D* and is connected across line *B* and *C* of 3ϕ side.
- It has primary *BC* and secondary $a_1 a_2$.
- The teaser transformer is connected between the line terminal *A* and centre taped *D* (Fig. 3.24)

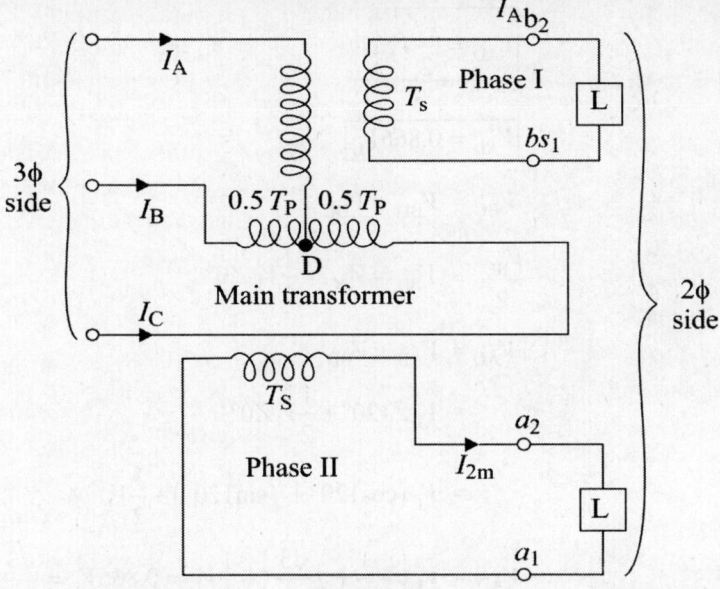

Fig. 3.24

- It has primary AD and secondary $b_1 b_2$.
- Frequently indentical interchangeable transformer are used for the scott connection in which each transformer has a primary winding T_p turns and it provided with tapping as $0.5\ T_p$ and $0.866\ T_p$.

Phasor diagram (Fig. 3.25)

$$V_{AB} = V_{BC} = V_{CA} = V_L \text{ (balanced)}$$

Let V_{BC} is reference

$$V_{BC} = V_L \angle -120°$$
$$V_{CA} = V_L \angle -120° \text{ or } V_L \angle -240°$$
$$V_{AB} = V_L \angle 120° \text{ or } V_L \angle -240°$$

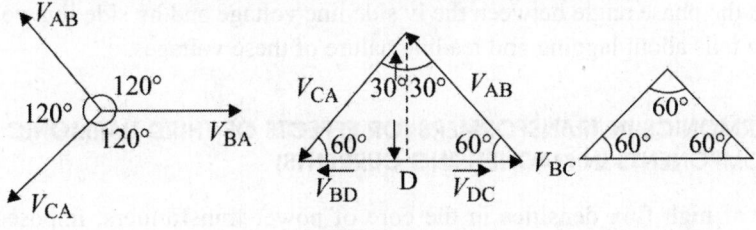

Fig. 3.25

$$= V_{AB} \sin 60°$$
$$V_{AD} = V_L \sin 60° \qquad \begin{vmatrix} r \angle \theta \\ \cos\theta + j\sin\theta \end{vmatrix}$$

$$\boxed{V_{AD} = \frac{V_3}{2} V_L}$$

$$\boxed{V_{AD} = 0.866 V_L}$$

In $\qquad V_{BC} = V_{BD} + V_{DC}$

$\therefore \qquad \dfrac{V_{BC}}{2} = V_{BD} = V_{DC} = \dfrac{1}{2} V_L \angle 0°$

$\qquad V_{AD} = V_{AB} + V_{BD}$

$\qquad\qquad = V_L \angle 120° + \dfrac{1}{2} V_1 \angle 0°$

$\qquad\qquad = V_L \left(\cos 120° + j \sin 120° \right) + \dfrac{1}{2} V_L$

$\qquad V_{AD} = V_L \left(-\dfrac{1}{2} + j \dfrac{\sqrt{3}}{2} \right) + \dfrac{1}{2} V_L = 0.866 V_L = \dfrac{\sqrt{3}}{2} V_L$

APPLICATION OF SCOTT CONNECTION

• Electric furnace installation.
• Electric train.
• Flow of power in either direction.
  ∴ 3φ to 2φ
  or 2φ to 3φ

Example 3.3: What is meant by three phase transformer groups? What are the possible connection for a 3φ transform bank?

Solution: Classification of a three-phase transformer is terms of phasor groups is based on the phase angle between the lv side line voltage and hv side line voltage.

It also tells about lagging and leading nature of these voltages.

3.6 HARMONICS IN TRANSFORMERS (OR EFFECTS OF THIRD HARMONIC COMPONENTS IN MAGNETISING CURRENTS)

The use of high flux densities in the core of power transformers, imposed by the requirements of an economical design and the reduction in size, results in high saturation level and the departure from rectilinearity of the flux-current relation of B–H (or B–AT) curve. Due to saturation effects, a sinusoidal flux and emf necessitate a pronounced third and less pronounced higher-order harmonic components in the magnetizing current. If for any reason the third harmonic current is not permitted to flow, i.e. when the magnetizing current is sinusoidal, the flux is flat-topped containing "depressing" third harmonic and as a consequence third harmonic voltages are present in the induced emf.

The above problem in single phase and three phase transformers is discussed below.

1. **Single phase transformers:** A sinusoidal supply voltage cannot itself give rise to any third harmonic current; it supplies fundamental active power, part of which is converted to the harmonic power in the transformer core by reason of its non-linear B–H characteristic. The flow of a third harmonic current implies the presence of a third harmonic emf, and, therefore, a corresponding harmonic in the the flux waveform, the harmonic currents flow through the effective third harmonic impedances of the primary winding and its supply network and also to a limited extent in the secondary winding and load on the secondary side (since the load impedance is usually much larger than that of the supply network, the third harmonic current flow principally in the primary circuit), The third harmonic flux is combined with the fundamental flux in such a way that the resultant flux has a waveform intermediate between a sinusoid and that waveform produced by a sinusoidal current, the exact degree of compromise being determined by the impedance of the circuits to third harmonic currents. If the impedance of the supply network is negligibly small in comparison to the third harmonic impedance Z_3 of the transformer, the third harmonic emf E_3 will be balanced by the corresponding impedance drop I_3Z_3 and no harmonic voltage appears across the primary terminals. On the other hand, if the impedance to third harmonics could be made entirely negligible, then only a vanishing small third harmonic emf would be needed to circulate a magnetizing current in addition to the fundamental magnetizing current so as to permit a substantially sinusoidal flux. The higher harmonic effects due to core saturation are similar to those described for the third harmonic.

Thus we see that the wave distortion caused by the influence of hysteresis loss and of saturation of the iron is not very serious, except for the resulting error computation of losses, in case of single plase transformers.

2. **Three-phase bank of single phase transformers:** While considering three phase transformers it is essential to distinguish between the cases where the phases are magnetically separate and where they are magnetically as well as electrically interlinked. In case of three phase banks of single phase transformers, the magnetic circuits are obviously separate, and each core must itself produce the flux demand by the conditions of its electrical connections. which then determine the flow of harmonic currents. In the following description both supply and load are considered to be balanced and star-connected,

(*a*) **Delta–delta (Dd) connection:** It is to be observed that the phase difference in third harmonic currents and voltages on a 3 phase system in $3 \times 120° = 360°$ or $0°$ which means that these are cophasal. Therefore, third harmonic (in general of the order 3 n called triplen) currents and voltages can not be present on lines of a 3 phase system as these do not add up to zero.

Delta connection provides a closed path to triplen emfs, which circulate currents in the closed mesh and are absorbed by the triplen leakage impedance drop: there is no resultant triplen voltage at the line terminals, The impedance to harmonic cunents is usually small (although it may be greater than fundamental impedance of the transformer), and very small emfs are sufficient to circulate considerable harmonic currents.

The action in delta/delta connection is as follows:

The supply voltage, being sinusoidal, provides only sinusoidal magnetizing current. Such a current produces a non-sinusoidal flat-topped flux wave, since there is no peak current to overcome the saturation at high flux densities. This flat-topped flux wave induces a peak emf with a strong third harmonic component, which does not balance the applied sinusoidal voltage, but leaves as resultant, third harmonic components (and also higher harmonics). The third harmonic components now circulate a triplen current in the mesh, which is in addition to the sinusoidal supply current restores the flux to almost sinusoidal. If the triplen mesh impedance is low, a very small departure of the flux from a perfect sine is different to give a large third harmonic magnetizing current. Thus the conditions are established whereby the primary voltage and primary current are sinusoidal. the flux is very nearly a sinusoidal, and the necessary triplen magnetizing current component is circulated round the closed mesh. Inspite of the presence of a third harmonic in the magnetizing current, of each phase the line currents do not, have any harmonic component, since the combination of the two-third harmonic components of the phase currents connected from any line cancel each other. In a $A-A$ connection the primaries carry all of the fundamental component of the exciting current while the third harmonic component of the exciting current can divide so that part of it exists in the primaries and the remaining portion in the secondaries.

(*b*) **Star–delta or delta–star (Yd or Dy) connection without neutral:** So long as there is no neutral wire, either of these connections operates substantially as in delta–delta connections but since there is a mesh connection on only one side, the triplen mesh impedance is now larger, and the compensation of the magnetizing current will necessitate a greater divergence of the flux wave from a true sinusoid. In star–delta (Δ–Y) connection the primaries carry the fundamental component of the exciting current and the secondaries carry the third harmonic component whereas in delta–star (Δ–Y) connection the secondary voltages do not contain any harmonic component.

(*c*) **Star–star (Yy) connection with neutral:** As in Δ connection of the primaries, each primary is directly connected to the respective phase of the supply. Again, there is nothing to interfere with the existence of a third harmonic in the magnetizing current, and the voltage induced in each transformer will be nearly sinusoidal. The line currents in this case will contain the third harmonic component of the magnetizing current. Under balanced conditions third harmonic components of the three phases will be in phase with each other and of the same magnitude. The current in the neutral wire of the supply system, therefore, will not be zero but will be equal to three times the third harmonic component of one phase.

(d) **Star–star (Yy) connection without neutral:** The triplen emfs are all directed simultaneously away from or towards the star point and, therefore, cancel between any pair of lines. Consequently no triplen currents can flow, and (apart from the fifth, seventh, ... harmonic components) the input magnetizing currents are sinusoidal. The flux waveform is flattened, its triplen harmonics providing the triplen emf. The balance between applied voltage and inuuced emf between line terminals is maintained since the triplen emf balance out, but the effect on the star–point voltage is to make it oscillate. The different conditions are illlustrated in Fig. 3.26. The voltages, e_{r1}, e_{y1}, e_{b1} represent the fundamental terms, mutually displaced by 120°. There is a third harmonic emf, e_3 co-phasal in all legs, additional to the fundamental, and the resultant voltages across the legs are shown dotted. The diagram illustrates four successive instants displaced by one-twelfth period. It is to be noted that the third harmonic phasor e_3 rotates thrice as the fundamental system. All the line-to-star point voltages fluctuate, causing "neutral oscillation". The non-triplen harmonics form balanced systems with a 120° phase displacement and appear across the lines to cause a self suppressing circulating current.

The phenomenon of oscillating neutral is highly undesirable and because of this star connection with isolated neutrals is not use in practice.

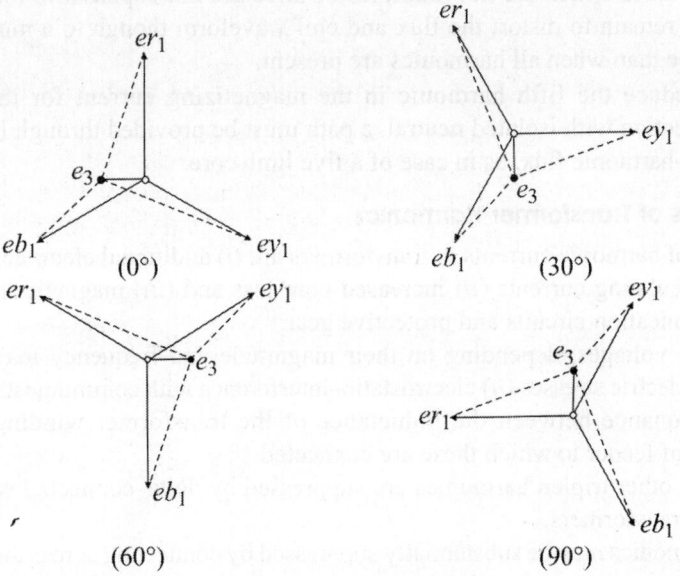

3.26: Neutral oscillation

(e) **Other star–delta (Yd) connections:** All the variants of the possible connections will fall under one or other of the cases *(a)* to *(d)* already dealt with a mesh connection always provides a closed path for triplen currents, while a neutral conductor on the supply side allows triplen currents to flow into the supply network. A neutral conductor is not operative on the secondary side on open-circuit.

3. **Three phase transformer units:** In the 3-phase shell-type transformer, the magnetic circuits are separate and so do not interact, and the discussion made above for the bank of single phase transformers will apply, with the very common three-limbed core type transformer, however, the phases are magnetically interlinked. Any triplen flux harmonic that exists will be direc- ted either all upwards or all down- wards in the limbs together at any instant. They must, therefore, find return paths outside the iron circuit (through the air or oil and walls of the tank, if any), as shown in Fig. 3.27. The magnetic reluctance of these paths is high and the third harmonic flux components are reduced to negligible proportions. Occasionally the third harmonic fluxes have been found to cause losses in the tank walls.

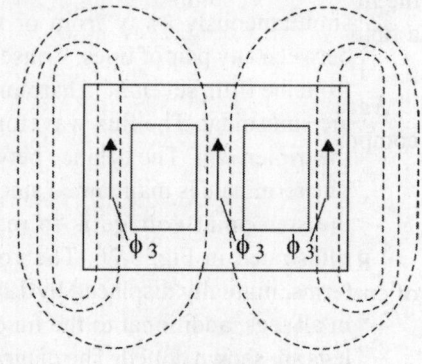

Fig. 3.27

Harmonics which are not multiples of three are not cophasal in the limbs, so these remain to distort the flux and emf waveform though to a much smaller degree than when all harmonics are present.

To reduce the fifth harmonic in the magnetizing current for the star/star connection with isolated neutral, a path must be provided through iron for the third–harmonic flux, as in case of a five limb core.

3.6.1 Effects of Transformer Harmonics

The effects of harmonic currents in transformers are *(i)* additional electrical (I^2R) loss owing to circulating currents *(ii)* increased core loss and *(iii)* magnetic interference with communication circuits and protective gear.

Harmonic voltages, depending on their magnitude and frequency may cause *(i)* increased dielectric stresses *(ii)* electrostatic–interference with communication circuits and *(iii)* resonance between the inductance of the transformer windings and the capacitance of feeder to which these are connected.

Third and other triplen harmonics are suppressed by delta–connected windings in three phase transformers.

Other harmonics may be substantially suppressed by connecting across the terminals of the transformer resonating shunts tuned to the frequencies which it is required to eliminate.

Example 3.4: The voltage applied to the primary of a single phase transformer at no–load is given by

$$v = 200 \cos \omega t + 80 \cos 3\, \omega t$$

If the primary has 300 turns, then calculate the maximum value of the flux in the core for a fundamental frequency of 50 Hz.

[Pb. Univ. Elec. Machines-I, Dec 1992]

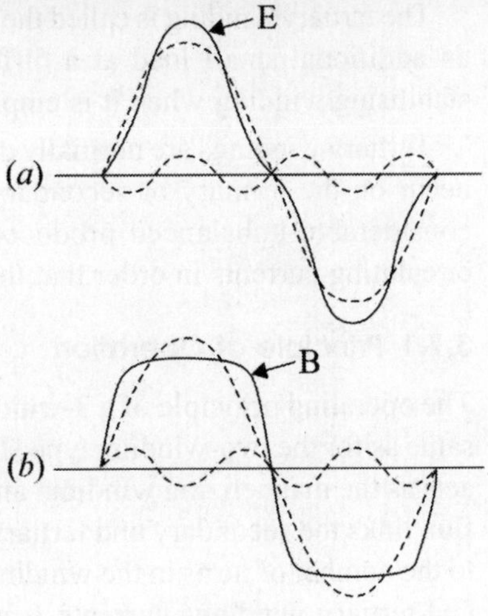

Solution: RMS value of fundamental component of emf

$$E_1 = \frac{200}{\sqrt{2}} = 100\sqrt{2} \text{ or } 141.42 \text{ V}$$

RMS value of third harmonic component of emf

$$E_3 = \frac{80}{\sqrt{2}} = 40\sqrt{2} \text{ or } 56.57 \text{ V}$$

Peak value of fundamental component of flux,

$$\phi_{1max} = \frac{E_1}{4.44\, f N_1} = \frac{141.42}{4.44 \times 50 \times 300} = 2.12 \text{ Wb}$$

Peak value of third harmonic component of flux,

$$\phi_{3max} = \frac{E_3}{3 \times 4.44\, f N_1} = \frac{56.57}{3 \times 4.44 \times 50 \times 300} = 0.283 \text{ Wb}$$

Maximum value of flux in the core

$$\phi_{max} = \phi_{1max} - \phi_{3max}, \text{ as obvious from Fig. 3.27.}$$
$$= 2.12 - 0.283 = 1.837 \text{ Wb}$$

Fig. 3.28

3.7 TERTIARY WINDINGS

Transformers may be built with a third winding, called the tertiary, in addition to the normal primary and secondary and such transformers are called the triple–wound (or 3–winding) transformers. The tertiary winding may serve any of the following purpose:

(*i*) To supply the substain auxiliaries at a voltage different from those of the primary and secondary windings.

(*ii*) To supply phase compensating device such as condensers operated at a voltage which is different form both primary and secondary voltage.

(*iii*) To interconnect three supply systems operating at different voltages.

(*iv*) To load large split-winding generators.

(*v*) To measure voltage of an HV testing transformer.

(*vi*) In star/star connected transformers, to allow sufficient earth fault current to flow for operation of protective equipment to suppress harmonic voltages and to limit voltage unbalance when the main load is unsymmetrical.

The tertiary winding is called the auxiliary winding when it is employed for supplying an additional small load at a different voltage. On the other hand, it is called the stabilising winding when it is employed to limit to short-circuit current.

Tertiary windings are normally delta-connected so that when faults and short-circuits occur on the primary or secondary sides (particularly between lines and earth), the considerable unbalanced produced in phase voltages may be compensated by the circulating currents in order that there is no over-heating of the windings.

3.7.1 Principle of Operation

The operating principle of a 3-winding (or triple wound) transformer is essentially the same as for the two-winding type. The primary winding of a three winding transformer acts as the magnetizing winding, and its current produces the main magnetic flux. The flux links the secondary and tertiary windings and induces emfs in them in proportion to the number of turns in the windings. When loads are connected across the secondary and tertiary windings currents I_2 and I_3 flow in them. The demagnetizing effect of these currents will be determined by the geometrical sum of the magnetizing forces of the secondary and tertiary windings. So the magnetizing force of the primary winding must balance this combined secondary magnetizing force and must have a magnetizing component.

i.e.
$$I_1 N_1 = (-I_2 N_2) + (-I_3 N_3)$$

or
$$I_1 = I_2' + I_3' + I_0$$

where
$$I_2' = -I_2 \frac{N_2}{N_1} \text{ and } I_3' = -I_3 \frac{N_3}{N_1}$$

Thus in a 3–winding transformer, there occurs a simultaneous transfer of electrical energy to two secondary circuits: the secondary and tertiary windings.

It is highly impossible that both secondary and tertiary windings will be fully loaded at the same time and also that the currents I_2' and I_3' will at that time, be in phase. For this reason, the primary winding is usually designed for a lower load than the sum of the rated powers of the secondary and tertiary windings.

3.7.2 Construction

As mentioned above, a three-winding transformer consists of three sets of windings (primary, secondary and tertiary). Three winding transformers may be either 3-phase units or 1-phase units connected in a 3-phase bank, such transformers have a large kVA rating. Since the three windings are operated at three different voltages, they may be termed as high voltage, medium voltage and low voltage windings.

Two alternative arrangements of windings of a 3-winding transformer are illustrated in Fig. 3.29(a) and 3.29(b) respectively. The hv winding in both cases is the outermost winding because of insulation consideration. The lv and mv windings are placed adjacent to each other with either of the two near to each other in order to reduce the leakage reactance between this pair of windings.

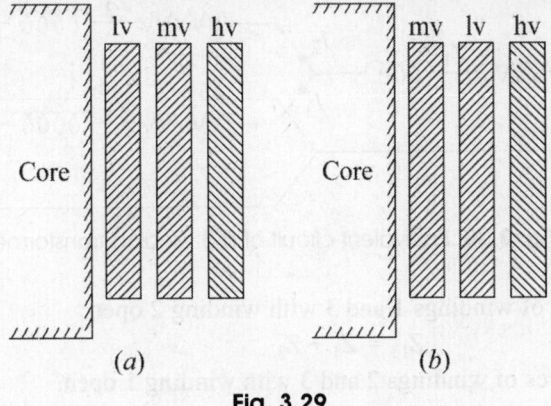

Fig. 3.29

Thus the two possible winding arrangements for windings starting from core outwards are: (*i*) lv, mv and hv.

In two-winding transformers, the kVA ratings of both the primary and the secondary windings are the same but it is not so in case of 3-winding transformers. The kVA rating of the 3-winding transformer is considered to be equal to the largest kVA rating of any of its windings.

The rating of tertiary winding depends upon the intended application, it is provided for supply of an additional load, the winding is designed and calculated on the same basis as the primary and secondary. When employed for balancing of loads and controlling short-circuit currents, the tertiary winding carries current only for short-periods, and its rating depends mainly on its heat capacity. In practice, the *x*-section of the winding wire is generally determined by the fault conditions irrespective of the fact for what application it is going to be used.

3.7.3 Equivalent Circuit

The equivalent circuit of a 3-winding transformer can be drawn with each winding represented by its own resistance and leakage reactance (all values are reduced to a common rating base and respective voltage basis. The subscripts 1, 2 and 3 indicate the primary, secondary and tertiary respectively. For simplicity, the effect of the exciting current is ignored.

It may be noted that the load division between the secondary and tertiary circuits is completely arbitrary. Three external circuits are connected between terminals 1, 2 and 3 respectively and the common terminal labelled O. Neglecting exciting current I_0 we have $I_1 + I_2 + I_3 = 0$.

The impedances for such an equivalent circuit can be readily determined from the data of three short circuit tests as follows. For circuit shown in Fig. 3.30.

SC impedance of windings 1 and 2 with winding 3 open (impedance of equivalent circuit when the terminals of circuit 2 are short-circuited and the terminals of circuit 3 are open)

$$Z_{12} = Z_1 + Z_2$$

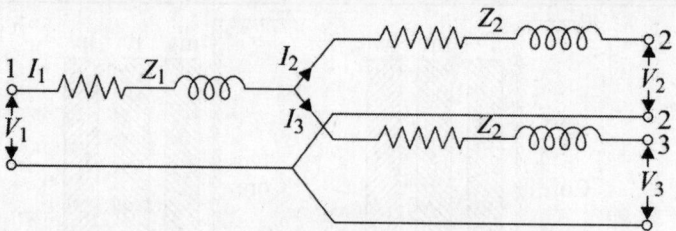

Fig. 3.30: Equivalent circuit of a 3-winding transformer

SC impedance of windings 1 and 3 with winding 2 open

$$Z_{13} = Z_1 + Z_3$$

and SC impedances of windings 2 and 3 with winding 1 open,

$$Z_{23} = Z_2 + Z_3$$

All the impedances are referred to a common base.

Solving equations, we have

$$Z_1 = \frac{1}{2}\left(Z_{12} + Z_{13} - Z_{23}\right)$$

$$Z_2 = \frac{1}{2}\left(Z_{23} + Z_{12} - Z_{13}\right)$$

and

$$Z_3 = \frac{1}{2}\left(Z_{13} + Z_{23} - Z_{12}\right)$$

The core loss exciting or magnetizing impedance and turn-ratio can be determined by performing open-circuit test on any of the three windings.

3.7.4 Stabilization by Tertiary Winding

The star-star connected transformer comprising single phase units, or three-phase shell or 5-limb core type units have the drawbacks (i) that they cannot readily supply unbalanced loads between line and neutral and (ii) that their phase emfs may get distorted by third harmonic emfs. The reason is that magnetic flux set up in the iron core due to third harmonic and zero sequence currents have an iron path. The zero phase sequence impedance is, therefore high (0.5 pu). The flow of earth fault is severely restricted, and may be insufficient to operate the protective gear. The zero-sequence impedance to out-of-balance or earth-fault currents include the very high impedance between the secondary winding of the loaded or faulty phase and the primary windings of the other two phases. Abnormal voltages, therefore, occur when zero phase sequence current flows. The output load or fault current flows through one phase only, whereas the corresponding input load current, in the absence of a primary neutral, has to return through the other two phases B and C (shown in Fig. 3.31) which have very high reactance owing to absence of their secondary currents. The voltages of phases B and C rise and become nearly equal to the primary line voltage and the voltage of the faulty or loaded phase A reduces to a very low value. Thus serious voltage unbalance is developed. By use of delta-connected tertiary winding, zero sequence currents are

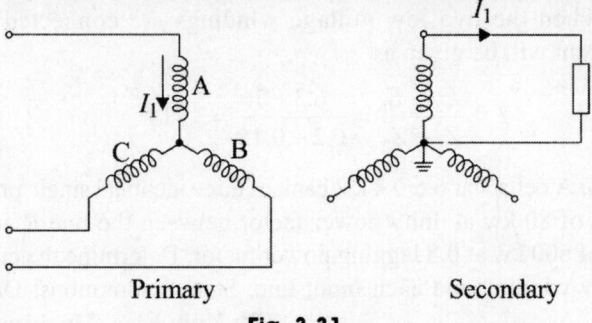

Primary   Secondary

Fig. 3.31

provided low reactance paths and voltage unbalance on unbalanced loads or ground faults is prevented. This is illustrated in Fig. 3.32. It has been assumed that each winding has equal turns per phase. This action is in brief, to interlink magnetically the separate magnetic circuits.

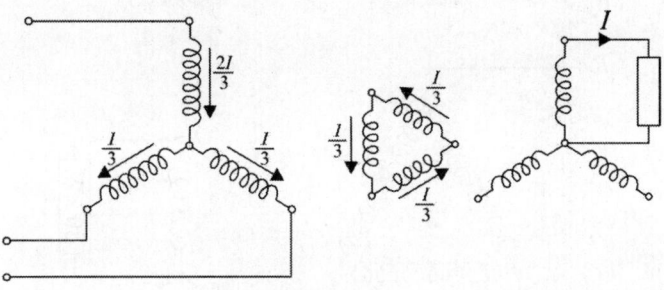

Fig. 3.32

The mesh connected to tertiary provides a path for the third harmonic currents required for magnetizing to the comparatively high saturation flux densities now employed. The circulation of the third harmonic currents then provided, with the primary current, the excitation necessary to secure a substantially sinusoidal flux and emf.

The two drawbacks, mentioned above are reduced to some extent in three limb core type transformers, becasue zero-sequence flux is forced out of the core limbs to high–reluctance air and tank leading to reduction in flux and zero-sequence impedance thereby reducing voltage unbalanced considerably. The third-harmonic flux also has similarly high reluctance path and so magnitude of the third harmonic flux is small. The distortion in voltage wave shape is small and the transformer has more or less sinusoidal flux and emf when the input voltage is sinusoidal. It is never the less usual to provide a delta-connected tertiary even to the three-limb core type transformer if it is star/star connected.

Example 3.5: A small 10 VA, 115 V primary transformer has two secondary windings of 6.3 V and 5 V respectively with resistive impedances of 0.2 Ω and 0.15 Ω respectively. Find the circulating current in secondary when low voltage windings are connected in parallel.   **[AMIE Sec B; Elec Machines; Winter 1993]**

Solution: When the two low voltage windings are connected in parallel, the circulating current will be given as

$$I = \frac{E_2 - E_1}{Z_1 + Z_2} = \frac{6.3 - 5}{0.2 + 0.15} = 3.714$$

Example 3.6: A delta/star 6.6/0.4 kV bank of three identical single phase transformers supplies a load of 80 kw at unity power factor between the line R and neutral and a balanced load of 600 kw at 0.8 lagging power factor. Determine the current magnitude in each primary winding and each input line. State assumptions. Draw the relevant diagrams. **[Pb Univ Elec. Machines-I; June 1991]**

Solution: Secondary phase voltage,

$$V_{p2} = \frac{0.4 \times 1000}{\sqrt{3}} = 231 \text{V}$$

Primary phase voltage, $V_{pl} = 6.6 \times 1000 = 6.600$ V

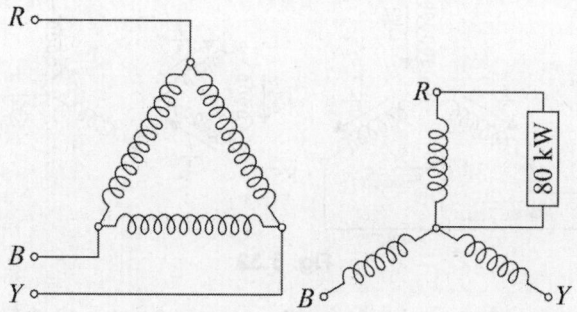

Fig. 3.33

Turn-ratio $K = \dfrac{V_{p_2}}{V_{p_1}} = \dfrac{231}{6600}$

Current in red phase of secondary due to single phase load or 80 kW at unity pf

$$= \frac{80 \times 1000}{231} = 346.32 \text{ A}$$

Current in all phases due to a balanced load of 600 kW at 0.8 pf lagging

$$= \frac{600 \times 1000}{3 \times 231 \times 0.8} = 1082.53 \text{ A}$$

lagging behind their respective phase voltages by an angle $\cos^{-1} 0.8$ (or 36.87°)

Assuming V_{RY} as the reference phasor for primary

$$V_{RY} = 6600 \angle 0° \text{ V}; \ V_{YB} = 6600 \angle -120° \text{ and } V_{BR} = 6600 \angle 120°$$

Current in phase RY due to single phase load of 80 kW at unity power

$$= 346.32 \times \frac{231}{6600} = 12.12\,A \text{ in phase with voltage } V_{RY}$$

Current in phase RY due to balanced load of 600 kW at 0.8 pf lagging

$$= 1082.53 \times \frac{231}{6600} = 37.88\,A \angle -36.87$$

Current in phase RY

$$I_{RY} = 12.12\angle 0 + 37.88\angle -36.87$$
$$= 12.12 + (30.3 - j\,22.73) = (42.42 - j22.73)\,A$$

Current in phase YB

$$I_{YB} = 37.88 \angle -120° - 36.87° = 37.88 \angle 156.87°$$
$$= (-34.835 - j14.88)\,A$$

Current in phase BR

$$I_{BR} = 37.88 \angle -120° - 36.87°$$
$$= 37.88 \angle 83.13° = (4.53 + j37.61)\,A$$

Line current, $I_R = (I_{RY} - I_{BR}) = (42.42 - j\,22.73) - (4.53 + j\,37.61)$
$$= (37.89 - j\,60.34)\,A$$

Line current, $I_Y = I_{YB} - I_{RY}$
$$= (-34.835 - j\,14.88) - (42.42 - j\,22.73)$$
$$= (-77.255 + j\,7.85)\,A$$

Line current, $I_B = I_{BR} - I_{YB}$
$$= (4.53 + j\,37.61) - (-34.835 - j\,14.88)$$
$$= (39.365 + j\,57.49)\,A$$

Currents in different primaries are as follows

$$I_{RY} = \sqrt{(42.42)^2 + (22.73)^2} = 48.126\,A$$
$$I_{YB} = 37.88\,A$$
$$I_{BR} = 37.88\,A$$

Current in line R, $I_R = \sqrt{(37.89)^2 + (60.34)^2} = 71.25\,A$

Current in line Y, $I_Y = \sqrt{(-77.255)^2 + (7.85)^2} = 77.65\,A$

Current in line B, $I_B = \sqrt{(39.365)^2 + (57.49)^2} = 65.61\,A$

Example 3.7: The following is the data pertaining to a 3-winding transformer:

hv in star; 10 MV A, 33 kV

mv in star; 7.5 MVA, 11 kV

lv in delta; 7.5 MVA, 3.3 kV

The per unit leakage reactances are:

$$x_{12} = j0.092$$
$$x_{13} = j0.062$$ measured on hv side

and $x_{23} = j \, 0.069$ measured on mv side

For the Y–equivalent circuit, find the coil leakage reactance in pu and also in ohms.

[Pb Univ; Elec. Machines I; June 1993]

Solution: $x_{12} = j \, 0.092$ measured on hv side

$x_{13} = j \, 0.062$ measure on hv side

$x_{23} = j \, 0.069$ measured on mv side

and $\quad x_{23} = j \, 0.069$ measured on mv side, i.e. referred to MVA of 7.5

$$= j0.069 \times \frac{10}{7.5} = j0.092$$

measured on hv side i.e. referred to base MVA of 10 coil leakage reactance referred to common base MVA of 10

$$x_1 = \frac{1}{2}(x_{12} + x_{13} - x_{23}) = \frac{1}{2}(j0.092 + j0.02 - j0.092) = j0.031$$

$$x_2 = \frac{1}{2}(x_{23} + x_{12} - x_{13}) = \frac{1}{2}(j0.092 + j0.092 - j0.062) = j0.061$$

$$x_3 = \frac{1}{2}(x_{13} + x_{23} - x_{12}) = \frac{1}{2}(j0.062 + j0.092 - j0.092) = j0.031$$

Leakage reactance of primary wind,

$$X_1 = \frac{x_1 \times (KV)^2}{MVA} = \frac{0.031 \times (33)^2}{10} = 3.3759\,\Omega$$

Leakage reactance of secondary winding,

$$X_2 = \frac{x_2 \times (KV)^2}{MVA} = \frac{0.061 \times (11)^2}{10} = 0.7381\,\Omega$$

Leakage reactance of tertiary winding,

$$X_3 = 3 \times \frac{x_3 \times (KV)^2}{MVA} = \frac{3 \times 0.031 \times (3.3)^2}{10} = 0.1013\,\Omega$$

Example 3.8: The short–circuit tests gave the following pu values of a-winding transformer $Z_{12} = (0.01 + j \, 0.06)$; $Z_{13} = (0.02 + j \, 0.06)$ and $Z_{31} = (0.016 + j \, 0.1)$.

The open-circuit test on the primary side gave the following values per unit $Y_0 = (0.02 - j \, 0.05)$. The secondary is supplying full rating at 0.8 pf lagging and tertiary open. The primary current and voltage at the terminals of the primary.

[Pb Univ; Elec. Machines-I; June 1991]

Solution: $\quad Z_{12} = 0.01 + j \, 0.06$

$Z_{23} = 0.02 + j \, 0.06$

$Z_{31} = 0.016 + j \, 0.1$

$$Z_1 = \frac{1}{2}(Z_{12} + Z_{13} - Z_{23})$$

$$= \frac{1}{2}[(0.01 + j0.06) + (0.016 + j0.1) - (0.02 + j0.06)]$$

$$= 0.003 + j0.05$$

$$Z_2 = \frac{1}{2}(Z_{23} + Z_{12} - Z_{13})$$

$$= \frac{1}{2}[(0.02 + j0.06) + (0.01 + j0.06) - (0.016 + j0.1)]$$

$$= 0.007 + j\,0.01$$

$$\frac{1}{2}(Z_{23} + Z_{12} - Z_{13}) = \frac{1}{2}[(0.02 + j0.06) + (0.01 + j0.06) - (0.016 + j0.1)]$$

$$Z_3 = \frac{1}{2}(Z_{13} + Z_{23} - Z_{12})$$

$$= \frac{1}{2}[(0.016 + j0.1) + (0.02 + j0.06) - (0.01 + j0.06)]$$

Secondary load current, $I_2 = 1\,(0.8 - j\,0.6)$

No–load current, $I_0 = 1\,(0.02 - j\,0.05)$

Primary current, $I_1 = I_2 + I_0$

$$= (0.8 - j\,0.6) + (0.02 - j\,0.05)$$

$$= (0.82 - j\,0.65) = 1.0464$$

Voltage at the terminals of the primary

$$= 1 + I_2 Z_2 + I_1 Z_1$$

$$= 1 + (0.8 - j0.6)(0.007 + j0.01) + (0.82 - j0.65)(0.003 + j0.05)$$

$$= 1.04656 + j\,0.006 = 1.04658$$

3.8 TRANSFORMER NOISE

Under no-load conditions, the 'hum' caused by energized power transformers originates in the core, where the laminations tend to vibrate by magnetic forces. The noise is transmitted through the oil to the tank side and thence to the surroundings. The essential factors for noise production in transformers are (i) magnetostriction; (ii) the mechanical vibrations caused by the laminations, depending upon the tightness of clamping, size, gauge, associated structural parts, etc (iii) the mechanical vibration of tank walls and (iv) the damping.

In general the total noise emission may be reduced by (i) preventing vibration of core-plate, which needs the use of a lower flux density and attention to constuctional features such as clamping bolts, proportions, and dimensions of the "steps" in plate width, tightness of clamping and uniformity at plates; (ii) sound insulating the transformer from the tank by cushions, padding, or oil–barriers; (iii) preventing tank wall vibration by suitable design of tank and stiffeners; and (iv) sound insulating the tank from the ground or surrounding air. There is no complete solution to the noise problem.

3.9 INITIAL RUSH OF CURRENT (OR SWITCHING-IN TRANSIENTS)

When the primary side of a transformer is switched on to normal voltage with secondary open-circuited, a transient rush of current may take place. This rush depends upon the point in the cycle at which the voltage wave is switched-on, the value and direction of the residual core flux, the shape of the saturation curve, and the normal flux density are used. The worst conditions may occur when the applied voltage has zero value at the instant of switching (Fig. 3.34).

The approximate value of the magnetizing inrush current may be determined as follows, neglecting the limiting effects of primary impedance drop and the effects of saturation.

Let the applied voltage be $v = V_{max} \cos(\omega t + \alpha)$ while the primary is switched on at time $t = 0$.

At every instant the voltage applied—must be balanced by the emf induced by the flux created in the core by the magnetizing current (neglecting the small drops is resistance and leakage reactance).

So:
$$e = -V_{max} \cos (\omega t + \alpha) \qquad ...(3.1)$$

Also
$$e = -N_1 \frac{d\phi}{dt} \qquad ...(3.2)$$

where N_1 is number of primary turns and ϕ is the flux in the core. Comparing Eqs (3.1) and (3.2) we have

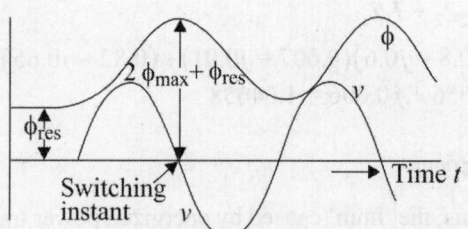

Fig. 3.34

$$\frac{d\phi}{dt} = \frac{V_{max}}{N_1} \cos(\omega t + \alpha)$$

or
$$\phi = \frac{V_{max}}{N_1} \int \cos(\omega t + \alpha) dt = \frac{V_{max}}{\omega N_1} \sin(\omega t + \alpha) + C$$

$$= \phi_{max} \sin(\omega t + \alpha) + C$$

where $\phi_{max} = \dfrac{V_{max}}{\omega N_1}$ and C is asymmetrical component of core flux.

Now the core is initially demagnetized and it may contain a small residual flux ϕ_{res}. So $\phi = \phi_{res}$ at $t = 0$. Thus asymmetrical component of core flux.

$$C = \phi_{res} - \phi_{max} \sin \alpha \qquad ...(3.3)$$

Equation 3.3 may be rewritten as

$$\phi = \phi_{max} \sin (\omega_t + \alpha) + \phi_{res} - \phi_{max} \sin \alpha$$

If the transformer is switched on at the instant that the applied voltage passes through a peak value, i.e. when $\alpha = 0$, a flux corresponding to steady state operation is immediately established in the transformer and also does the magnetizing current. Such conditions are those of normal no-load operation and there is no transient.

The worst conditions for any connection are when $\alpha = \dfrac{-\pi}{2}$, i.e. when the switch is closed at the instant the applied voltage passes through a zero value. In this case $\phi = -\phi_{max} \cos \omega t + \phi_{res} + \phi_{max}$, thus the flux attains nearly double the normal flux after half a cycle, as illustrated in Fig. 3.34. This is referred to as the *doubling effect*. The corresponding magnetizing current will be very large as the core goes into deep saturation region of magnetization; it may indeed be as large as 100 times the normal magnetizing current. If the normal magnetizing current is 5% of the normal rated current, the doubling effect may result it initially to several times rated current. The double flux–density will not actually be reached owing to considerable resistance and leakage reactance drops with large currents. Further, the current gradually falls to the normal magnetizing currents. Because of the low time constant of the transformer circuit, the distortion effects of the transients may last several seconds. The transformer switching transient is referred to as the *inrush current*. Never less the high initial current may cause intense mechanical stress in the transformer windings. This is why the windings of large power transformers are strongly braced. A typical oscillogram of the in rush current is illustrated in Fig. 3.35.

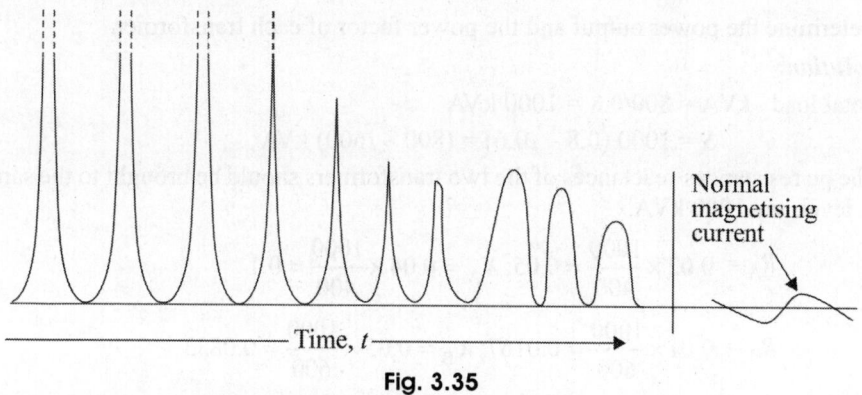

Fig. 3.35

The in rush current does not present a direct danger to the transformer but it may lead to its being tripped out of the power circuit. Therefore, the protection apparatus must be so designed as to avoid the undue tripping of transformer out of power circuit.

The above discussion pertains to a single-phase transformer. When a three-phase transformer is switched on a more or less considerable current inrush should be expected. since there will always be a phase whose voltage at the switching instant is nearly zero.

3.10 PARALLEL OPERATION OF 3–PHASE TRANSFORMERS

All the conditions for the successful parallel operation of single-phase transformers also apply to the parallel running of 3-phase transformers but with following additions:

 (i) The secondaries of all transformers mast have the same phase sequence.

 (ii) The phase displacement between primary and secondary line voltages must be the same for all transformers which are to be operated in parallel.

 (iii) The secondaries of oil transformers must have the same magnitude of line voltage.

The above three conditions must be strictly observed. If these conditions are not compiled with, the secondaries will simply short-circuit one another and no output will be possible. The main difficulty arising from the parallel connection of 3-phase transformers is to ensure that condition (ii) is satisfied. For this, transformers in the same group should be connected in parallel.

Note: In dealing with 3-phase transformers, we consider balanced loading conditions. Therefore, calculations are made for one phase only.

Example 3.9: A three-phase transformer which have the same turns ratio are connected in parallel and supply to a total load of 800 kW at 0.8 power factor lagging. Their ratings are as follows:

Transformer	Rating	pu resistance	pu reactance
A	400 kVA	0.02	0.04
B	600 kVA	0.01	0.05

Determine the power output and the power factor of each transformer.

Solution:

Total load   kVA = 800/0.8 = 1000 kVA

$$S = 1000\,(0.8 - j0.6) = (800 - j600)\ \text{kVA}$$

The pu resistances/reactances of the two transformers should be brought to the same kVA level, say 1000 kVA.

$$R_A = 0.02 \times \frac{1000}{400} = 0.05;\ X_A = 0.04 \times \frac{1000}{400} = 0.1$$

$$R_B = 0.01 \times \frac{1000}{600} = 0.0167;\ X_B = 0.05 \times \frac{1000}{600} = 0.0833$$

$$S_B = S\frac{Z_A}{Z_A + Z_B} = (800 - j600) \times \frac{0.05 + j0.1}{0.05 + 0.01 + 0.0167 + j0.833}$$

$$= (800 - j600) \times \frac{0.05 + j0.1}{(0.0667 + 0.1833)} = (414 - j392)\text{kVA}$$

$$S_A = S - S_B = (800 - j600) - (414 - j392) = (386 - j208)\ \text{kVA}$$

Transformer A : 386 kW, 440 kVA, 0.878 pf lagging

Transformer B : 414 kW, 570 kVA, 0.726 pf lagging

EXERCISE

1. State the type of three phase transformer.
2. What are the advantages and disadvantages of a delta–delta connections?
3. What is open delta or V–V connection? Why is it used?
4. Discuss the advantages and limitations of three single phase transformer connected in a bank over one three phase transformer. Also explain the principal features of any four connection on three phase transformer. **[UPTU 2004–05]**
5. What is an autotransformer. Also show that for the same capacity and voltage ratio, the autotransformer requires less copper than a two-winding transformer.

   [UPTU 2004–05]
6. Discuss the advantages, disadvantages and applications of autotransformer compare the conductor saving of autotransfonner with a two-winding transformer. **[UPTU 2005–06]**
7. Draw the Scott connection of transformers and mark the terminal. What are the applicalions of scott connection. **[UPTU 2007–08]**

4

Three Phase Induction Motors

4.1 THREE PHASE INDUCTION MOTOR

4.1.1 Introduction

- The induction motors are basically AC motors, i.e. they need an alternating voltage for their operation.
- They can operate on their single phase (or) three-phase AC supply, however, the single phase induction motors find very limited area of applications.
- In almost 85% applications the three-phase induction motors are preferred.
- Depending on the type of rotor, the induction motors are classified into two types:
  1. Squirrel cage induction motor.
  2. Slip ring induction motor.

4.1.2 Advantages of Induction Motors over DC Motor

- Low maintenance.
- Ruggedness, smaller size and weight.
- Cost effective.
- They can operate in dusty and explosive environments, because the brushes are not being used.
- They can operate at higher speed of the order of 12,000 rpm.
- It has self-starting torque.

4.1.3 Disadvantages of Induction Motors

- Low starting torque
- Lagging power factor.
- It is essentially a constant speed motor and its speed cannot be changed easily.

4.1.4 Applications of Induction Motors

(i) Fans (ii) Pumps (iii) Extruders (iv) Conveyors (v) Chemical industries (vi) Paper and sugar industries etc.

4.2 CONSTRUCTION OF INDUCTION MACHINES

Similar to other rotating electrical machines, a three-phase induction motor also consists of two main parts: the stator and the rotor (the stator is the stationary part and the rotor is the rotating part), apart from these two main parts, a three-phase induction motor also requires bearings, bearing covers, end plates, etc. for its assembly.

The stator of a three-phase induction motor has three main parts namely, stator frame, stator core and stator windings. The stator frame can either be casted or can be fabricated from rolled steel plates. The stator core is built up of high silicon sheet steel laminations of thickness 0.4 to 0.5 mm. Each lamination is separated from the other by means of either varnish, paper or oxide coating. Each lamination is slotted on the inner periphery so as to house the winding. The laminations for small machines are in the form of complete rings, but for large machines these may be made in sections. The insulated stator conductors are connected to form a three-phase winding, the stator phase windings may be either star- or delta-connected.

The rotor is also built up of their laminations of the same material as the stator. The laminated cylindrical core is mounted directly on the shaft or a spider carried by the shaft. These laminations are slotted on their outer periphery to house the rotor conductors. There are two types of induction motor rotors:

1. Squirrel cage or simply cage rotor
2. Phase wound or wound rotor or slip ring rotors.

In either case, the rotor windings are contained in slots in a laminated iron core which is mounted on on the shaft. In small machines, the lamination stack is pressed directly on the shaft. In larger machines, the core is mechanically connected to the shaft through a set of spokes called a 'spider'.

The motor having the first type of rotor is known as a squirrel cage induction motor. This type of rotor is cheap and has a simple and rugged construction. It is cylindrical in shape and is made of sheet steel laminations. Here the slots provided to accommodate the rotor conductors, are not made parallel to the shaft but they are skewed. The purpose of skewing is (*a*) to reduce the magnetic hum and (*b*) to reduce the magnetic locking. The rotor conductors are short-circuited at the ends by brazing the copper rings, resembling the cage of a squirrel and hence the name squirrel cage rotor.

In present days, *die–cast rotors* have become very popular. The assembled rotor laminations are placed in a mould. The molten aluminium is forced under pressure to form thc bars. Figure 4.1 (*a* to *c*) shows a typical stator and rotor (both squirrel cage type and slip ring type) assembly. Figure 4.1(*d*) shows the schematic of a cage rotor separately.

The motor having the second type rotor, i.e. wound type rotor, is named as a slip ring induction motor. In this motor, the rotor is wound for three-phase, similar to stator winding using open type slots in the rotor lamination. Rotor winding is always star connected and thus only three remaining ends of the windings are brought out and connected to the slip rings as shown in Fig. 4.2. With the help of these slip rings and brushes, additional resistances can also be connected in series with each rotor phase.

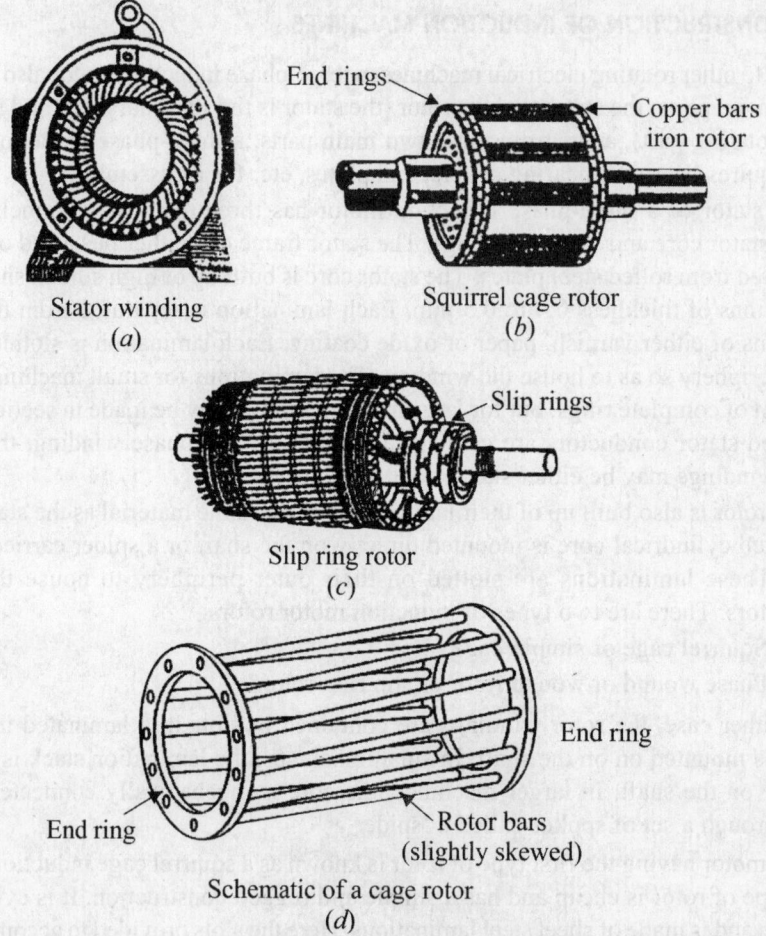

Fig. 4.1: Stator and rotor parts

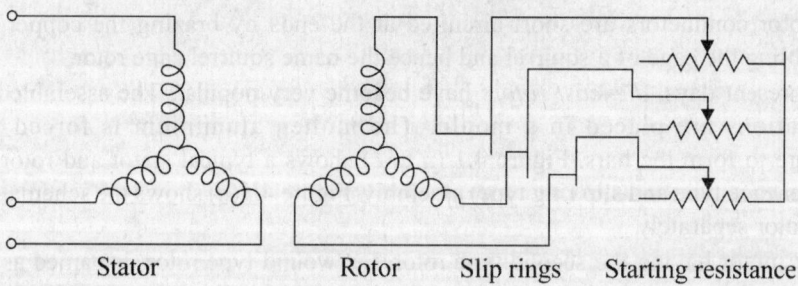

Fig. 4.2: Addition of external resistances to the rotor of wound rotor induction motor

This will increase the starting torque provided by the motor and will also help in reducing, the starting current. When running under normal condition, the external resistances are removed completely from the rotor by short circuiting these additional resistances from the rotor circuit and rotor behaves just like a squirrel cage rotor (Table 4.1).

Table 4.1: Comparison between squirrel cage and slip ring induction motor

S. no.	Squirrel cage	Slip ring
1.	Low starting torque	High starting torque
2.	Starting current in 5 to 6 times of full load current	Starting current is 2 to 3 times of full load current
3.	Higher efficiency	Lower efficiency
4.	Starter is required	Can be started directly on line with the help of external resistance
5.	Minimum maintenance	High degree of of maintenance
6.	No slip rings, brush gears, etc.	Slip rings, brush, gears are required
7.	No speed cotrol	Speed control is possible
8.	It is economical	It has higher cost

4.3 PRINCIPLE OF OPERATION

A three-phase induction motor has a stator winding which is supplied by three–phase alternating balanced voltage and has balanced three-phase currents in the winding. The rotor is not excited from any source and has only magnetic coupling with the stator. Under normal running conditions, the rotor winding (cage or slipring) is always short circuuited to allow induced currents to flow in the rotor winding. The flow of three-phase currents in the stator winding produces a rotating magnetic field of constant amplitude and rotates at a synchronous speed. Let us assume that the rotor is at standstill initially; the rotating stator field induces an emf in the rotor conductor by transformer action. Since the rotor circuit is a closed set of conductors, a current flows in the rotor circuit. This rotor current then produces a rotor field. The interaction of stator and rotor field produces a torque which causes the rotation of the rotor in the direction of the stator rotating field.

As per Lenz's law, the rotor field will try to oppose the very cause of its production. Thus it speed up in the direction of the stator field so that relative speed difference between these two fields is zero. In this way, the three-phase induction motor catches up the speed.

When the rotor is at standstill, the relative motion between the stator field and rotor feild is maximum. Therefore, the emf induced in the rotor and rotor current are reduced. However, the rotor cannot attain the speed of the stator field which is equal to the synchronous speed. This is evidently due to the reason that if the rotor is moving at synchronous speed, there is no relative motion between the stator field and rotor field. Hence the rotor induced emf and current become zero and the torque becomes zero. This would cause the rotor speed to decrease. As the rotor speed falls below the synchronous speed, the rotor emf and current continue to increase. Therefore, the electromagnetic torque continues to increase.

Finally, the rotor speed becomes constant at a value at speed slightly less than that of the stator field, the torque developed equals the sum of load torque and the mechanical losses.

4.4 ROTATING MAGNETIC FIELD DUE TO 3-PHASE CURRENTS

When a 3-phase winding is energised from a 3-phase supply, a rotating magnetic field is produced. This field is such that its poles do not remain in a fixed position on the stator but go on shifting their positions around the stator. For this reason, it is called a *rotating field*. It can be shown that magnitude of this rotating field is constant and is equal to 1.5 ϕ_m where ϕ_m is the maximum flux due to any phase.

To see how rotating field is produced, consider a 2-pole, 3-phase winding as shown in Fig. 4.4(*a*). The three phases *x, y* and *z* are energised from a 3-phase source and currents in these phases are indicated as I_x and I_z. Referring to Fig. 4.4(*b*), the fluxes produced by these currents are given by;

$$\phi_x = \phi_m \sin \omega t$$
$$\phi_y = \phi_m \sin (\omega t - 120°)$$
$$\phi_z = \phi_m \sin (\omega t - 240°)$$

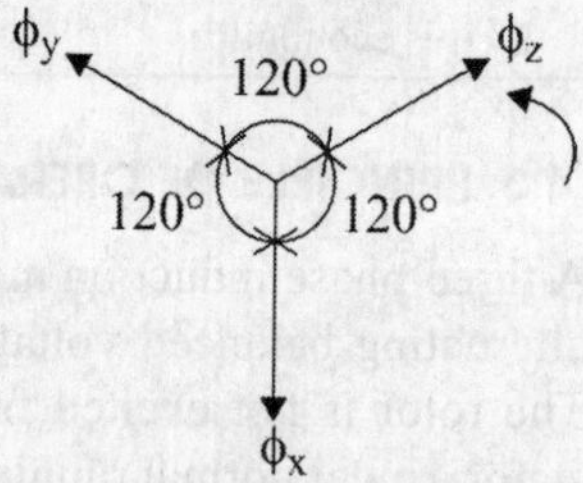

Here ϕ_m is the maximum flux due to any phase. Figure 4.3 shows the phasor diagram of the three fluxes. We shall now prove that this 3-phase supply produces a rotating field of constant magnitude equal to 1.5 ϕ_m.

(*i*) At instant [see Fig. 4.4(*b*) and (*c*)], the current in phase *x* is zero and currents in phases *y* and *z* are equal and opposite. The currents are flowing outward in the top conductors and inward in the bottom conductors. This establishes a resultant flux towards right. The magnitude of the resultant flux is constant and is equal to 1.5 ϕ_m as proved under.

At instant 1, $\omega t = 0°$. Therefore, the three fluxes are given by;

$$\phi_x = 0; \ \phi_y = \phi_m \sin(-120) = \frac{\sqrt{3}}{2} \phi_m$$

$$\phi_z = \phi_x = \phi_m \sin(-240°) = \frac{\sqrt{3}}{2} \phi_m$$

The phasor sum of $-\phi_y$ and $-\phi_z$ is the resultant flux ϕ_r. It is clear that:

$$\text{Resultant flux, } \phi_r = 2 \times \frac{\sqrt{3}}{2} \phi_m \cos \frac{60°}{2} = 2 \times \frac{\sqrt{3}}{2} \phi_m \times \frac{\sqrt{3}}{2} = 1.5 \phi_m$$

(*ii*) At instant 2, the current is maximum (negative) in phase *Y* and 0.5 maximum (positive) in phases *x* and *z*. The magnitude *y* of resultant flux is 1.5 + as proved under:

At instant 2, $\omega t = 30°$. Therefore, the three fluxes are given by

$$\phi_x = \phi_m \sin 30° = \frac{\phi_m}{2}$$

Fig. 4.3

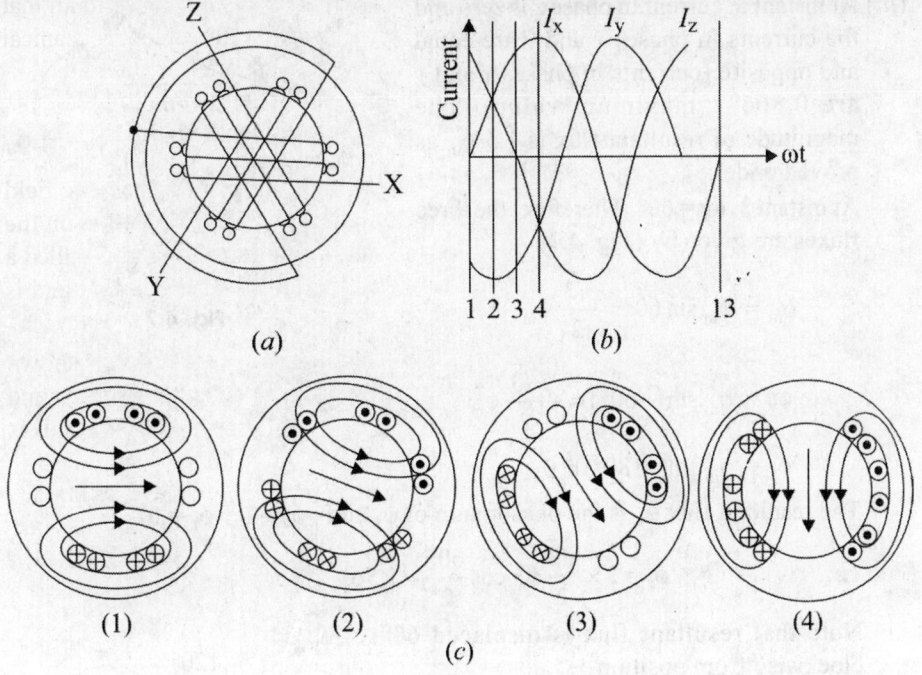

(a)

(b)

(1)   (2)   (3)   (4)

(c)

Fig. 4.4

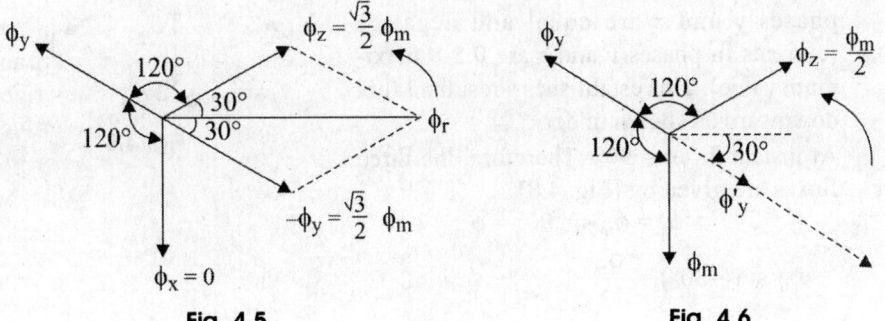

$\phi_z = \dfrac{\sqrt{3}}{2}\phi_m$

$-\phi_y = \dfrac{\sqrt{3}}{2}\phi_m$

$\phi_x = 0$

Fig. 4.5

$\phi_z = \dfrac{\phi_m}{2}$

ϕ'_y

ϕ_m

Fig. 4.6

$$\phi_y = \phi_m \sin-(90°) = -\phi_m$$

$$\phi_z = \phi_m \sin(-210°) = \frac{\phi_m}{2}$$

The phasor sum of $-\phi_z$ and ϕ_z is the resultant flux ϕ_r (Figs 4.5 and 4.6).

Phasor sum of ϕ_x and ϕ_z, $\phi'_r = 2 \times \dfrac{\phi_m}{2}\cos\dfrac{120°}{2} = \dfrac{\phi_m}{2}$

Phasor sum of ϕ'_r and $-\phi_y$, $\phi_r = \dfrac{\phi_m}{2} + \phi_m = 1.5\phi_m$

Note that resultant flux is displaced 30° clockwise from position I.

(*iii*) At instant 3, current in phase z is zero and the currents in phases x and y are equal and opposite (currents in phases x and y are $0.866 \times$ maximum value). The magnitude of resultant flux is 1.5 ϕ_m as proved under:

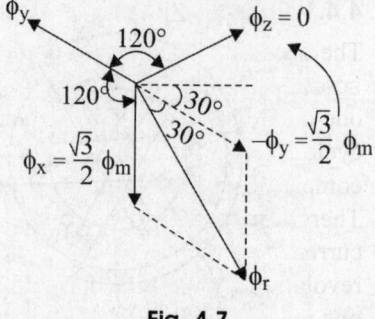

At instant 3, $\omega t = 60°$. Therefore, the three fluxes are given by (Fig. 4.7)

$$\phi_m = \phi_m \sin 60° = \frac{\sqrt{3}}{2}\phi_m$$

Fig. 4.7

$$\phi_y = \phi_m \sin(-60°) = -\frac{\sqrt{3}}{2}\phi_m$$

$$\phi_z = \phi_m \sin(-180°) = 0$$

The resultant flux ϕ_x is the phasor sum of ϕ_x and $-\phi_y = (\therefore \phi_z = 0)$.

$$\therefore \qquad \phi_r = 2 \times \frac{\sqrt{3}}{2}\phi_m \cos\frac{60°}{2} = 1.5\phi_x$$

Note that resultant flux is displaced 60° clockwise from position 1.

(*iv*) At instant 4, the current in phase x is maximum (positive) and the currents in phases y and z are equal and negative (currents in phases y and z are $0.5 \times$ maximum value). This establishes a resultant flux downward as shown under.

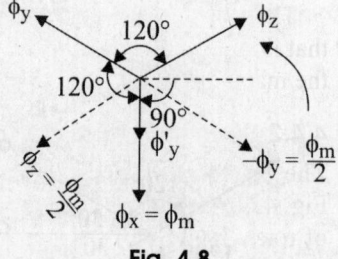

At instant 4, $\omega t = 90°$. Therefore, the three fluxes are given by (Fig. 4.8)

$$\phi_x = \phi_m \sin 90° = \phi_m$$

Fig. 4.8

$$\phi_m \sin(-30°) = \frac{\phi_m}{2}$$

$$= \phi_m \sin(-150°) = \frac{\phi_m}{2}$$

The phasor sum of ϕ_x, $-\phi_y$ and $-\phi_z$ is the resultant flux ϕ_r,

Phasor sum of $-\phi_z$ and $-\phi_y$, $\phi_r' = 2 \times \frac{\phi_m}{2}\cos\frac{120°}{2} = \frac{\phi_m}{2}$

Phasor sum of ϕ_r' and ϕ_x', $\quad \phi_r = \frac{\phi_m}{2} + \phi_m = 1.5\phi_m$

Note that the resultant flux is downward, i.e. it is displaced 90° clockwise from position 1.

It follows from the above discussion that a 3-phase supply produces a rotating field of constant value ($= 1.5\phi_m$ where m is the maximum flux due to any phase).

4.4.1 Speed of Rotating Magnetic Field

The speed at which the rotating magnetic field revolves is called the *synchronous speed* (N_S). Referring to Fig. 4.4(b), the time instant 4 represents the completion of one-quarter cycle of alternating current from the time instant I_x. During this one quarter cycle, the field has rotated through 90°. At a time instant represented by 13 or one complete cycle of current I_x from the origin, the field has completed one revolution. Therefore, for a 2-pole stator winding, the field makes one revolution in one cycle of current. In a 4-pole stator winding, it can be shown that the rotating field makes one revolution in two cycles of current. In general, for P poles, the rotating field makes one revolution in $P/2$ cycles of current.

$$\therefore \qquad \text{Cycles of current} = \frac{P}{2} \times \text{revolutions of field}$$

or   cycles of current per second $= \dfrac{P}{2} \times$ revolutions of field per second

Since revolutions per second is equal to the revolutions per minute (N_S) divided by 60 and the number of cycles per second is the frequency f

$$f = \frac{P}{2} \times \frac{N}{60} = \frac{N_S P}{120} \quad \text{or} \quad N_S = \frac{120 f}{P}$$

The speed of the rotating magnetic field is the same as the speed of the alternator that is supplying power to the motor if the two have the same number of poles. Hence the magnetic flux is said to rotate at synchronous speed.

4.4.2 Direction of Rotating Magnetic Field

The phase sequence of the three-phase voltage applied to the stator winding in Fig. 4.4(b) is x–y–z. If this sequence is changed to x–z–y, it is observed that direction of rotation of the field is reversed, i.e. the field rotates counterclockwise rather than clockwise. However, the number of poles and speed at which the magnetic field rotates remain unchanged. Thus, it is necessary only to change the phase sequence in order to change the direction of rotation of the magnetic field. For a three-phase supply, this can be done by interchanging any two of the three lines. As the rotor in a three-phase induction motor runs in the same direction as the rotating magnetic field. Therefore, the direction of rotation of a phase induction motor can be reversed by interchanging any two of the three motor supply lines.

4.5 ALTERNATE ANALYSIS FOR ROTATING MAGNETIC FIELD

We shall now use another useful method to find the magnitude and speed of the resultant flux due to three-phase currents. The three-phase sinusoidal currents produce fluxes ϕ_1, ϕ_2 and ϕ_3 which vary sinusoidally. The resultant flux at any instant will be the vector sum of all the three at that instant. The fluxes are represented by three variable magnitude vectors. In Fig. 4.9, the individual flux

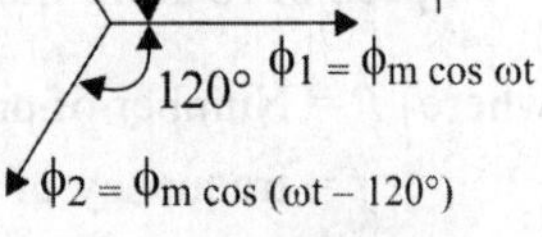

$\phi_3 = \phi_m \cos(\omega t - 240°)$

$\phi_1 = \phi_m \cos \omega t$

$\phi_2 = \phi_m \cos(\omega t - 120°)$

120° 120° 120°

Fig. 4.9

directions are fixed but their magnitudes vary sinusoidally as does the current that produces them. To find the magnitude of the resultant flux, resolve each flux into horizontal and vertical components and then find their vector sum (Fig. 4.10).

$$\phi_h = \phi_m \cos t\, \omega t - \phi_m \cos(\omega t - 120°)\cos 60° - \phi_m \cos(\omega t - 240°)\cos 60°$$

$$= \frac{3}{2}\phi_m \cos \omega t$$

$$\phi_v = 0 - \phi_m \cos(\omega t - 120°)\sin 60° + \phi_m \cos(\omega t - 240°)\sin 60° = -\frac{3}{2}\phi_m \sin \omega t$$

The resultant flux is given by (see Fig. 4.10)

$$\phi_r = \sqrt{\phi_h^2 + \phi_v^2} = \frac{3}{2}\phi_m \left[\cos^2 \omega t + (-\sin \omega t)^2\right]^{1/2} = \frac{3}{2}\phi_m = 1.5\phi_m = \text{constant}$$

Thus the resultant flux has constant magnitude $(1.5\,\phi_m)$ and does not change with time.

The angular displacement of ϕ_r relative to the *ox* axis is

$$\tan \theta = \frac{\phi_v}{\phi_h} = \frac{\frac{3}{2}\phi_m \sin \omega t}{\frac{3}{2}\phi_m \cos \omega t} = \tan \omega t$$

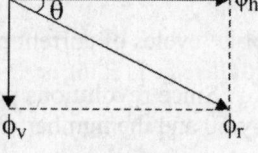

Fig. 4.10

$$\theta = \omega t$$

Thus the resultant magnetic field rotates at constant angular velocity $\omega (= 2\pi f)$ rad/sec. For a P pole machine, the rotation speed (ω_m) is

$$\omega_m = \frac{2}{P}\omega\, \text{rad/sec}$$

or

$$\frac{2\pi N_S}{60} = \frac{2}{P}\omega\, \text{rad/sec}$$

$$N_S = \frac{120\, f}{P}$$

Thus the resultant flux due to three-phase current is of constant value $(= 1.5\,\phi_m$ where ϕ_m is the maximum flux in any phase) and this flux rotates around the stator winding at a synchronous speed of $120\,f/P$ rpm.

For example, for a 6-pole, 50 Hz, 3-phase induction motor, $N_S = 120 \times 50/6 = 1000$ rpm It means that flux rotates around the stator at a speed of 1000 rpm.

4.6 SYNCHRONOUS SPEED

Speed of rotating flux $N_S = \dfrac{120\, f}{P}$ (Fig. 4.11)

where P = Number of poles

f = frequency of supply

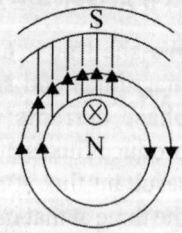

Fig. 4.11

4.7 SLIP (s)

- The rotor speed of an induction machine is different from the speed of rotating magnetic field.
- The percentage difference of speed is called slip.

$$s = \frac{N_S - N_r}{N_S} \text{ or } \boxed{N_r = (1-s)N_S}$$

- where N_S = synchronous speed (rpm)

  N_r = mechanical speed of rotor (rpm)
- Under normal operating conditions, $s = 0.01$ to 0.05, which is very small and the actual speed is very close to synchronous speed. But s is not negligible.

Case 1: If rotor is stationary then $N_r = 0$ and $s = 1$ at start. This is the maximum value of slip possible for induction motor which occurs at start.

Case 2: If rotor is running with synchronous speed, i.e. $N_r = N_S$ then $s = 0$. This is not possible for an induction motor. Slip of induction motor can not be zero under any circumstances. So, the value of slip varies from $s = 0$ to 1.

4.8 EFFECT OF SLIP ON MOTOR PARAMETERS

- In case of a transformer, frequency of the induced emf in the secondary is same as the voltage applied to primary.
- Now in case of induction motor at start $N_r = 0$ and slip $s = 1$.
- Under this condition as long as $s = 1$, the frequency in induced emf in rotor is same as the voltage applied to the stator.
- But as motor gathers speed, induction motor has some slip corresponding to speed N_r.
- In such case, the frequency of induced emf in rotor is no longer same as that of stator voltage.
- Slip effects the frequency of rotor induced emf due to this some other rotor parameters also get affected.
- Let us study the effect the slip on the following rotor parameters:

  Rotor frequency

  Magnitude of rotor induced emf

  Rotor reactance

  Rotor power factor

  Rotor current

4.8.1 Rotor Frequency

In case of induction motor, the speed of rotating magnetic field given by

$$N_S = \frac{120f}{P}$$

- When the rotor is stationary, rotor conductor get cut by rotating flux at the synchronous speed.
- So frequency of rotor current or rotor emf is same as the supply frequency.

$$N_S = \frac{120f}{P} \qquad \qquad ...(4.1)$$

- Now when rotor rotate at speed N, then rotor conductors get cut by $N_S - N_r$ speed.

$$N_S - N_r = \frac{120f}{P} \qquad \qquad ...(4.2)$$

- Divide Eq. (4.2) by Eq. (4.1)

$$\frac{N_S - N_r}{N_S} = \frac{120f_r}{P} \times \frac{P}{120f}$$

$$\frac{f_r}{f} = \frac{N_S - N_r}{N_S} \qquad \qquad \therefore \quad \frac{N_S - N_r}{N_S} = s$$

$$\frac{f_r}{f} = s$$

Rotor frequency = Slip × supply frequency

$$\therefore \qquad \boxed{f_r = sf}$$

- Thus frequency of rotor induced emf in running condition (f_r) is slip times the supply frequency (f).

4.8.2 Effect of Slip on Rotor Induced Emf

- The magnitude of rotor induced emf depends on the relative speed ($N_S - N_r$) the motor.
- If the relative speed is maximum at stand still condition. So emf (E_2) at standstill condition is also maximum.
- If the relative speed is minimum at the running condition. So emf (E_2) at standstill condition is also minimum.
- So at standstill condition of rotor, the induced emf E_2 will be

$$E_2 \propto N_S - N_r; \qquad N_r = 0 \text{ at standstill condition}$$
$$E_2 \propto N_S \qquad \qquad ...(4.3)$$

- In running condition of rotor, the induced emf in rotor circuit E_{2r} will be

$$E_{2r} \propto N_S - N_r \qquad \qquad ...(4.4)$$

- From Eqs (4.3) and (4.4)

$$\frac{E_{2r}}{E_2} = \frac{N_S - N_r}{N_S}$$

$$\frac{E_{2r}}{E_2} = s$$

$$\boxed{E_{2r} = sE_2}$$

- The magnitude of the induced emf in the rotor also reduces by slip times magnitude of induced emf at standstill condition.

4.8.3 Effect of Slip on Rotor Resistance and Reactance

- Let R_2 be the rotor resistance at standstill. Then R_2 will not depend on the frequency $R_2 = R_{2r}$.
- But x_2 depends on the frequency

$$x_2 = 2\pi rfL$$

where
$$f = \text{supply frequency}$$
$$x_2 = \text{reactance at standstill}$$

- Now, let x_2 be rotor reactance in running condition. Then

$$x_2 = 2\pi fr \cdot L \quad \therefore \ f_r = s \cdot f$$

Since,
$$f_r = s \cdot f$$
$$x_{2r} = 2\pi rsf \cdot L$$
$$= S \cdot 2\pi fL$$
$$x_{2r} = sx_2$$

Hence, rotor impedance in standstill condition,

$$z_2 = R_2 + jx_2 \Omega/ph \quad [z_2 = \text{impedance at standstill condition}]$$

$$z_2 = \sqrt{R_2^2 + x_2^2} \ \Omega/\text{ph}$$

z_{2r} = Rotor impedance in running condition
$$= R_2 + jx_{2x} = R_2 + j(sx_2)\Omega/\text{ph}$$

$$\boxed{Z_{2r} = \sqrt{R_2^2 + (sx_2)^2}\,\Omega/\text{ph}}$$

4.8.4 Effect of Slip on Rotor Power Factor

Rotor power factor on standstill condition

$$\cos\phi_2 = \frac{R_2}{z_2} = \frac{R_2}{\sqrt{R_2^2 + x_2^2}}$$

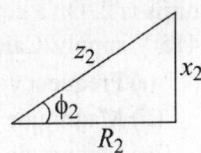

Fig. 4.12

- Rotor power factor at running condition

$$\cos\phi_2 = \frac{R_2}{2z_r} = \frac{R_2}{\sqrt{R_2^2 + (sx_2)^2}}$$

- As the rotor winding is inductive, the rotor power factor is always lagging in nature (Figs 4.12 and 4.13).

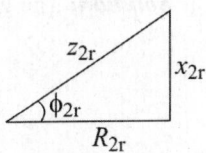

Fig. 4.13

4.8.5 Effect of Slip on Rotor Current

- The simplified equivalent circuit of rotor at standstill condition of rotors (Fig. 4.14).

- Then rotor current $I_2 = \dfrac{E_2}{\sqrt{(R_2)^2 + x_2^2}}$

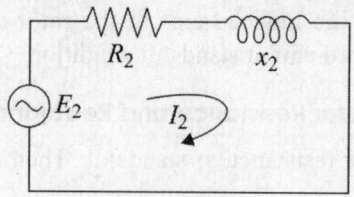

Fig. 4.14(a)

• Now the equivalent circuit of rotor circuit running condition (Fig. 4.14b)

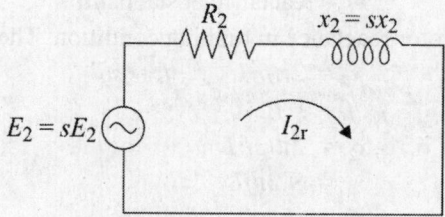

Fig. 4.14(b)

• Then the rotor current

$$I_{2r} = \frac{sE_2}{\sqrt{R_2^2 + (sx_2^2)}}$$

SOLVED EXAMPLES

Example 4.1: For a 4-pole, 3-phase, 50 Hz induction motor ratio of stator to rotor turns is 2. On a certain load, its speed is observed to be 1455 rpm when connected to 415 V supply. Calculate:

(*i*) Frequency of rotor emf in running condition.

(*ii*) Magnitude of induced emf in the rotor at standstill.

(*iii*) Magnitude of induced emf in the rotor under running condition. Assume star connected stator. **(UPTU 2007â08)**

Solution: The given values are

$$K = \text{Rotor turns/stator turns}$$

$$P = 4f = 50 \text{ Hz}, N_r = 1455 \text{ rpm}, E_{sin} = 415 \text{ V}$$

$$N_S = \frac{120f}{P} = \frac{120 \times 50}{4} = 1500 \text{ rpm}$$

• For a given load $N_r = 1455$ rpm

$$s = \frac{N_S - N_r}{N_S} = \frac{1500 - 1455}{1500}$$

$$= 0.03 \text{ or } 3\% \textbf{ Ans.}$$

- $f_r = s \cdot f = 0.3 \times 50 = 1.5$ Hz
- At standstill, induction motor acts as a transformer

$$\therefore \qquad \frac{E_2(ph)}{E_1(ph)} = \frac{N_2}{N_1} = \frac{\text{Rotor turns}}{\text{Stator turns}} = K$$

But ratio of stator to rotor turns is given as 2,

$$\frac{N_1}{N_2} = 2 \qquad \therefore \frac{N_2}{N_1} = \frac{1}{2} = K \text{ and } E_{1\,\text{line}} = 415 \text{ V}$$

- The given values are always line values unless and until specifically stated as per phase.

$$\therefore \quad E_{1ph} = \frac{E_1}{\sqrt{3}} = \frac{415}{\sqrt{3}} = 239.6 \text{ V}; \text{ As star connection } E_{\text{line}} = \sqrt{3}\, E_{ph}$$

$$\therefore \qquad\qquad E_{1ph} = 239.6 \text{ V}$$

$$\therefore \qquad\qquad \frac{E_{2ph}}{E_{1ph}} = \frac{1}{2} \quad \therefore E_{2ph} = \frac{1}{2} \times 239.6 = 119.8 \text{ V}$$

Rotor induced emf on standstill is 119.7 V.

- In running condition,

$$E_{2r} = sE_2 = 0.03 \times 119.8 = 3.594 \text{ V}$$

The value of rotor induced emf in the running condition is also very very small.

Example 4.2: The voltage applied to stator of 3ϕ, 4 pole IM has a frequency of 50 Hz. The frequency of emf. induced in rotor is 2 Hz. Calculate (*i*) slip (*ii*) actual speed at which the motor is running.

Solution: The given data

$$P = 4, f = 50 \text{ Hz}$$

$$= \frac{120 \times 0.5}{4} = 15 \text{ rpm at full load}$$

$$= 15 \text{ rpm}$$

Speed of rotor wrt stator,

$$N = 1500 \text{ rpm}$$

Example 4.3: A 3ϕ, 440 V, 50 Hz, lm runs at speed of 1450 rpm when it delivers rated output power. Determine

(*i*) No of poles in machine

(*ii*) Speed of rotating magnetic field

(*iii*) Rotor induced emf if stator to rotor turn ratio is 1:0.8. Assume the winding factors are same

(*iv*) Frequency of rotor current

Solution: As synchronous and rotor speed are nearly same

(*i*) Number of poles

$$P = \frac{120 f}{N_s} = \frac{120 \times 50}{1450} \approx 4$$

(*ii*) Speed of rotating magnetic field,

$$N_S = \frac{120 \times 50}{4} = 1500 \, \text{rpm}$$

(*iii*) Since

∴ Produced emf in rotor at standslill ≈ KE, $0.8 \times 400 = 352$ V.

$$s = \frac{N_S - N_r}{N_S} = \frac{1500 - 1450}{1500} \times 100 = 3.33\%$$

Rotor induced voltage = $S \times 352$

$$= \frac{3.33}{100} \times 352 = 11.74 \, \text{volts}$$

(*iv*) Frequency of rotor current,

$$f_r = s \cdot f = 0.0333 \times 50$$
$$= 1.66 \, \text{Hz}$$

4.9 LOSSES IN INDUCTION MOTOR

- The input electric power fed to the stator of the motor is converted into mechanical power at the shaft of the motor.
- The various losses during energy conversion are shown in the Fig. 4.15.

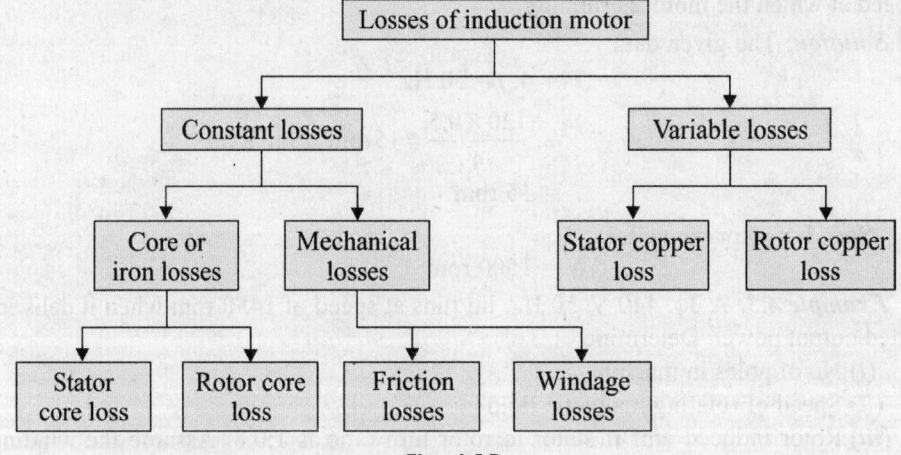

Fig. 4.15

4.9.1 Constant Losses

- Core losses occur in stator core and rotor core. These are also called *iron losses*. There losses include eddy current losses and hystersis losses.
- Eddy current losses are minimised by using laminated construction.

- While hysteresis losses are minimized by selecting high grade silicon steel as the material for stator and rotor.
- The iron losses depends on the frequency. The stator frequency is always supply frequency hence stator iron losses and hystersis losses.
- As against this in rotor circuit, the frequency is very very small which is slip supply frequency.
- Hence rotor iron losses are very small and hence generally neglected, in the running condition.
- The mechanical losses include frictional losses at the bearing and windage losses. The friction changes with speed but practically the drop in speed is very small hence there losses are assumed to be part of constant losses.

4.9.2 Variable Losses

- This include the copper losses in stator and rotor windings due to current flowing in the winding. As current changes as load changes, these losses are said to be variable losses.
- Generally stator iron variable are combined with stator copper losses at a particular load to specify total stator losses at particular load condition.

Total Rotor copper losses = $3I_{2r}^2 R_2$

where I_{2r} = Rotor current per phase at a particular load

R_2 = Rotor resistance per phase.

4.10 POWER FLOW DIAGRAM FOR AN INDUCTION MOTOR

- Induction motor converts an electrical power supplied to it into mechanical power.
- The various stages in this conversion is called *power flow in an induction motor*.
- The three-phase supply given to the stator is the net electrical input to the motor. If motor power factor is cos ϕ and $V_L I_L$ are link values of supply voltage and current drawn, then net input electrical power supplied to the motor can be calculated as

$$P_{in} = \sqrt{3} V_L I_L \cos \phi$$

where P_{in} = Net input electrical power

- This part of this power is utilized to the supply the losses will the stator which are stator core loss as well as copper losses.
- The remaining power is delivered to the rotor magnetically through the air gap with the help of rotating magnetic field. This is called *rotor input* denoted as P_2

So $\qquad P_2 = P_{in} -$ stator losses (core loss + copper loss)

- The rotor is not able to convert its entire input to the mechanical drive as it has to supply rotor losses.

- The rotor losses are dominantly copper losses as rotor Iron losses are very small and hence generally neglected, so rotor losses are rotor copper losses denoted as P_C.

So

$$P_C = 3I_{2r}^2 \times R_2$$

where, I_{2r} = Rotor current per phase in running condition

R_2 = Rotor resistance per phase

- After supplying these losses, the remaining part of P_2 is converted into mechanical power developed by the motor denoted as P_m

$$P_m = P_2 - P_C$$

- Now this power, motor tries to deliver to the load connected to the shaft. But during this mechanical transmission, part of P_m is utilized to provide mechanical losses like friction and windage losses.
- Finally the power is available to the load at shaft. This is called net output of the motor denoted as P_{out}. This is also called *shaft power*.

$$P_{out} = P_m - \text{mechanical losses}$$

- The rating of the motor is specified in terms of value of P_{out} when load is in full load condition. This is expressed in horse power and called *horse power rating of the motor*.
- The stages can be shown diagrammatically called power flow diagram of an induction motor. This is shown in the Fig. 4.16.

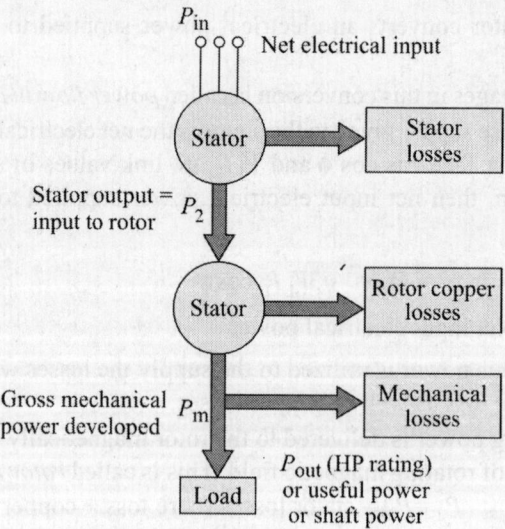

Fig. 4.16: Power flow diagram

4.11 EFFICIENCY OF AN INDUCTOR MOTOR

- The ratio of net power available at the shaft (P_{out}) and the net electrical power input (P_{in}) to the motor is called *overall efficiency of an induction motor*.

$$\% \, \eta = \frac{P_{out}}{P_{in}} \times 100$$

- The maximum efficiency occurs when variable losses equal to constant losses. When motor is at no load condition, current drawn by the motor is small.
- Hence efficiency is low. As load increases, current drawn by the motor increases so copper losses also increases.
- When such variable losses achieve the same value as that of constant losses, efficiency attains its maximum value.
- If load is increased further, variable losses become greater than constant losses hence deviating from condition for maximum, efficiency starts decreasing.
- Hence the nature of the curve of efficiency against output power of the motor is shown in Fig. 4.17.

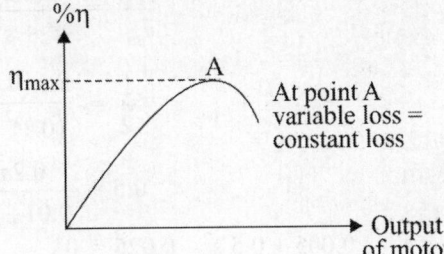

Fig. 4.17: Efficiency curve for an induction motor

SOLVED EXAMPLES

Example 4.4: A 400 V, 4 pole, 3-phase, 50 Hz induction motor has a rotor resistance and reactance per phase of 0.01 Ω and 0.1 Ω respectively, determine

(*i*) Maximum torque in N·m and the corresponding slip.

(*ii*) The full load slip and power output in watts if maximum torque is twice the full load torque and the ratio of stator to rotor turns is 4.   [UPTU 2003–04]

Solution: $V_2 = 400$ V, $f = 50$ Hz, $P = 4$, $R_2 = 0.01$ Ω/phases
$x_2 = 0.1$ Ω/phase.

Step 1: To find the maximum torque:

$$\frac{N_2}{N_1} = \frac{E_2}{E_1} \qquad \therefore E_2 = \frac{N_2}{N_1} \times E_1 = \frac{1}{4} \times 400$$

$$E_2 = 100 \text{ V}$$

$$E_{ph} = \frac{100}{\sqrt{3}} = 57.73 \text{ V}$$

$$\therefore \qquad T_m = \frac{KE_2^2}{2x_2} = \frac{3}{2\pi n_s} \times \frac{E_2^2}{2x_2}, \text{ where } K = \frac{3}{2\pi n_s}$$

But
$$N_s = \frac{120 f}{P} = \frac{120 \times 50}{4} = 1500\,\text{rpm}$$

$$n_s = \frac{1500}{60} = 25\,\text{rps}$$

$$T_m = \frac{3}{2\pi \times 25} \times \frac{(57.73)^2}{2 \times 0.1} = 318.25\,\text{Nm}$$

Step II: To find corresponding slip

$$a = s_m = \frac{R_2}{x_2} = \frac{0.01}{0.1} = 0.1$$

Step III: To find full load slip

$$\frac{T_{FL}}{T_m} = \frac{2a \cdot s}{a^2 + s^2}$$

$$\frac{1}{2} = \frac{2 \times (0.1) \times s}{(0.1)^2 \times s^2}$$

$$\therefore \qquad 0.5 = \frac{0.2s}{0.01 + s^2}$$

$$0.005 + 0.5\,s^2 - 0.025 = 0$$

$$\therefore \qquad 0.5\,s^2 + 0.2s - 0.005 = 0$$

$$\therefore \text{ using} \qquad s = \frac{-b \pm \sqrt{b^2 - 4ac}}{2a}$$

$$s = \frac{0.2 \pm \sqrt{(-0.02)^2 - 4 \times 0.5 \times 0.005}}{2 \times 0.5}$$

$$s = \frac{0.2 \pm \sqrt{-0.03}}{1} = 0.2 \pm 0.173$$

4.12 ROTOR TORQUE

The torque T developed by the rotor is directly proportional to:

   (*i*)   rotor current

  (*ii*)   rotor emf

 (*iii*)   power factor of the rotor circuit

$$\therefore \qquad T \propto E_2 I_2 \cos\phi_2$$

or
$$T = K E_2 I_2 \cos f_2$$

where I_2 = rotor current at standstill

       E_2 = rotor emf at standstill

    Note: The values of rotor emf, rotor current and rotor power factor are taken for the given conditions.

4.13 STARTING TORQUE (T_s)

Let E_2 = rotor emf, per phase at standstill
$\quad X_2$ = rotor reactance per phase at standstill
$\quad R_2$ = rotor resistance per phase

Rotor impedance/phase, $\quad Z_2 = \sqrt{R_2^2 + X_2^2} \quad$ (at standstill)

Rotor current/phase, $\quad I_2 = \dfrac{E_2}{Z_2} = \dfrac{E}{\sqrt{R_2^2 + X_2^2}} \quad$ (at standstill)

Rotor pf, $\quad \cos\phi_2 = \dfrac{R_2}{Z_2} = \dfrac{R}{\sqrt{R_2^2 + X_2^2}}$

$\therefore$ Starting torque, $\quad T_s = KE_2 I_2 \cos\phi_2$

$$= KE_2 \times \frac{E_2}{\sqrt{R_2^2 + X_2^2}} \times \frac{R_2}{\sqrt{R_2^2 + X_2^2}}$$

$$= \frac{KE_2^2 R_2}{R_2^2 + X_2^2}$$

Generally, the stator supply voltage is constant so that flux per pole set up by the stator is also fixed. This in turn means that emf E_2 induced in the rotor will be constant.

$$\therefore \qquad T_s = \frac{K_1 R_2}{R_2^2 + X_2^2} = \frac{K_1 R_2}{Z_2^2}$$

where K_1 is another constant.

- It is clear that the magnitude of starting torque would depend upon the relative values of R_2 and X_2, i.e. rotor resistance/phase and standstill rotor reactance/phase.

It can be shown that $K = 3/2\ \pi N_s$.

$$\therefore \qquad T_s = \frac{3}{2\pi N_s} \cdot \frac{E_2^2 R_2}{R_2^2 + X_2^2}$$

Note that here N_s is in rps.

4.14 CONDITION FOR MAXIMUM STARTING TORQUE

It can be proved that starting torque will be maximum when rotor resistance/phase is equal to standstill rotor reactance/phase.

Now $\qquad T_s = \dfrac{K_1 R_2}{R_2^2 + X_2^2} \qquad\qquad$ ...(4.5)

Differentiating Eq. (4.5) wrt R_2 and equating the result to zero, we get

$$\frac{dT_s}{dR_2} = K_1 \left[\frac{1}{R_2^2 + X_2^2} - \frac{R_2 (2R_2)}{\left(R_2 + X_2^2\right)^2} \right] = 0$$

$$R_2 + X_2 = 2R_2$$
or
$$R_2 = X_2$$

Hence starting torque will be maximum when:

Rotor resistance/phase = Standstill rotor reactance/phase

Under the condition of maximum starting torque, $\phi_2 = 45°$ and rotor power factor is 0.707 lagging [see Fig. 4.18(a)].

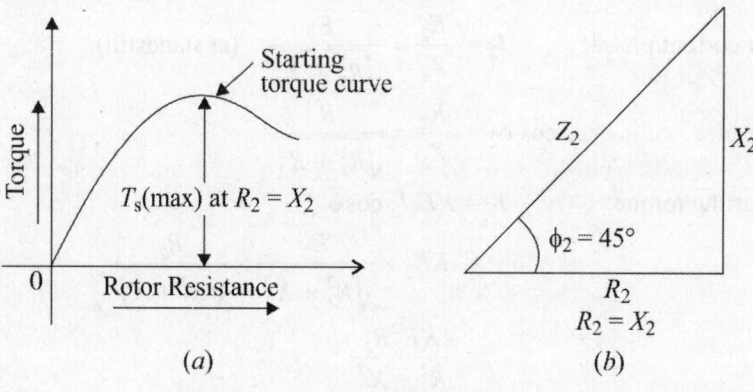

Fig. 4.18

Figure 4.18(b) shows the variation of starting torque with rotor resistance. As the rotor resistance is increased from a relatively low value, the starting torque increases until it becomes maximum when $R_2 = X_2$. If the rotor resistance is increased beyond this optimum value, the starting torque will decrease.

4.15 EFFECT OF CHANGE OF SUPPLY VOLTAGE

$$T_s = \frac{KE_2^2 R_2}{R_2^2 + X_2^2}$$

Since E_2 have supply voltage V

$$T_s = \frac{K_2 V^2 R_2}{R_2^2 + X_2^2}$$

where K_2 is another constant.

∴ $$T_s \propto V^2$$

Therefore, the starting torque is very sensitive to changes in the value of supply voltage. For example, a drop of 10% in supply voltage will decrease the starting torque by about 20%. This could mean the motor failing to start if it cannot produce a torque greater than the load torque plus frictional torque.

4.16 STARTING TORQUE OF 3-PHASE INDUCTION MOTORS

The rotor circuit of an induction motor has low resistance and high inductance. At starting, the rotor frequency is equal to the stator frequency (i.e. 50 Hz) so that rotor reactance is large compared with rotor resistance. Therefore, rotor current lags the rotor emf by a large angle, the power factor is low and consequently the starting torque is small. When resistance is added to the rotor circuit, the rotor power factor is improved which results in improved starting torque. This, of course, increases the rotor impedance and, therefore, decreases the value of rotor current but the effect of improved power factor predominates and the starting torque is increased.

(*i*) **Squirrel-cage motors:** Since the rotor bars are permanently short-circuited, it is not possible to add any external resistance in the rotor circuit at starting. Consequently, the starting torque of such motors is low. Squirrel cage motors have starting torque of 1.5 to 2 times the full-load value, the starting current is 5 to 9 times the full-load current.

(*ii*) **Wound rotor motors:** The resistance of the rotor circuit of such motors can be increased through the addition of external resistance. By inserting the proper value of external resistance (so that $R_2 = X_2$), maximum starting torque can be obtained. As the motor accelerates, the external resistance is gradually cut out until the rotor circuit is short-circuited on itself for running conditions.

Example 4.5: The 3-phase induction motor having star-connected rotor has an induced emf of 50 V between the slip rings at standstill on open-circuit. The rotor has a resistance and reactance per phase of 0.5 Ω and 4.5 Ω respectively. Find the current per phase and the power factor at starting when (*i*) the slip rings are short-circuited (*ii*) the slip rings are connected to a star-connected rheostat of 4 Ω per phase.

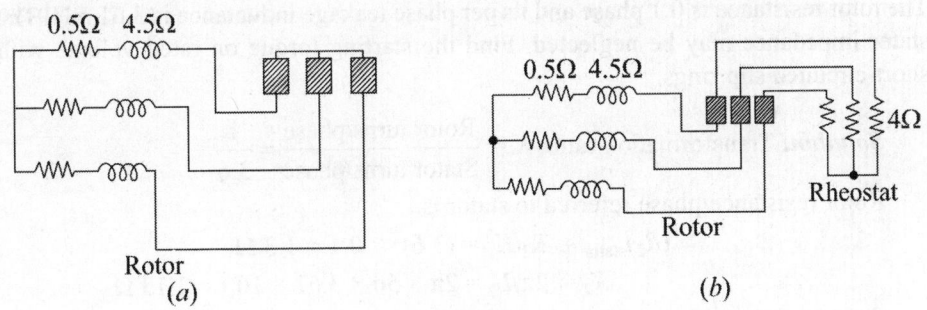

Fig. 4.19

Solution: Standstill rotor emf/phase, $E_2 = 50/\sqrt{3} = 28.87\,\text{V}$

(*i*) When slip rings are short-circuited [Fig. 4.19]

Rotor impedance/phase = $\sqrt{(0.5)^2 + (4.5)^2} = 4.53\,\Omega$

Rotor current/phase = $28.87/4.53 = 6.38\,\text{A}$

Rotor power factor = $0.5/4.53 = 0.11$ lag

(*ii*) When slip rings connected to rheostat [Fig 4.19(*b*)]

Rotor resistance/phase = $4 + 0.5 = 4.5\,\Omega$

Rotor impedance/phase = $(4.5)^2 + (4.5)^2 = 6.360$

Rotor current/phase = 28.87/6.36 = 4.54 A

Rotor power factor = 4.5/6.36 = 0.707 lag

It is clear that by inserting additional resistance in the rotor circuit, the pf of the rotor circuit is improved, though the rotor current is decreased. However, the effect of improved pf predominates the decrease in rotor current. Hence, starting torque is increased.

Example 4.6: A star-connected rotor winding of a 3-phase induction motor has a resistance of 0.1 Ω per phase and a standstill reactance of 1 $\sqrt{(0.5)^2 + (4.5)^2} = 4.53\Omega$ per phase. Find the value of external resistance per phase to be added to give maximum starting torque. What is the power factor of the rotor circuit then?

Solution: Let R ohm be the external resistance per phase to be added to the rotor circuit to give maximum starting torque. For maximum starting torque

rotor resistance/phase = Standstill rotor reactance/phase

or $\qquad 0.2 + R = 1$

∴ $\qquad R = 1 - 0.2 = 0.8\ \Omega$

Rotor impedance/phase = $\sqrt{1^2 + 1^2} = \sqrt{2}$

Rotor power factor = $1/\sqrt{2} = 0.707\ \text{lag}$

Example 4.7: A 150 kW; 3000 Y, 50 Hz, 6-pole, star-connected induction motor has a star-connected slip-ring rotor with a transformation ratio of 3.6 (stator/rotor). The rotor resistance is 0.1 phase and its per phase leakage inductance is 3.61 mH. The stator impedance may be neglected. Find the starting torque on rated voltage with short-circuited slip rings.

Solution: Transformation ratio, $K = \dfrac{\text{Rotor turns/phase}}{\text{Stator turns/phase}} = \dfrac{1}{3.6}$

Rotor resistance/phase referred to stator is

$$(R_2)_{\text{stator}} = R_2/K^2 = (3.6)^2 \times 0.1 = 1.3\ \Omega$$
$$X_2 = 2\pi f L_2 = 2\pi \times 50 \times 3.61 \times 10^{-3} = 1.13\ \Omega$$

Rotor reactance/phase referred to stator is

$$(X_2)_{\text{stator}} = X_2/K^2 = 1.13 \times (3.6)^2 = 14.7\ \Omega$$
$$N_S\,z = \frac{120f}{P} = \frac{120 \times 50}{6} = 1000\ \text{rpm} = \frac{50}{3}\ \text{rps}$$

Supply voltage/phase, $\quad E_1 = 3000\sqrt{3}\ \text{volts}$

Starting torque, $\qquad T_s = \dfrac{3}{2\pi N_s} \times \dfrac{E_1^2\left(R_2\right)_{\text{stator}}}{(R^2)_{\text{stator}}^2 + (X_2)_{\text{stator}}^2}$

$$= \frac{3}{2\pi(50/3)} \times \frac{(3000\sqrt{3})^2 \times 1.3}{(1.3)^2 \times (14.7)^2} = 513\ \text{Nm}$$

Example 4.8: The rotor resistance and standstill reactance per phase of a 3-phase induction motor are 0.02 Ω and 0.1 Ω respectively. What should be the value of the external resistance per phase to be inserted in the rotor circuit to give maximum torque at starting?

Solution: Let R_xΩ/phase be the external resistance added to the rotor circuit to obtain maximum starting torque.

Rotor resistance/phase, $R_2 = (0.02 + R_x)$Ω

Rotor reactance/phase at standstill, $X_2 = 0.1$ Ω

The starting torque will be maximum when:

rotor resistance/phase = Standstill rotor reactance/phase

or $\qquad\qquad\qquad 0.02 + R_x = 0.1$ Ω

$\qquad\qquad\qquad\qquad R_x = 0.1 - 0.02 = 0.08$ Ω

Example 4.9: The rotor resistance and standstill reactance of a 3-phase induction motor are 0.2 Ω and 1.0 Ω per phase. The voltage between the slip rings with rotor locked and a full voltage on the stator is 110 volt.

(*i*) Find the starting rotor current/phase when the rotor slip rings are short-circuited to form the normal running condition.

(*ii*) What should be the value of the external resistance per phase to be inserted in the rotor circuit to give maximum torque at starting? Also find the rotor current/phase under this condition.

Solution: (*i*) Figure 4.20 shows the conditions of the problem.

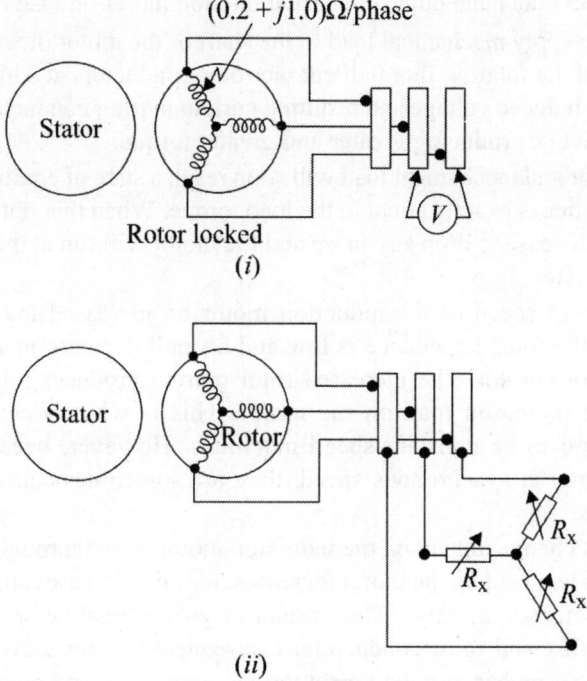

Fig. 4.20

Rotor emf/phase at standstill, $E_2 = 110/\sqrt{3} = 63.5\,V$

Rotor impedance/phase at standstill, $Z_2 = \sqrt{R_2^2 + X_2^2} = \sqrt{(0.2)^2 + (1)^2} = 1.02\,\Omega$

Rotor phase current at standstill, $I_2 = E_2/Z_2 = 63.5/1.02 = 62.3\,A$

(*ii*) Let $R_x\Omega$/phase be the external resistance added to the rotor circuit [see Fig. 4.20] to obtain the maximum starting torque.

The starting torque will be maximwn when:

Rotor resistance/phase = Standstill rotor reactance/phase

or $\qquad\qquad 0.2 + R_x = 1$

or $\qquad\qquad R_x = 1 - 0.2 = 0.8\ \Omega$/phase

$\qquad$ Rotor impedance/phase = $\sqrt{(0.8 + 0.2)^2 + (1)^2} = 1.414\,\Omega$

$\qquad\qquad$ Rotor current/phase = $63.5/1.414 = 44.7\,A$

Note that rotor current/phase is reduced by about 30%. This will produce more than double the torque by improving the power factor of the rotor circuit. The efficiency has reduced considerably since the losses in the rotor have changed from $3I^2R = 3 \times (62.3)^2 \times 0.2 = 2328.5\ \Omega$ to $3 \times (44.7)^2 \times 1 = 5994.3\ W$. Therefore, the external resistances must be removed as soon as the motor has reached the working speed.

4.17 MOTOR UNDER LOAD

Let us now discuss the behaviour on-phase induction motor on load (Fig. 4.21).

(*i*) When we apply mechanical load to the shaft of the motor, it will begin to slow down and the rotating flux will cut the rotor conductors at a higher and higher rate. The induced voltage and resulting current in rotor conductors will increase progressively, producing greater and greater torque.

(*ii*) The motor and mechanical load will soon reach a state of equilibrium when the motor torque is exactly equal to the load torque. When this state is reached, the speed will cease to drop any more and the motor will run at the new speed at a constant rate.

(*iii*) The drop in speed of the induction motor on increased load is small. It is because the rotor impedance is low and a small decrease in speed produce a large rotor current. The increased rotor current produces a higher torque to meet the increased load on the motor. This is why induction motors are considered to be constant-speed machines. However, because they never actually run at synchronous speed, they are sometimes called *asynchronous machines*.

Note that change in load on the induction motor is met through the adjustment of slip. When load on the motor increases, the slip increases slightly (i.e. motor speed decreases slightly). This results in greater relative speed between the rotating flux and rotor conductors. Consequently, rotor current is increased, producing a higher torque to meet the increased load and reverse happens the load on the motor decrease.

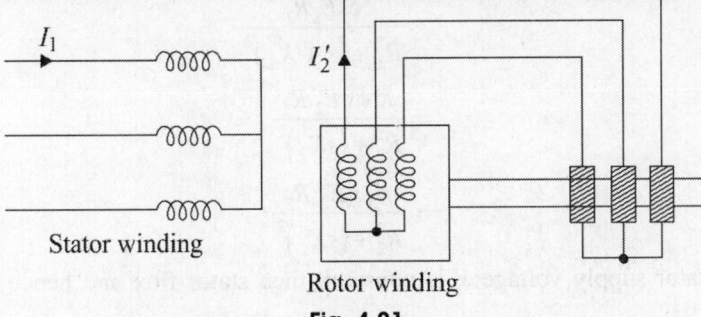

Stator winding

Rotor winding

Fig. 4.21

(iv) With increasing load, the increased load currents I_2 are in such a direction so as to decrease the stator flux (Lenz's law), thereby decreasing the counter emf in the stator windings. The decreased counter emf allows motor stator currents (I_1) to increase, thereby increasing the power input to the motor. It may be noted that action of the induction motor in adjusting its stator or primary current with changes of current in the rotor or secondary is very much similar to the changes occurring in transformer with changes in load.

4.18 TORQUE UNDER RUNNING CONDITIONS

Let the rotor at standstill have per phase induced emf E_2, reactance X_2 and resistance R_2. Then under running conditions (Fig. 4.22) at slip s,

Rotor emf/phase, $\qquad E_2' = s \cdot E_2$

Rotor reactance/phase, $\quad X_2' = sX_2$

Rotor impedance/phase, $I_2' = \dfrac{E_2'}{Z_2} = \dfrac{s \cdot E_2}{\sqrt{R_2^2 + (s \cdot X_2)^2}}$

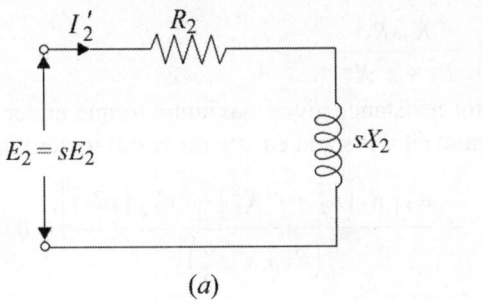

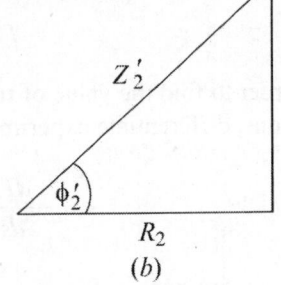

(a) $\qquad\qquad\qquad\qquad\qquad\qquad$ *(b)*

Fig. 4.22

Running torque, $\qquad\qquad T_r \propto E_2 I_2' \cos\phi_2'$

$$\propto \phi I_2' \cos\phi_2' \qquad (\because E_2' \propto \phi)$$

$$\propto \phi \times \frac{sE_2}{\sqrt{R_2^2 + (sX_2)^2}} \times \frac{R_2}{\sqrt{R_2^2 + (sX_2)^2}}$$

$$\propto \frac{\phi s E_2 R_2}{R_2^2 + (s + X_2)^2}$$

$$= \frac{K \phi s E_2 R_2}{R_2^2 + (sX_2)^2}$$

$$= \frac{K_1 s E_2^2 R_2}{R_2^2 + (sX_2)^2}$$

If the stator supply voltage V is constant, then stator flux and hence E_2 will be constant.

$$\therefore \qquad T = \frac{K_2 sR_2}{R_2^2 + (sX_2)^2}$$

where K_2 is another constant.

It may be seen that running torque is:

(i) directly proportional to slip, i.e. if slip increases (i.e. motor speed decreases), the torque will increase and vice-versa.

(ii) directly proportional to square of supply voltage ($\therefore E_2 \propto V$).

It can be shown that value of $K_1 = 3/2\pi N_s$ where N_s is in rps

$$\therefore \qquad T_r = \frac{3}{2\pi N_s} \cdot \frac{s E_2^2 R_2}{R_2^2 + (sX_2)} = \frac{3}{2\pi N_s} \cdot \frac{s E_2^2 R_2}{(Z_2')^2}$$

At starting, $s = 1$ so that starting torque is

$$T_s = \frac{3}{2\pi N_s} \cdot \frac{E_2^2 R_2}{R_2^2 + X_2^2}$$

4.19 MAXIMUM TORQUE UNDER RUNNING CONDITIONS

$$T = \frac{K_2 sR_2}{R_2^2 + s^2 X_2^2} \qquad \qquad ...(4.6)$$

In order to find the value of rotor resistance gives maximum torque under running conditions, differentiate experimental (i) wrt s and equate the result to zero, i.e.

$$\frac{dT_r}{ds} = \frac{K_2 \left[R_2 \left(R_2^2 + s^2 X_2^2 \right) - 2s X_2^2 (sR_2) \right]}{\left(R_2 + s^2 X_2^2 \right)^2} = 0$$

or $\qquad \left(R_2^2 + s^2 X_2^2 \right) - 2s^2 X_2^2 = 0$

or $\qquad R_2^2 = s^2 X_2^2$

or $\qquad R_2 = sX_2$

Thus for maximum torque (T_m) under running conditions:

Rotor resistance/phase = Fractional slip × Standstill rotor reactance/phase

Now $$T_r \propto \frac{sR^2}{R_2^2 + s^2 X^2} \qquad \text{...from Eq. (4.6)}$$

For maximum torque, $R_2 = sX_2$, putting $R_2 = sX_2$ in the above expression, the maximum torque is given by

$$T_m \propto \frac{1}{2X_2}$$

Slip corresponding to maximum torque, $s = R_2/X_2$

It can be shown that:

$$T_m = \frac{3}{2\pi N_s} \times \frac{E_2^2}{R_2^2 + s_2^2 X_2^2}$$

It is evident from the above relation that:

(i) The value of rotor resistance does not alter the value of the maximum torque but it only changes the value of the slip at which it occurs.

(ii) The maximum torque varies inversely as the standstill reactance. Therefore, it should be kept as small as possible.

(iii) The maximum torque varies directly with the square of the applied voltage.

(iv) To obtain maximum torque at starting ($s = 1$), the rotor resistance must be made equal to rotor reactance at standstill.

Example 4.10: A 4-pole, 50 Hz, 3-phase induction motor has a rotor resistance of 0.024 Ω per phase and standstill reactance of 0.6 Ω per phase. Determine the speed at which maximum torque is developed.

Solution:

Rotor resistance/phase, $R_2 = 0.024 \ \Omega$

Standstill rotor reactance/phase, $X_2 = 0.6 \ \Omega$

Slip corresponding to max. torque, $s = \dfrac{R_2}{X_2} = \dfrac{0.024}{0.6\Omega} = 0.04$

Speed corresponding to max. torque

$$N = N_s(1 - s)$$

$$= \frac{120 f}{P}(1 - s) = \frac{120 \times 50}{4}(1 - 0.04) = 1440 \, \text{rpm}$$

Example 4.11: A 440 V, 3-phase, 50 Hz, 4-pole, Y–connected induction motor has a full-load speed of 1425 rpm. The rotor has an impedance of $(0.4 + j4)$ ohm per phase and rotor/stator turn ratio is 0.8. Calculate (i) full-load torque (ii) rotor current and (iii) full-load rotor Cu loss.

Solution: $N_s = 120 \, f/P = 120 \times 50/4 = 1500 \, \text{rpm} = 25 \, \text{rps}$

$$s = \frac{N_s - N}{N_s} = \frac{1500 - 1425}{1500} = \frac{75}{1500} = 0.05$$

$$E_1 = 440\sqrt{3} = 254 \, \text{V/phase}$$

$$E_2 = KE = 0.8 \times 254 = 203.2 \, \text{V}$$

(i) Full–load torque, $T_f = \dfrac{3}{2\pi N_s} \times \dfrac{sE_2^2 R_2}{R_2^2 + (sX_2)^2}$

$$= \dfrac{3}{2\pi \times 25} \times \dfrac{0.05 \times (203.2)^2 \times 0.4}{(0.4)^2 + (0.05 \times 4)^2} = 78.87\,\text{Nm}$$

(ii) Rotor current, $I_2' = \dfrac{sE_2}{\sqrt{(R_2)^2 + (sx_2)^2}} = \dfrac{0.05 \times 203.2}{\sqrt{(0.04)^2 + (0.05 \times 4)^2}} = 22.73\,\text{A}$

(iii) Total rotor Cu loss = $3I_2^2 R_2 = 3 \times (22.73)^2 \times 0.4 = 620\,\text{W}$

Example 4.12: A 3-phase induction motor has a 4-pole star-connected stator winding. The motor runs on a 50 Hz supply with 200 V between lines. The rotor resistance and standstill reactance per phase are 0.1Ω and 0.9 Ω respectively. The ratio of rotor to stator turns is 0.67. Calculate (i) torque at 4% slip (ii) maximum torque (iii) speed at maximum torque.

Solution: $E_1 = 200\sqrt{3} = 115.5\,\text{V}; E_2 = KE_1 = 115.5 \times 0.67 = 77.4\,\text{V}$
$N_s = 120\,f/P = 120 \times 50/4 = 1500\,\text{rpm} = 25\,\text{rps}$

(i) At 4% slip, the torque is given by

$$T_r = \dfrac{3}{2\pi N_s} \times \dfrac{sE_2^2 R_2}{R_2^2 + (sX_2)^2} = \dfrac{3}{2\pi \times 25} \times \dfrac{0.04 \times (77.4)^2 \times 0.1}{(0.1)^2 + (0.04 \times 0.9)^2} = 40.5\,\text{Nm}$$

(ii) The maximum torque occurs when $R_2 = sX_2$ or $s = R/X_2 = 1/9$.

$$\therefore \quad T_m = \dfrac{3}{2\pi N_s} \times \dfrac{sE_2^2 R_2}{R_2^2 + (sX_2)^2} = \dfrac{3}{2\pi \times 25} \times \dfrac{(1/9) \times (77.4)^2 \times 0.1}{(0.1)^2 + (0.1)^2} = 63.6\,\text{Nm}$$

(iii) Slip at maximum torque, $s = R_2/X_2 = 1/9$

Speed at maximum torque = $(1-s)N_s = \left(1 - \dfrac{1}{9}\right) \times 1500 = 1333\,\text{rpm}$

Example 4.13: The resistance and standstill reactance of each phase of a 3-phase induction motor with star-connected rotor are 0.06 Ω and 0.4 Ω respectively. The full-load slip is 4%. Calculate the resistance per phase of a star-connected rheostat which when connected to the rotor circuit will give a pull-out torque at one half of the full-load speed. What is then the power factor?

Solution: Let N_s be the synchronous speed.

Full-load speed, $N = (1 - s_f)\,N_s = (1 - 0.04)N_s = 0.96\,N_s$

One–half of full-load speed, $v = \dfrac{0.96\,N_s}{2} = 0.48\,N_s$

Slip at half of full load speed, $s = \dfrac{N_s - 0.48\,N_s}{N_s} = 0.52$

Rotor resistance/phase, $R_2 = 0.06\,\Omega$
Rotor reactance/phase at standstill, $X_2 = 0.4\,\Omega$

Rotor reactance at slip $s = sX_2 = 0.52 \times 0.4 = 0.208 \ \Omega$

Let R_x ohm/phase be the resistance of the star-connected rheostat added to rotor circuit that will give pull-out (or maximum) torque at one-half of the full-load speed. Then

$$0.06 + R_x = 0.208$$

or

$$R_x = 0.208 - 0.06 = 0.148 \text{ W}$$

$$\text{Power factor} = \frac{R_2 + R_x}{\sqrt{(R_2 + R_x)^2 + (sX_2)^2}} = \frac{0.06 + 0.148}{\sqrt{(0.06 + 0.148)^2}} = 0.707 \text{ lag}$$

Full-load, starting and maximum torque

$$T_f \propto \frac{sR^2}{R_2^2 + (sX_2)^2}$$

$$T_s \propto \frac{R_2}{R_2^2 + X_2^2}$$

$$T_m \propto \frac{1}{2X_2}$$

Note that s corresponds to full-load slip.

(i)

$$\frac{T_m}{T_f} = \frac{R_2^2 + (sX_2)^2}{2sR_2X_2}$$

Dividing the numerator and denominator on RHS by X_2^2, we get,

$$\frac{T_m}{T_f} = \frac{(R_2 / X_2)^2 + s^2}{2s(R_2 / X_2)} = \frac{a^2 + s^2}{2as}$$

where

$$a = \frac{R_2}{X_2} = \frac{\text{Rotor resistance/phase}}{\text{Standstill rotor reactance/phase}}$$

(ii)

$$\frac{T_m}{T_s} = \frac{R_2^2 + X_2^2}{2R_2X_2}$$

Dividing the numerator and denominator on RHS by X_2^2, we get

$$\frac{T_m}{T_s} = \frac{(R_2 / X_2)^2 + 1}{2(R_2 / X_2)} = \frac{a^2 + 1}{2a}$$

where

$$a = \frac{R_2}{X_2} = \frac{\text{Rotor resistance/phase}}{\text{Standstill rotor reactance/phase}}$$

4.20 EQUIVALENT CIRCUIT OF INDUCTION MOTOR

- We have already seen that the induction motor can be treated as generalized transformer.
- Transformer works on the principle of electromagnetic induction. The induction motor also works on the same principle.

- The energy transfer from stator to rotor of the induction motor takes place entirely with the help of a flux mutually linking the two.
- Thus stator acts as a primary while the rotor acts as a rotating secondary when induction motor is treated as a transformer.

If E_1 = Induced voltage in stator per phase.

E_2 = Rotor induced emf per phase on standstill.

$$K = \frac{N_2}{N_1} = \text{Rotor winding turns}$$

Then, $K = \dfrac{E_2}{E_1}$

- Thus if V_1 is the supply voltage per phase to stator, it produces the flux which links with both stator and rotor.
- Due to self induction E_1 is the induced emf in stator per phase while E_2 is the induced emf. in rotor due to mutual induction at standstill.
- In running condition the induced emf. in rotor be comes E_2 which is sE_2.

Now E_{2r} = Rotor induced emf in running condition per phase

R_2 = Rotor resistance per phase

x_{2r} = Stator reactance per phase in running condition

R_1 = Stator resistance per phase

x_1 = Stator reactance per phase

- So indution motor can be represented as a transformer as shown in Fig. 4.23.

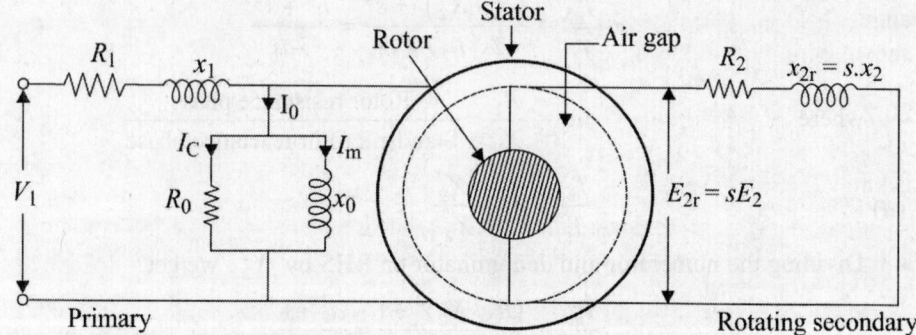

Fig. 4.23: Induction motor as a transformer

4.20.1 Basic Equivalent Circuit

When induction motor is on no load, it draws a current from the supply to produce the flux in air gap and to supply iron losses.

This current I_0 has two components.

- active component which supplies no load losses.
- I_m = magnetizing component which sets up flux in core and air gap.

These two currents give us the elements of an exciting branch as

$$R_0 = \text{Representing no load losses} = \frac{V_1}{I_C}$$

$$X_0 = \text{Representing flux set up} = \frac{V_1}{I_m}$$

Thus, $\qquad I_0 = I_C + I_m$

The equivalent circuit of induction motor can be represented as shown in Fig. 4.24.

- The stator and rotor sides are shown separated by air gap.

$\qquad I_{2r} = $ Rotor current in running condition

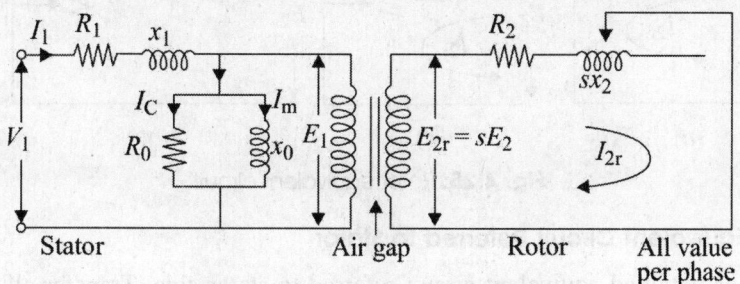

Fig. 4.24: Basic equivalent circuit

Note: It is important to note that as load on the motor changes, the motor speed changes. Thus slip changes. As slip changes the reactance x_{2r} changes. Hence $x_{2r} = s_{X2}$ is shown variable.

4.20.2 Rotor Equivalent Circuit

It is know that, $\qquad I_{2r} = \dfrac{sE_2}{\sqrt{R_2^2 + (sx_2)^2}} = \dfrac{E_2}{\sqrt{\left(\dfrac{R_2}{s}\right)^2 + x_2^2}}$

- So it can be assumed that equivalent rotor circuit in the running conditions has a fixed reactance X_2 and voltage E_2, but a variable resistance $\dfrac{R_2}{s}$ as indicated in the above equation.

Now $\qquad \dfrac{R_2}{s} = R_2 + \dfrac{R_2}{s} - R_2$

$$= R_2 + R_2\left(\frac{1}{s} - 1\right) = R_2 + R_2\left(\frac{1-s}{s}\right)$$

- So the variable rotor resistance R_2 has two parts.
- Rotor resistance R_2, itself represents copper loss.

- $R_2\left(\dfrac{1}{s}-1\right)$ which represents load resistance RL. So it is electrical equivalent of mechanical load on the motor.

  [**Note:** Thus mechanical load on the motor is represented by the pure resistance of value $R_2\left(\dfrac{1}{s}-1\right)$.]

- So rotor equivalent circuit can be shown as in Fig. 4.25(a)–(c).

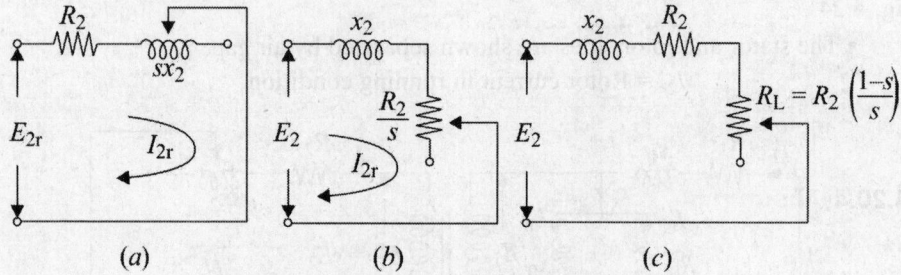

$$(a) \qquad\qquad (b) \qquad\qquad (c)$$

Fig. 4.25: Rotor equivalent circuit

4.20.3 Equivalent Circuit Referred to Stator

Now let us obtained equivalent circuit referred to stator side. Transfer all the rotor parameters to stator.

$$K = \frac{E_2}{E_1} = \text{transformation ratio}$$

$$E_2' = \frac{E_2}{K}$$

- The rotor current I_{2r} has its referred component on the stator side which is I_{2r}'

$$I_{2r}' = KI_{2r} = \frac{KsE_2}{\sqrt{R_2^2 + (sx_2)^2}}$$

$$x_2' = \frac{x_2}{K_2} = \text{Referred rotor reactance}$$

$$R_2' = \frac{R_2}{K^2} = \text{Referred rotor resistance}$$

$$= \frac{R_L}{K^2} = \frac{R_2}{K^2}\left(\frac{1-s}{s}\right)$$

$$R_L' = R_2'\left(\frac{1-s}{s}\right)$$

Thus R_L' is referred to mechanical load on stator. So equivalent circuit referred to stator can be shown as in the Fig. 4.26.

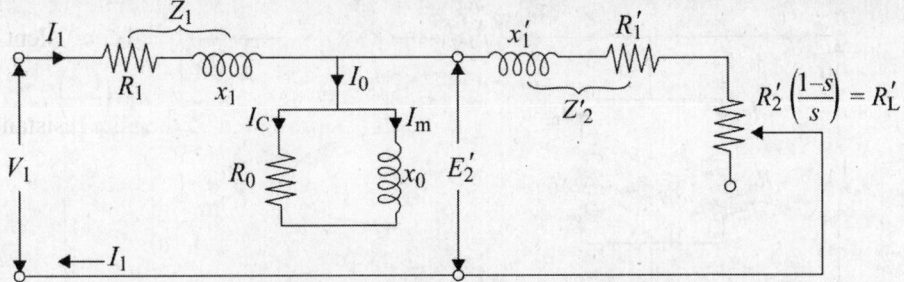

Fig. 4.26: Equivalent circuit referred to stator

- The resistance is fictitious resistance representing the mechanical load on the motor.

4.20.4 Approximate Equivalent Circuit of Induction Motor

- Similar to the transformer, the equivalent circuit can be modified by shifting the exciting circuit (R_0 and x_0) purely across the supply, to the left of R_1 and x_1.
- Due to this we are neglecting the drop across R_1 and x_1 due to I_0 which is very small.
- Hence the circuit is called *approximate equivalent circuit*. The circuit is shown in the Fig. 4.27.

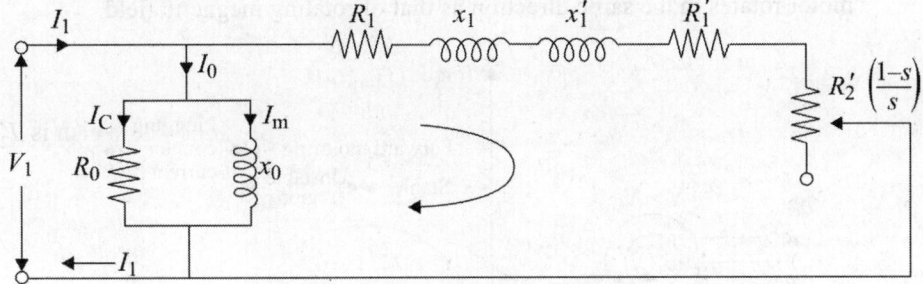

Fig. 4.27: Approximate equivalent circuit

- Now the resistances R_1 and R'_2 while reactances x_1 and x_2 can be combined so we get,

$$R_{1e} = \text{Equivalent resistance referred to stator} = R_1 + R'_2$$
$$X_{1e} = \text{Equivalent reactance referred to stator} = x_1 + x'_2$$

while
$$\bar{I}_1 = \bar{I}_0 + \bar{I}'_{2e}$$

and
$$\bar{I}_1 = \bar{I}_C + \bar{I}_m \quad \text{(phasor sum)}$$

- Thus the equivalent circuit can be shown in Fig. 4.28.

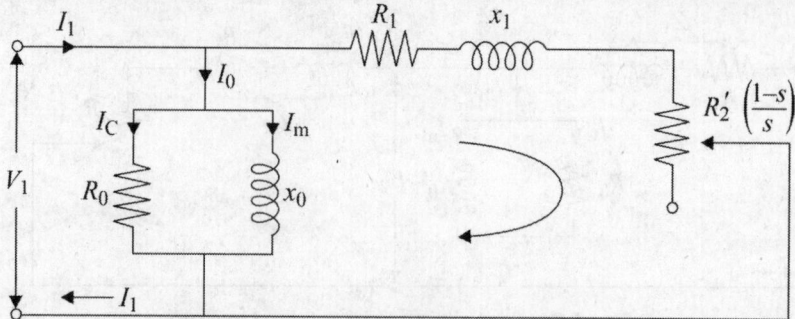

Fig. 4.28

4.21 TORQUE–SLIP CHARACTERISTICS OF 3ϕ INDUCTION MOTOR

- The torque slip or torque speed characteristics of induction motor is shown in Fig. 4.29.
- The characteristics can be divided into:
- Forward Motoring
- Plugging/Regeneration

Forward Motoring

- The forward motoring region corresponds to the values of slip between 0 and 1.
- In the forward motoring region of the characteristics shown in Fig 4.29, the motor rotates in the same direction as that of rotating megnetic field.

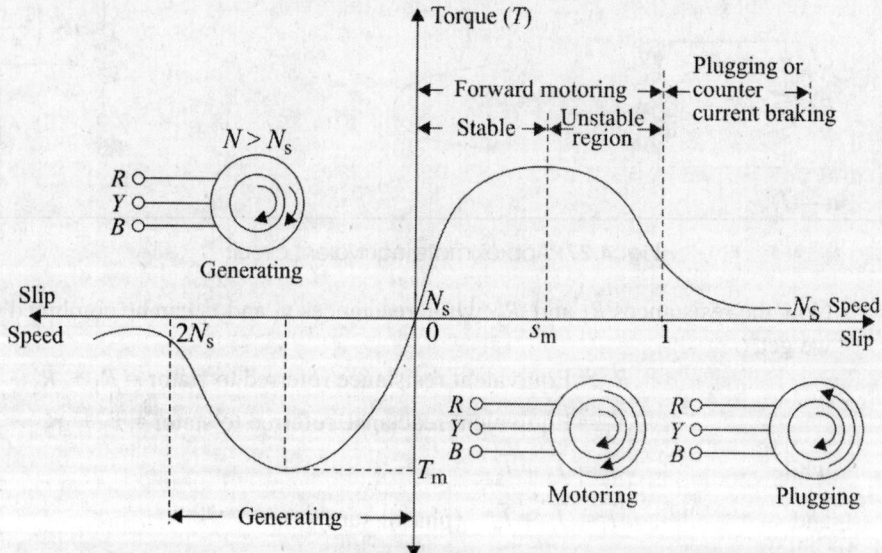

Fig. 4.29: Torque-speed characteristics of 3-phase induction motor

- The torque produce by motor is zero at synchronous speed or for $s = 0$. This is because the induced voltage in rotor is zero when $N_r = N_s$.
- The torque increases as the slip increases while the air gap flux remains constant.
- Once torque reaches its maximum value T max at the critical slip $s = s'_m$, the torque decreases, with increase in slip due to reduction in air gap flux.

Stable Region of Operation

- In the forward motoring region of $(0 \leq s \leq 1)$, the torque speed characteristics as shown in Fig. 4.29, the region of $(0 \leq s \leq s_m)$ is said to be stable region of operation and the operating point of the motor should be in this region of the charactesistics.
- This is stable region because in this region with increase in the torque demand, the motor speed decrease.
- The region of $(s_m \leq s \leq 1)$ is unstable region as in this region with increase in torque the speed of motor increases.

4.21.1 Analysis of the Torque-Slip Characteristics

- Let us analyze only the forward motoring-portion of the torque-slip characteristics. This portion has been drawn in Fig. 4.30.
- For convenience divide this characteristics into two portions as follows:
  - Stable region extending from $s = 0$ to $s = s_m$
  - Unstable region extending from $s = s_m$ to $s = 1$.

Stable Region ($s = 0$ to $s = s_m$)

- The general expression for the motor torque is as follows

$$T \propto \frac{sR_2}{R_2^2 + (sx_2)^2}$$

- In the stable region, the value of slip is small. Hence this region is also called as the low slip region.
- As 's' is small, $(sx_2)^2$ becomes negligible as compared to R_2. Hence the expression for torque get modified as

$$T \propto \frac{sR_2}{R_2^2} \qquad \therefore\ T \propto \frac{s}{R_2}$$

- But R_2 is constant, Hence,

$$T \propto s \text{ for low slip region.}$$

- Thus in low slip region, the torque is directly proportional to the slip 's'. Hence as load increase speed decreases slip increases which in turn increases the torque in a linear manner.
- Therefore in low slip region the torque increases linearly with increase in slip.
- At $s = 0$, $T = 0$ as shown in Fig. 4.30.

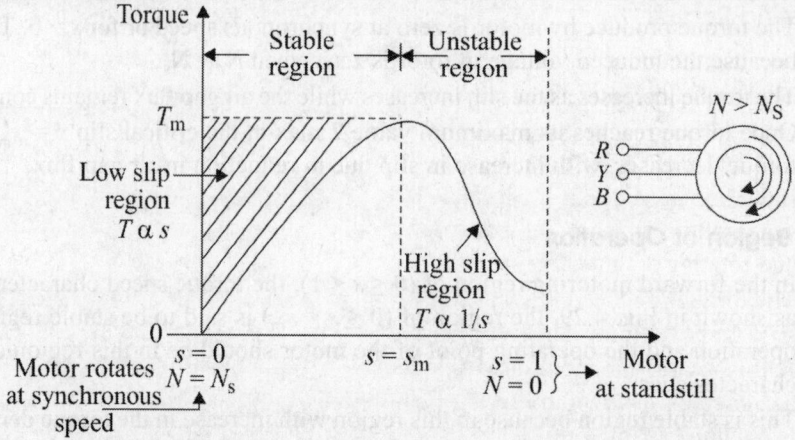

Fig. 4.30: Forward motoring part of torque-ship characteristics.

UNSTABLE REGION ($S_M < S < 1$)

- In the unstable region extending from $s = s_m$ to $s = 1$, the value of 's' is large so this region is also called *high slip region*.

- As 's' is large, $(sx_2)^2$ is much larger than R_2^2. Hence we can neglect R_2^2 to get the modified expression for torque.

$$T \propto \frac{sR_2}{sx_2} \propto \frac{R_2}{sx_2^2}$$

- But $\left(\dfrac{R_2^2}{x_2^2}\right)$ is a constant. Hence

$$\boxed{T \propto \frac{1}{s}}$$

- Thus in the high slip region the motor torque is inversely proportional the slip 's'. Hence torque decreases with increase in slip as shown in Fig. 4.30, and the reduction in torque is nonlinear.

4.21.2 Pullout Torque or Breakdown Torque

- The maximum torque T_m obtained at $s = s_m$ is also called as the *pull on torque* or *breakdown torque*.
- If the load torque increases beyond the pull out torque then the induction motor will be pushed into the unstable region and will finally come to a standstill condition.

Other Regions of Operation

- There are two more regions on the torque-slip characteristics namely the 'plugging' region and the 'generator operation' region which are devoted to induction motor braking.

Generating Region

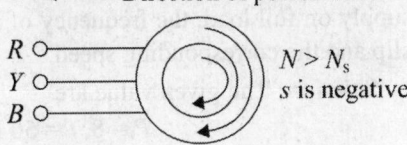

— Direction of power flow

- For the generating region, the slip needs to be negative and between 0 and -1 as shown in Fig. 4.31(a).

$N > N_s$
s is negative

Fig. 4.31(a): Generating mode

- The slip will be negative if and only. If the rotor speed N_r is higher than the synchronous speed N_s, however rotor and rotating magnetic field both rotate in the same direction.
- The torque produced is in the opposite direction to that of the motoring mode so it is shown to be negative.
- In this region, the motor acts as a generator, and returns the power back to the a.c. source as shown in Fig. 4.31(a), maximum torque in the generating mode is obtained at slip $s = -s_m$.

4.21.3 Plugging or Counter Current Braking

- As shown in Fig. 4.31(b), the motor operates in the plugging or counter current braking mode for values of $s > 1$.
- But $s = \dfrac{N_s - N_r}{N_s}$, therefore to get values of $s > 1$, N_r must be negative, i.e. N_s and N_r must have opposite direction, i.e. the RMP (rotating magnetic field) and rotor should rotate in opposite directions.
- This is achieved by interchanging any two phases of the stator supply as shown in Fig. 4.31(b).
- The motor is already rotaling therefore due to inertia it continue to rotate in the same direction.

Y and B have been interchanged
RMF
Rotor
Rotating in counter-clockwise direction and rotating in clockwise direction braking torque

Fig. 4.31(b): Plugging or counter current breaking

- The motor current is high and motor heating takes place in this mode of operation. This is a major disadvantage of plugging.
- The advantage of plugging is the high braking torque produced by the rotor current.

4.21.4 Why the Name Counter Current Braking

- The name of this type of braking is because the braking takes place due to a torque which is produced by a current which flows in the opposite direction to that in the normal motoring operation.

Example 4.14: A 8 pole, three-phase induction motor is supplied from 50 Hz, AC supply on full load, the frequency of induced emf in rotor is 2 Hz. Find the full load slip and the corresponding speed.

Solution: The given value are

$$P = 8, f = 50 \text{ Hz}, f_r = 2 \text{ Hz}$$

Now $$f_r = s \cdot f$$

$$\therefore \qquad 2 = s \times 50 \quad \therefore \quad s = \frac{2}{50} = 0.04$$

$$\therefore \qquad \% s = 0.04 \times 100 = 4\%$$

The corresponding speed is given by

$$N_r = N_s (1 - s) \quad \text{from } s = \frac{N_s - N_r}{N_r}$$

where $$N_s = \frac{120 f}{P} = \frac{120 \times 50}{8} = 750 \text{ rpm}$$

$$N_r = 750 (1 - 0.04)$$
$$= 720 \text{ rpm (full load speed)}$$

Example 4.15: A 6 pole, 50 Hz three-phase induction motor has rotor resistance of 0.4 ohm/phase maximum torque is 200 N·m at 850 rpm. Find.

(*i*) Torque at 4% slip

(*ii*) Additional rotor resistance to get 2/3rd of maximum torque at starting.
                                                      [UPTU 2005–06]

Solution: Given: $P = 6 f = 50$ Hz, $R_2 = 0.4$ W/ph, $T_m = 200$ N·m at $N = 850$ rpm.

To find (*i*) Torque at 4% slip. (*ii*) Additional resistance to get 2/3rd of T_{max} at start.

As maximum torque occurs at $N = 850$ rpm, the slip corresponding to this speed is slip at maximum torque, i.e.

$$s_m = \frac{R_2}{x_2}$$

(*i*) Synchronous speed $$N_s = \frac{120 f}{P}$$

$$= \frac{120 \times 50}{6} = 1000 \text{ rpm}$$

$$\therefore \qquad s = \frac{N_s - N_r}{N_s} = \frac{1000 - 850}{1000} = 0.15$$

$$s_m = \frac{R_2}{x_2}$$

$$\therefore \qquad x_2 = \frac{R_2}{s_m} = \frac{0.4}{0.15} = 2.67 \,\Omega / \text{phase}$$

$$= \frac{T_L}{T_m} = \frac{2as}{a^2 + s^2} \quad \therefore \quad T_L = T_m \times \frac{2as}{a^2 + s^2}$$

Considering the relation $\quad T_L = \dfrac{200(2 \times 0.15 \times 0.04)}{(0.15)^2 \times (0.04)^2} = 99.585\text{N-m}$

To find additional resistance to get 2/3rd of maximum torque at start

$$T_{st} = \frac{2}{3}(T_m)$$

$$\frac{T_{st}}{T_m} = \frac{\left(\dfrac{E_2^2 R_2}{R_2^2 + x_2^2} \right)}{\left(\dfrac{E_2^2}{2x_2} \right)^2}$$

$$\frac{2}{3} = \frac{2R_2 x_2}{\left(R_2^2 + x_2^2 \right)}$$

Substituting the value we get

$$0.667 = \frac{2 \times R_2 \times 2.67}{R_2^2 + (2.67)^2}$$

$$R_2^2 + 7.1289 = 8R_2^2$$

$$= R_2^2 + 8R_2' + 7.1289$$

$$\therefore \qquad R_2' = \frac{8 \pm \sqrt{-(8)^2 - 4 \times 1 \times 7.289}}{2 \times 1} = \frac{8 \pm 5.96}{2}$$

$$= 4 \pm 2.97 = 6.97 \ \Omega \ \text{or} \ 1.02 \ \Omega$$

Neglect the larger value, since there is no point in increasing R_2 beyond x_2.

4.22 SPEED CONTROL OF INDUCTION MOTOR

- A 3-phase induction motor is practically a constant speed motor like a DC shunt motor. But the speed of DC shunt motor can be varied easily just by using simple rheostats.
- In 3-phase induction motor it is very dificult to achieve smooth speed control and if the speed control is achieved by some means, the performance of the induction motor in terms of its power factor, efficiency, etc. gets adversely affected.
- We know that the rotor speed of an induction motor is given by

$$N_r = N_s(1 - s)$$

$$N_s = \frac{120f}{P}$$

$$N_r = \frac{120f}{P}(1 - s)$$

- From the above expression, it is seen that the motor speed can be changed by:
  - A change in frequency (f)
  - Number of pole (P)
  - Slip (s)
- Another way to change the speed is to change the torque produced by the motor which is given by

$$T = \frac{sE_2^2 R_2}{R_2^2 + (sx_2)^2}$$

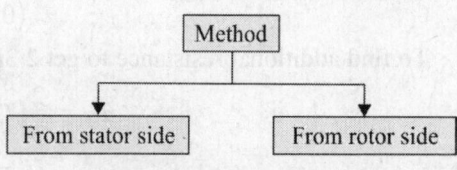

Fig. 4.32: Block diagram

Thus speed of the IM can be controlled basically by two methods as shown in Fig. 4.32.

4.22.1 Speed Control by Adding Rheostats in Stator Circuit

- We have seen that the reduced voltage can be applied to the stator by adding the rheostats in the stator circuit.
- The arrangement is shown in the Fig. 4.33. The part of the voltage get dropped across the resistances and reduced voltage gels applied across the stator.

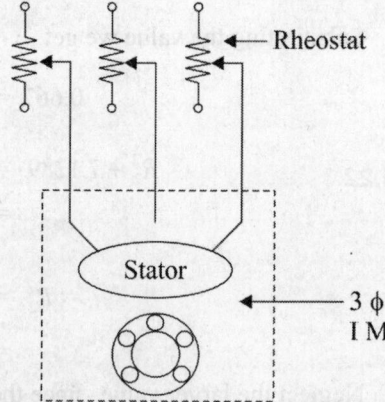

Disadvantages

1. Power loss in the external stator rheostats so efficiency reduces.
2. Starting torque is low.

Fig. 4.33: Stator Resistance Control

Applications

1. Speed control of small motors.
2. For the speed control of blowers and fans.

4.22.2 Speed Control by Multiple Stator Winding

- This method is easily applicable to squirrel cage motors because a cage winding automatically reacts to create the same number of poles as the stator.
- In this method the stator is provided with two seperate windings which are wound for two different number of poles.
- One winding is energized at a time suppose that a motor has two winding for 4 pole and 8 poles. For 50 Hz supply the synchronous speed will be 150 and 750 rpm respectively.
- The speed torque characteristics of multiple stator winding are shown in the Fig. 4.34.

• In this method, the two stator winding are insulated from each other. When anyone winding is used, the other winding should be kept open circuited by a switch to avoid any circulating current to flow in that winding due to induced emf in the winding. By this method, smooth control of speed over a wide range is not possible.

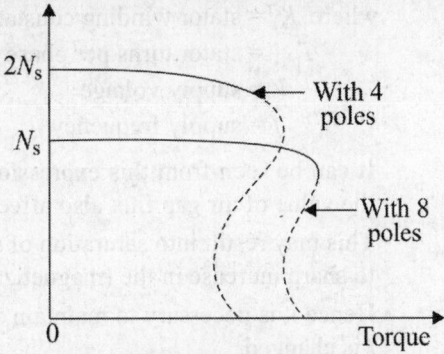

Fig. 4.34: Speed torque curves

Note: Multiple stator winding method is also called pole changing method of controlling the speed.

Disadvantages

1. The speed change takes place in steps.
2. These are only a limited number of speed values that are achievable by this method.
3. This method is ideally suitable for the squirrel cage motors but not so suitable for the slip ring induction motors.

4.22.3 Supply Frequency Control or V/f Control

• This method of speed control provides wide speed control range with gradual variation of the speed through this range.

• The major difficulty with this method is to get the variable frequency supply.

• The auxiliary equipment required for this pupose results in high cost, increased maintenace and lowering of the overall efficiency. That is why this method is not employed for general purpose speed control applications.

• In spite of the fact that this scheme is complicated, there are certain applications in which it is wide, continuously variable, speed range and good speed regulation makes its use highly desirable.

• We know that the synchronous speed is given by

$$N_s = \frac{120f}{P}$$

• Thus the controlling the supply frequency smoothly the synchronous speed can be controlled over a wide range.

• This gives smooth speed control of an induction motor. But the expression for the air gap flux is given by

$$\phi_g = \frac{1}{4.44K_1T_{ph1}}\left(\frac{V}{f}\right)$$

where K_1 = stator winding constant

T_{ph1} = stator turns per phase

V = supply voltage

f = supply frequency

It can be seen from this expression that if the supply frequency f is changed, the value of air gap flux also affected.

- This may result into saturation of stator and rotor cores such a saturation leads to sharp increase in the (magnetization) no load current of the motor.

- Hence it is necessary to maintain air gap flux constant when supply frequency f is changed.

- To achieve this, it is necessary to keep V/f ratio constant.

- This ensures constant air gap flux giving speed control without affecting the perfomance of the of the motor. Hence this method is called V/f control.

- Hence, in this method, the supply to the induction motor required is variable voltage. Variable frequency supply can be achieved by an electronic scheme using converter and inverter circuitry. The scheme is shown in the Fig. 4.35.

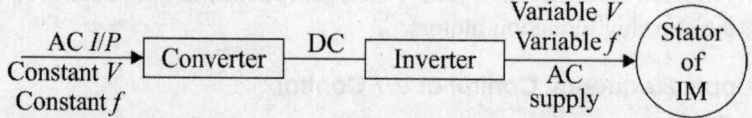

Fig. 4.35: Electronic scheme for V/f control

- The normal supply available is constant voltage constant frequency AC supply.

- The converter converts this supply into a DC supply. This DC supply is then given to the inverter.

- The inverter is a device which converts DC supply, to varible frequency AC supply which is required to keep V/f ratio constant. But selecting the proper frequency and maintaining V/f constant, smooth speed control of the induction motor is possible.

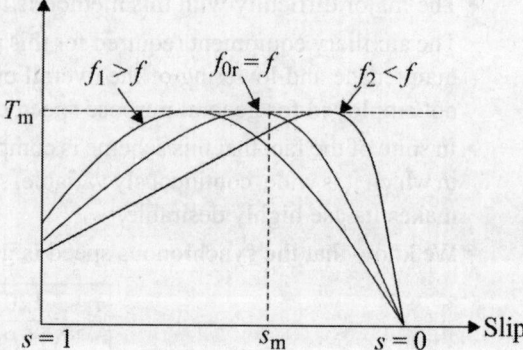

Fig. 4.36: Torque–slip characteristics with variable f and constant (V/f)

- If f is the normal working frequency then Fig. 4.36, shows the torque-slip characteristics for the frequency $f_1 > f$ and $f_2 > f$, i.e. for frequencies above and below the normal frequency.

Advantages

- Smooth control of speed
- Air gap flux remains constant so no possibility of core saturation.
- It is possible to get a constant maximum torque at all the speed.

Disadvantages

- Control circuit is complicated
- It is not possible to use the output of this inverter for any other applications.

Applications

- This type of control produces very low starting torque and therefore is preferred in the applications such as fans, blowers and centrifugal pumps where the starting torque requirement is not high.

4.22.4 Speed Control by Variation of Supply Voltage

- This is a slip-control method with constant frequency variable supply voltage. In this method of speed control of induction motors, the voltage applied to the stator is varied for varying the speed.

- We know that, $T \propto \dfrac{sE_2^2 R_2}{R_2^2 + (sx_2)^2}$.

- Now E_2, the rotor induced emf at standstill depends on the supply voltage V.
- Also for low slip region, which is operating region of the induction motor, $(sx_2)^2$ R_2 and hence can be neglected.

$$T \propto \frac{sV^2 R_2}{R_2^2 +} \propto sV^2$$

- Now if supply voltage is reduced below rated valued, as per above equation, torque produced also decreases.
- But to supply the same load it is necessary to develop same torque hence value of slip increases so that torque produced remains same.
- Slip increases means motor reacts by running at lower speed. The speed-torque

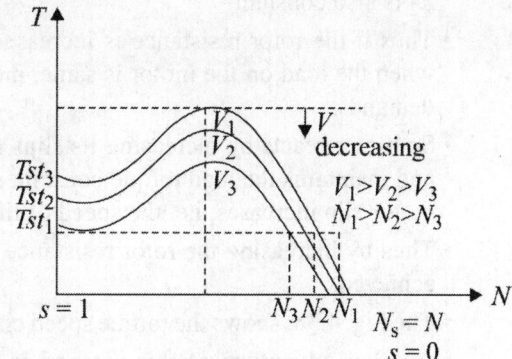

Fig. 4.37

characteristics for the motor using supply voltage control are shown in the Fig. 4.37.

- But in this method, due to reduction in voltage. Current drawn by the motor increases.

- Large change in voltage for small change in speed is required is biggest disadvantage.
- Due to increased current, the motor may get overheated, additional voltage changing equipment is necessary.
- Hence this method is rarely used in practice motor driving type.
- Due to reduced voltage and DC supply, this DC supply is then decreases, decreasing the value of maximum torque as well.

4.23 SPEED CONTROL BY VARIATION OF ROTOR RESISTANCE

- As the name implies, this type of control is only possible with wound rotor or slip ring induction motor, i.e. this method cannot applied to squirel cage motors.
- We know that wound rotor or slip ring motors are usually started by connecting starting resistances in the secondary circuit, which are shorted out as the motor speed up.
- If the ohmic values of these resistors are properly chosen and if these resistors are designed for continuous operation, they can serve dual purpose starting and speed control.
- This method of speed control has characteristics similar to those of DC shunt motor speed control by means of resistance in series with the armature we know $T \propto \dfrac{sV^2 R_2}{R_2^2 + (sx_2)^2}$.
- For low slip region $(sx_2)^2 \ll R_2$ and can be neglected for constant supply voltage, E_2 is also constant.
- Thus if the rotor resistance is increased, the torque produced decreases. But when the load on the motor is same, motor has to supply same torque as load demands.
- So motor reacts by increasing its slips to compensate decrease in T due to R_2 and maintains the load torque constant so due to additional rotor resistance R_2 motor slip increases, i.e. the speed of the motor decreases.
- Thus by increasing the rotor resistance R_2 speeds below normal value can be achieved.
- The Fig. 4.38, shows the torque speed characterstics for rotor resistance control.
- Another advantage of this method is that the starting torque of the motor increases proportional to rotor resistance.

Disadvantages

- A lot of power is wasted in the external rotor resistance.
- This method cannot be used for the squirel cage induction motor.

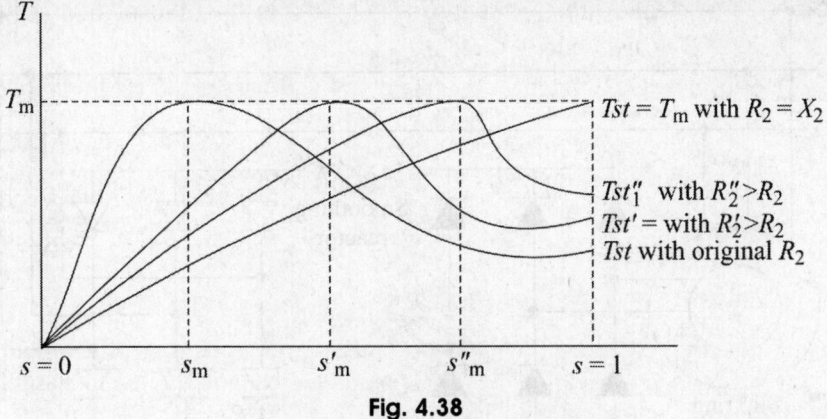

Fig. 4.38

- The speed above the normal value cannot be obtained.
- The large speed changes are not possible.

4.24 SPEED CONTROL BY SLIP ENERGY RECOVERY METHOD

- In the rotor resistance control, the slip power in the rotor circuit wasted as I^2R loss during the low speed operation.
- The efficiency of the drive system by this method of speed control is, therefore, reduced.
- The slip power from the rotor circuit can be recovered and fed back to the AC source so as to utilize it outside the motor.
- Thus, the overall efficiency of the drive system can be increased. The basic principle of slip power recovery is to connect an external source of emf of slip frequency to the rotor circuit. A method for recovering the slip power is shown in Fig. 4.39. This method is known as static scherbius drive. It provides the speed control of a slip ring induction motor below synchronous speed.
- A portion of rotor AC power (slip power) is converted into DC by a diode bridge.
- The rectified current is smoothed by the smoothing reactor. The output of the rectifier is then connected to the DC terminals of the inverter which inverts this DC power to AC power and feed it back to the AC source.
- The inverter is a controlled rectifier operated in the inversion mode.

SOLVED EXAMPLES

Example 4.16: A three phase induction motor is wound for 4 poles and is supplied from 50 Hz system, calculate

    (i) N_s (ii) Rotor speed when slip is 4% (iii) Rotor frequency when rotor runs at 600 rpm.   **[UPTU 2003-04]**

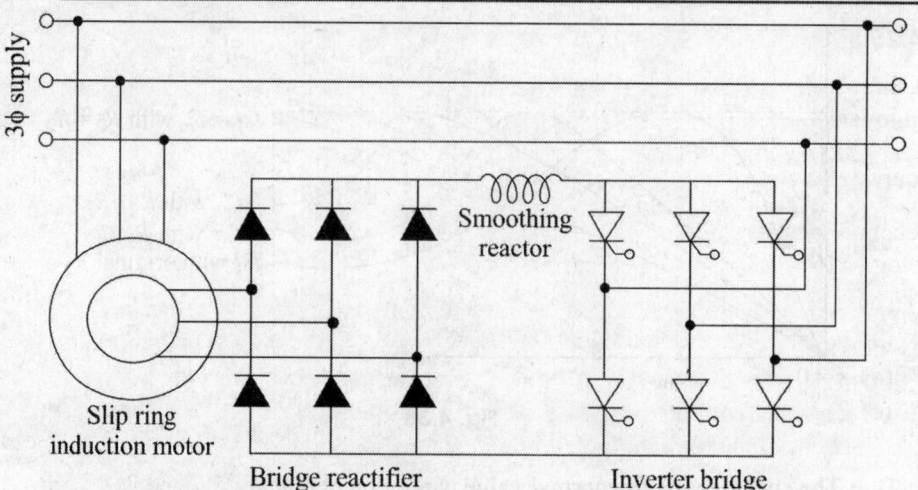

Fig. 4.39: Static Scherbius drive for speed control of slip ring induction motor

Solution: Given: $P = 4$, s = 4%, i.e. 0.04, $f = 50$ Hz

(i) $N_s = \dfrac{120f}{P} = \dfrac{120 \times 50}{4} = 1500$ rpm

(ii) $N_r = N_s(1-s) = 1500(1-0.04) = 1440$ rpm

(iii) $N_1 = 600$ rpm

$\therefore \qquad s_1 = \dfrac{N_s - N_1}{N_s} \times 100 = \dfrac{1500 - 600}{1500} \times 100 = 60\%$

$f_r = s_1 f = 0.6 \times 50 = 30$ Hz

Example 4.17: A 5 Hp, 230 V, 50 Hz induction motor has a rated full load speed of 950 rpm. The induced voltage per phase of rotor at standstill is 100 V, calculate

(i) number of poles and % full load slip.

(ii) rotor induced voltage and its frequency at full load. **[UPTU 2004–05]**

Solution: Given: $V_L = 230$ V, $f = 50$ Hz, $N = 950$ rpm, $E_2 = 100$ V.

(i) The practical value of full load slip is about 4 to 6%. Hence the nearest synchronous speed to $N_r = 950$ rpm and $N_s = 1000$ rpm

But $\qquad N_s = \dfrac{120f}{P}$, i.e. $1000 = \dfrac{120 \times 50}{P}$ or $P = 6$ poles

$s = \dfrac{N_s - N}{N_s} \times 100 = \dfrac{1000 - 950}{1000} \times 100 - 5\%$

(ii) $E_{2r} = sE_2 - 0.05 \times 100 = 5$V

$f_r = 2 \cdot f$

$= 0.05 \times 50 = 2.5$ Hz

4.25 STARTING OF THREE PHASE INDUCTION MOTORS

A three phase induction motor has a definite positive starting torque. When switched on to supply it starts itself but draws a high starting current. This is evident from the equivalent circuit. At the time of starting, slip $s = 1$ and hence the resistance $\left[\dfrac{R_2 (1-s)}{s} \right]$ becomes zero (the motor behaves as a short circuited transformer). The current in the rotor and the stator windings may be about five times more than full load values.

These high rotor and stator currents cause many problems:

(a) High electromagnetic forces between the conductors on the same part.

(b) High heat generation causing high temperature may damage the insulation.

(c) High current (at low power factor) may cause an appreciable drop in supply voltage causing undesirable effects on other equipments.

Therefore suitable means must be provided with the motor at start, to limit the starting current up to safe value.

The device which is used to start the three-phase induction motor is termed as starter. The function of the starter is to limit the initial rush of current to a predetermined safe value.

The various methods of starting the three-phase induction motor are:

1. By direct on line (DOL) starter
2. By star-delta starter
3. By autotransformer starter

4.25.1 Direct On Line Starting (DOL)

For small size squirrel cage (less than 2 HP) motor or for motors in power system where inrush of high-starting current is permissible, direct start may be used. For these small motors, the starting torque is about twice the full-load torque and the starting period lasts only a few seconds.

Figure 4.40 shows a starter for direct starting with in-built short circuit, over-load and under voltage protection. When the motor is to be started, the main switch is put on and start button is pressed. This energy the relay coil S induces normally to open contacts S_1, S_2, S_3 to close. Power is supplied to the motor and it starts. The contact S_4 also shuts, thus shorting out the starting switch allowing the operator to release it without removing power from the S relay. When the stop button is pressed, the S relay is de-energised and the S contacts open, thus stopping the motor. Short circuit protection is provided by fuses F_1, F_2 and F_3. Thermal overload relay (OLC) protects the motor from sustained overloads opening the contact D.

4.25.2 Star–Delta Starter

Figure 4.41 shows the diagram of the star–delta starter. Star–delta starter can be used only for those three-phase induction motors whose stator winding has been designed

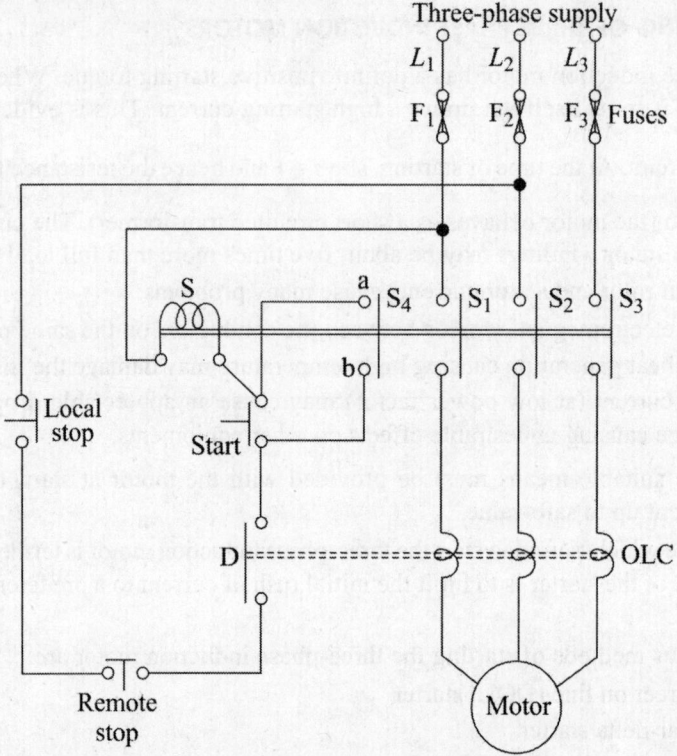

Fig. 4.40: Direct on-line starter

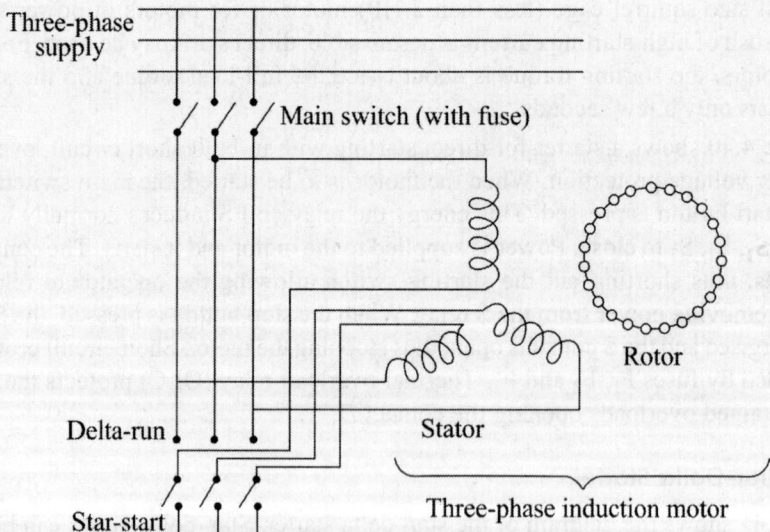

Fig. 4.41: Star–delta starter

for delta connection. All the six terminals (of the three phases) are brought out. For starting, the phases are connected in star thereby reducing the voltage of each phase to $\frac{1}{\sqrt{3}}$ of its normal value.

We have $\qquad P_{ag} = \dfrac{I_2'^2 R_2'}{s}$

$\therefore \qquad\qquad T = \dfrac{P_{ag}}{\omega_s} = \dfrac{1}{\omega_s} \cdot I_2' R_2' \times \dfrac{1}{s}$

At $s = 1$ (at starting)

$\qquad\qquad T_s = \dfrac{1}{\omega_s} \cdot I_2''^2 R_2' \; [I_2''$ is the rotor current reflected at primary at starting]

$\therefore \qquad\qquad \dfrac{T_s}{T} = \dfrac{I_2''^2}{I_2'^2} \cdot s$

If T represents full load torque, I_2' the full load rotor current reflected to primary, we have $I_{f1} = I_2'$, neglecting the magnetizing branch current. Similarly I_2'' represents the starting current (I_s) at stator, the magnetizing branch being neglected.

$\therefore$ We have $\dfrac{T_s}{T} = \left(\dfrac{I_s}{I_{fl}}\right)^2 \times S_{fl}$ ...(4.7)

The starting line current of the motor with star–delta starter is thus also reduced to $\frac{1}{\sqrt{3}}$ full voltage starting line current. The starting torque which is proportional to $\left(\dfrac{E_1}{\sqrt{3}}\right)^2$ is reduced to 1/3 of the full load torque. Thus, for star-delta starter though we are able to reduce the starting current, we sacrifice the torque and the starting torque reduces to 1/3 of the full load torque.

Let us analyse the star delta starting method to find the torque. We assume that the motor first operates with star connection Fig. 4.42(a) and when speeds up it operates with delta connection of the stator as shown in Fig. 4.42(b).

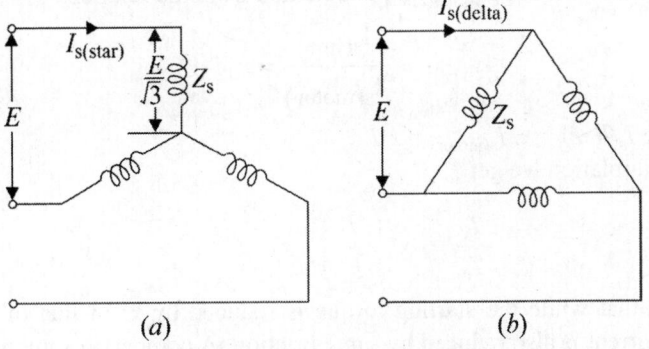

(a) $\qquad\qquad\qquad\qquad\qquad\qquad (b)$

Fig. 4.42: Star–delta starting

In Fig. 4.42(*a*) starting line (phase) current is $(I_{s(star)})$.

In Fig. 4.42(*b*) starting phase current $I_{P(start)} = \dfrac{E}{Z_s}$

Starting line current $\quad I_{s(delta)} = \sqrt{3}\, I_{P(star)}$

$\therefore\quad \dfrac{I_{s(star)}}{I_{s(delta)}} = \left(\dfrac{1}{\sqrt{3}} \cdot I_{P(star)}\right)\Big/\left(\sqrt{3}\, I_{P(star)}\right) = \dfrac{1}{3}$

Using equation (*i*) we can write

$$T_{s(star)/T_{fl}} = \frac{1}{3}\left(I_{p(star)/I_{fl}}\right)^2 \times s_{fl}.$$

Thus starting torque is 1/3rd of that obtained in DOL starting.

This method is a bit economical one but for motors rated beyond 3 kV, this method is not applicable. Like other three-phase motor starters, in this starter also overload coil and no-voltage coils are provided for the protection of the motor (not shown in the star–delta figure). An automatic star–delta starter can also be made by using push button, contactors, time delay relay (TDR), etc.

4.25.3 Autotransformer Starter

In this method, reduced voltage is obtained by some fixed tappings on the three-phase auto transformer. Generally 60% to 65% tappings can be used to obtain a safe value of starting current. The full rated voltage is applied to the motor by taking the auto-transformer out of the motor circuit when motor has picked up the speed up to 85% of its normal speed. Figure 4.43 shows the circuit.

Let us assume that the input voltage E is reduced to xE using auto-transformer tappings.

$\therefore$ the motor starting current is

$$I_s = xI$$

where I is the motor starting current when full voltage E is applied. However, the current drawn from the supply $[I_{s(line)}]$ is obtained from the relation:

$$\frac{I_{s(line)}}{I_{s(motor)}} = x$$

$\therefore$ We have $I_s(line) = x$, $I_{s(motor)} = x^2 I$.

Hence from relation we get

$$\frac{T_s}{T_{fl}} = x^2\left(\frac{I}{I_{fl}}\right)^2 \cdot s_{fl}$$

It is found that while the starting torque is reduced by x^2 of that of DOL start, starting line current is also reduced by same fraction. A comparison among direct on-line starter, star–delta and autotransformer starter is given in Table 4.2.

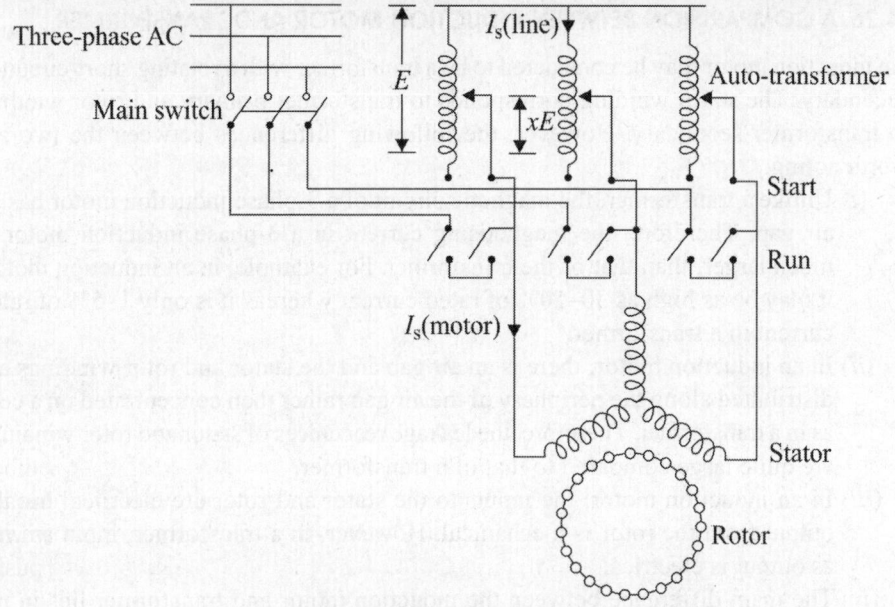

Fig. 4.43: Autotransformer starter

Table 4.2: Comparison among direct on-line starter, star–delta and autotransformer starter

DOL starter	Star–delta starter	Autotransformer starter
1. Full voltage is applied to the motor at the time of starting	1. Each winding gets 58% of the rated line voltage at the time of starting	1. The starting voltage can be adjusted according to the requirement
2. The starting current is 5–6 times of the full load current	2. The starting current is reduced to 1/3 that of direct on line starting	2. The starting current can be reduced as desired
3. The three windings are connected generally in star	3. The three windings are connected in star at the time of starting, and then in delta at the time of running	3. The three windings are generally connected in delta
4. Only three wires are to be brought out from the motor	4. Six wires to be brought out from the motor.	4. Only three wires are to be brought out from the motor
5. Easy to connect motor with direct on line	5. Identification of three starting leads and three end leads is not so easy.	5. Input and output connections of the autotransformers are to be made properly.
6. Very easy operation	6. It is required that connections are first to be made made in star, and then in delta either manually or automatically	6. Skilled operator is needed for connection and starting
7. Low cost	7. Reasonably more expensive	7. High cost
8. Less space required for installation	8. More space required	8. More space required
9. Used fot motor up to 5 HP	9. Up to 10 HP	9. Large motors

4.26 A COMPARISION BETWEEN INDUCTION MOTOR AND TRANSFORMER

An induction motor may be considered to be a transformer with a rotating short-circuited secondary. The stator winding corresponds to transformer primary and rotor winding to transformer secondary. However, the following differences between the two are worth noting:

(*i*) Unlike a transformer, the magnetic circuit of a 3-phase induction motor has an air gap. Therefore, the magnetising current in a 3-phase induction motor is much larger, than that of the transformer. For example, in an induction motor, it may be as high as 30–50% of rated current whereas it is only 1–5% of rated current in a transformer.

(*ii*) In an induction motor, there is an air gap and the stator and rotor windings are distributed along the periphery of the air gap rather than concentrated on a core as in a transformer. Therefore, the leakage reactances of stator and rotor windings are quite large compared to that of a transformer.

(*iii*) In an induction motor, the inputs to the stator and rotor are electrical but the output from the rotor is mechanical. However, in a transformer, input as well as output is electrical.

(*iv*) The main difference between the induction motor and transformer lies in the fact that the rolor voltage and its frequency are both proportional to slip *s*. If *f* is the stater frequency, E_2 is the per phase rotor emf at standstill and X_2 is the standstill rotor reactance/phase, then at any slip *s*, these values are:

Rotor emf phase, $E_2' = sE_2$, rotor reactance/phase, $X_2' = sX_2$

Rotor frequency, $f' = sf$

4.27 CRAWLING OF INDUCTION MOTORS

It has been found that induction motor, particularly the squirrel-cage type, sometimes show a tendency to run at speeds as low as one-seventh of their synchronous speed N_s. This peculiar behaviour of the cage motor at starting is known as *crawling of an induction motor*.

The above action is due to the fact that the AC winding of the stator produces a flux wave which is not a pure sine wave. It consists of a fundamental wave, which revolves synchronously and odd harmonics like 3rd, 5th, 7th, etc. which rotate either in the forward or backward direction at $N_s/3$, $N_s/5$ and $N_s/7$ respectively. These harmonic fields produce torques in the same way as the fundamental. The torque-speed curves of these harmonic torques have the same general shape as that due to the fundamental torque but with synchronous speed $= N_s/n$, where *n* is the order of the harmonic. The magnitude of the harmonic torque is $1/n^2$ of the fundamental torque.

Since third harmonic currents are absent in a balanced 3-phase system. the value of third harmonic torque is zero. Therefore, the total motor torque has the following three components:

(*i*) The fundamental torque T_f due to the fundamental flux rotating at a speed N_s.

(*ii*) The fifth harmonic torque T_5 due to the 5th harmonic flux rotating at a speed $N_s/5$.

(*iii*) The seventh harmonic torque T_7 due to the 7th harmonic flux rotating at a speed $N_s/7$.

Now the 5th harmonic currents have a phase difference of $-120°$ ($5 × 120° = 600° = -120°$) in the three stator windings. Therefore. the 5th harmonic flux rotates at a speed $N_s/5$ in a direction opposite to that produced by the funda- mental. The reverse torque pro- duced by 5th harmonic is small and may be neglected.

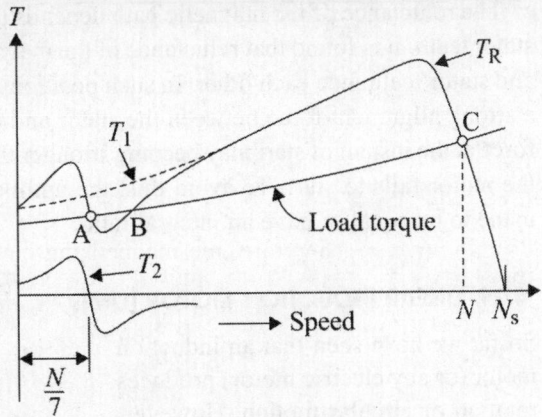

Fig. 4.44

Neglecting all higher harmonic, the resultant torque T_R is the sum of the fundamental torque (T_f) and the 7th harmonic torque (T_7). Figure 4.44 shows the torque speed characteristics due to the fundamental flux and the 7th harmonic flux as well as the resultant torque-speed characteristic. Note that the presence of the 7th harmonic flux has producced a dip in the normal torque-speed characteristic. The resultant curve and the load torque curve meet at point A which represents the stable operating point. Now point A represents a speed of $N/7$, where N is the normal speed of the motor. Therefore, the motor would not run up to full speed (N) but merely drive the load at a reduced speed of $N/7$. We say that the motor crawls at the speed $N/7$. The following points may be roted:

(*a*) If the motor torque developed is due to the fundamental flux alone, the motor will accelerate to the point C as the final operating speed (N) for the given load. However, the presence of the 7th harmonic flux has caused the motor to crawl at a speed corresponding to point A (i.e. speed equal to $N/7$).

(*b*) To make the motor to run up to full speed, the load would have to be reduced to a value less than that of the minimum occurring between A and B [see Fig. 4.44].

(*c*) By proper choice of coil pitch and distribution of coils of stator winding, the harmonic fluxes in the air-gap of the motor can be reduced to a minimum. This eliminates the crawling effect.

4.28 COGGING

A squirrel-cage rotor may show a peculiar behaviour in starting for certain relationship between the number of stator slots (S_1) and rotor slots (S_2). If S_1 is equal to or an integral multiple of S_2, the motor may refuse to start. This phenomenon is known as *cogging* and is due to the magnetic locking between the stator and rotor teeth.

The reluctance of the magnetic path depends upon the positions of rotor teeth w.r.t. stator teeth. It is found that reluctance of the magnetic path is minimum when the rotor and stator teeth face each other. In such positions of minimum reluctance, there exists a strong alignment force between the stator and the rotor at standstill. The alignment force at the instant of start may become stronger than the starting torque. Consequently, the motor fails to start. To avoid this, the number of stator and rotor slots are never made to be equal or have an integral ratio.

4.29 LINEAR INDUCTION MOTOR (LIM)

So far we have seen that an induction motor (or any electric motor) produces rotation or circular motion. However, it is possible to design an induction motor in such a way that it produces linear motion. Such a motor is called linear induction motor (LIM). Figure 4.45 shows the cross-sectional view of the rotary induction motor. Instead of a squirrel cage motor, a cylinder of conductors (usually made of metal) enclosing the rotor's ferro-magnetic core is considered. If the rotary motor of Fig. 4.46 cut along the line XY and laid flat, we get the linear induction motor as shown in Fig. 4.46. The stator is called primary and rotor the secondary of the linear induction motor. These terms are appropriate because any member (primary or secondary) can be made movable. Note that the primary side consists of magnetic core with a 3-phase winding and the secondary side is just a metal

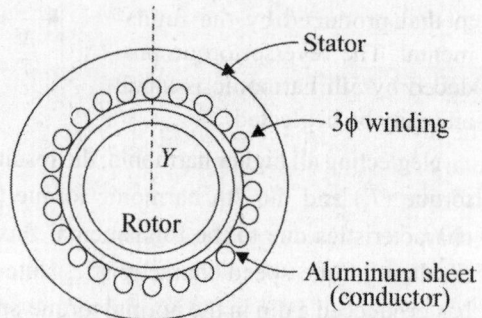

Fig. 4.45: Rotary induction motor

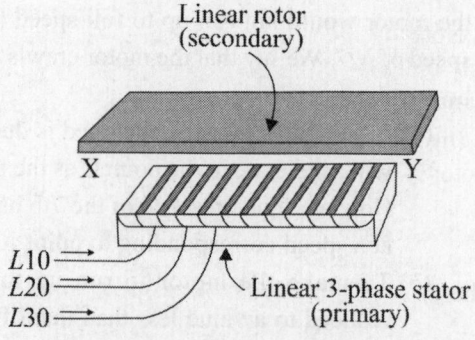

Fig. 4.46: Linear induction motor

sheet. Thus a linear induction motor consists of a linear motor and linear 3-phase stator with airgap between them.

Operation: When a 3-phase supply is connected to the stator of a rotary induction motor, a revolving flux is produced in the air gap of the machine. Similarly, when a 3-phase supply is connected to the primary of a linear induction motor, the flux produced moves at a constant speed in a straight line along the length of the primary. The speed of the travelling field is called the *linear synchronous speed* v_s. The travelling field will induce currents in the secondary. The induced currents will interact with the travelling field to produce a linear force F (or thrust). If one member is fixed and the other is free to move, the force will make the movable member to move. For example, suppose that the primary in Fig. 4.46 is fixed and the secondary is free to move. If the travelling flux moves from left to right, the secondary will also move to the right with linear velocity v_5 following the travelling field. Note that P (actual linear speed of rotor) will be less than v_5 (linear synchronous speed of the travelling field).

The basic difference between a linear induction motor and a rotary induction motor is that the latter has endless air gap while the former is open-ended due to the finite lengths of the primary and seconary sides. Also the angular velocity becomes linear velocity and the torque becomes the thrust or force in a linear induction motor. In order to maintain a constant thrust (force) over a considerable distance, one member is kept shorter than the other as shown in Fig. 4.46.

4.30 PROPERTIES OF LINEAR INDUCTION MOTOR

The properties of a linear induction motor are almost similar to those of a standard rotary induction motor. Consequently, the equations for slip, thrust, power, etc. are also very similar.

(*i*) **Linear synchronous speed (v_s):** The linear speed of the travelling wave of flux is called linear synchronous speed. It is given by:

$$v_s = 2wf$$

where v_s = Linear synchronous speed (m/s)

w = Width of pole-pitch (m)

f = Supply frequency (Hz)

Note that the synchronous speed (v_s) does not depend upon the number of poles but only on the width of pole-pitch. Moreover, the number of poles need not be an even number.

(*ii*) **Slip (s):** Similar to the rotary induction motor, the slip in a linear induction motor is defined as:

$$s = \frac{v_s - v}{v_s}$$

where v_s = Linear synchronous speed (m/s)

v = Linear speed of rotor (m/s)

(*iii*) **Thrust of force (F):** The thrust or force developed by the induction motor is given by:

$$F = \frac{P_r}{v_s}$$

and F = Thrust or foce (N)

P_r = Power supplied to the rotor (W)

v_s = Linear synchronous speed (m/s)

(*iv*) **Active power flow:** The active power flows through a linear induction motor in the same way it does through a rotary induction motor. Thus, we have:

(*a*) Mechanical power P_m developed by motor is

$$P_m = P_r(1 - s)$$

(*b*) Rotor copper losses $= sP_r$

(*c*) Efficiency, $\eta = \dfrac{\text{Output power}}{\text{Input power}}$

The linear induciton motor (LIM) requires a large air gap (15–30 mm) whereas the air gap for a rotary induction motor is small (1–1.5 mm). Therefore, the magnetising current is large and the power factor is low. For this reason, the efficiency of a linear induction motor is low.

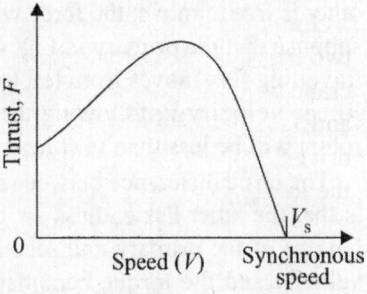

Fig. 4.47

(*v*) **Thrust-speed chracteristics:** The thrust–speed characteristic of the linear induction motor has the same form as the torque-speed characteristics of a rotary induction motor as shown in Fig. 4.47. Note that velocity in a linear induction motor decreases rapidly with the increasing thrust.

The LIM shown in Fig. 4.46 is called a single-sided LIM or SLIM. Another version is used in which primary is on both sides of the secondary as shown in Fig. 4.48. This is known as *double-sided LIM* or DLIM.

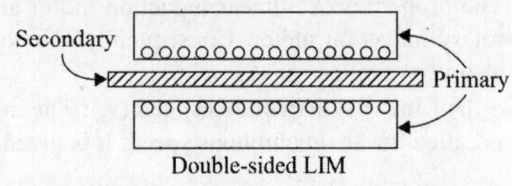

Double-sided LIM

Fig. 4.48

Applications: Theoretically, each type of rotating machine may find a linear counterpart. However, it is the induction motor that is being used in such industrial applications as high-speed ground transportation, sliding door systems, curtain pullers and conveyors.

Example 4.18: A linear induction motor has 98-poles and a pole–pitch of 50 cm. Determine the synchronous speed and motor speed in km/h if the frequency is 50 Hz and slip is 0.25.

Solution: Here, $w = 50$ cm $= 0.5$ m; $f = 50$ Hz; $s = 0.25$

Linear synchronous speed,
$$v_s = 2wf = 2 \times 0.5 \times 50$$
$$= 50\text{m/s} = \frac{50 \times 60 \times 60}{1000} \text{ km/h} = 180 \text{km/h}$$

Motor speed, $v = 180 - 180 \times 0.25 = 135$ km/h

Example 4.19: The stator of a linear induction motor is excited from a 75 Hz electronic source. If the distance between consecutive phase groups of phase *R* is 300 mm, calculate the linear speed of the magnetic field.

Solution: Pole–pitch, $w = 300$ mm $= 0.3$ m; supply frequency, $f = 75$ Hz

Linear synchronous speed, $v_s = 2, wf = 2 \times 0.3 \times 75 = 45$ m/s

$$= \frac{45 \times 60 \times 60}{1000} \text{ km/h} = 162 \text{ km/h}$$

Example 4.20: An overhead crane in a factory is driven horizontally by means of two linear induction motors whose "rotors" are the two steel I-beams upon which the crane rolls. The 3-phase, 4-pole linear stators (mounted on opposite sides of the crane and facing the respective webs of the I-beams) have a pole pitch of 8 cm and are driven bya variable-frequency electronic source. During a test on one of the motors, the following results were obtained:

Stator frequency	15Hz
Power to stator	5 kW
Copper loss and in)n loss in stator	1 kW
Crane speed	1.8 m/s

Calculate (*i*) synchronous speed and slip (*ii*) power to the rotor (*iii*) Cu loss in rotor (*iv*) mechanical power and thrust.

Solution: Here, Pole pitch, $w = 8$ cm $= 0.08$ m; Supply frequency, $f = 15$ Hz

(*i*) Linear synchronous speed, $v = 2wf = 2 \times 0.08 \times 15 = 2.4$ m/s

(*ii*) Input power to rotor, P_r = Power to stator – Cu loss and iron loss in stator

$$= 5 - 1 = 4 \text{ kW}$$

Rotor Cu loss $\quad sP_r = 0.25 \times 4 = 1$ kW

(*iv*) Mechanical cal power developed, $P_m = P_r -$ Rotor Cu loss $= 4 - 1 = 3$ kW

Thrust, $\qquad F = \dfrac{P_r}{V_s} = \dfrac{\left(4 \times 10^3\right)}{2.4} = 1.67 \times 10^3$ N

Example 4.21: The pole pitch of a linear induction motor is 0.5 m and the frequency of the applied 3-phase voltage is 60 Hz. The speed of the primary side of the motor is 200 km/h and the developed thrust is 100 kN. Calcuate the power developed by the motor and the copper loss in the secondary side.

Solution:

Speed of the motor, $\qquad v = 200 \text{ km/h} = \dfrac{200 \times 10^3}{3600} = 55.55 \text{m/s}$

Power developed by motor, P_m = Thrust × velocity $\times f \times v$
$$= (100 \times 10^3) \times 55.55 = 5555 \times 10^3 \text{ W}$$
$$= 5555 \text{ kW}$$

Synchronous speed of motor, $v_s = 2wf = 2 \times 0.5 \times 60 = 60$ m/s

Slip for operating condition, $\quad s = \dfrac{v_s - v}{v_s} = \dfrac{60 - 55.55}{60} = 0.074$

Input power to secondary, $\quad P_r = \dfrac{P_m}{1-s}$

$\therefore$ Cu loss in the secondary $= sP_r = s \times \dfrac{P_m}{1-s} = 0.074 \times \dfrac{5555 \times 10^3}{1 - 0.074}$
$$= 444000 \text{ W} = 444 \text{ kW}$$

SOLVED UNIVERSITY PROBLEMS FOR PRACTICE

Example 4.22: A 3-ϕ, 4 pole, 50 Hz induction motor runs at 1460 rpm. Determine the slip. **[UPTU 2003-04]**

Solution:
$$N_s = \frac{120f}{P} = \frac{120 \times 50}{1500} = 3 \times 50 = 1500 \text{ rpm}$$

$$s = \frac{N_s - N_r}{N_s} = \frac{1500 - 1460}{1500} \times 100$$

$$s\% = 2.667\% \text{ Ans.}$$

Example 4.23: A 12-pole, 3ϕ alternator driven at a speed of 500 rpm supplier power to an 8-pole 3ϕ induction motor. If the slip of the motor is 0.03 pu, calculate the speed. **[UPTU 2002-03]**

Solution:
$$N_s = \frac{120f}{P} = f = \frac{500 \times 12}{120} = 50 \text{ Hz}$$

$$= \frac{120f}{P} = \frac{120 \times 50}{8} = 750 \text{ rpm}$$

$$N_r = N_s(1-s) = 750(1-0.03)$$
$$= 727.5 \text{ rpm Ans.}$$

Example 4.24: A 3-ϕ, 4 pole IM is supplied from diesel-generator set running of 600 rpm. The generator has 10 poles, find the synchronous speed of the IM and also the actual speed for a slip of 4%.

Solution:
$$f = \frac{P \times N}{120} = \frac{10 \times 600}{120} = 50 \text{ Hz}$$

$$N_s = \frac{120f}{P} = \frac{120 \times 50}{4} = 1500 \text{ rpm}$$

$$N_r = N_s(1-s) = 1500(1-0.04)$$
$$= 1440 \text{ rpm Ans.}$$

Example 4.25: A motor generator set used for providing variable frequency AC supply consists of a 3ϕ. 10 pole synchronous motor and a 24-pole, 3ϕ synchronous generator. The motor-generator set is fed from a 25 Hz, 3ϕ AC supply. A 6 pole, 3ϕ induction motor is electrically connected to the terminals of the synchronous generator and runs at a slip of 5%. Determine (*i*) the frequency of the generator voltage of the synchronous generator, (*ii*) The speed at which the induction motor is running.

Solution: Speed of motor generator set

$$N_S = \frac{120 \times \text{Supply frequency } f_1}{\text{No. of poles on synchornous motor}}$$

$$= \frac{120 \times 2.5}{10} = 300 \text{ rpm}$$

(*i*) Frequency of generated value voltages

$$f_2 = \frac{\text{Speed of motor gen set} \times \text{No. of pole of syn gen.}}{120}$$

$$= \frac{300 \times 24}{120} = 60 \text{ Hz}$$

(*ii*) Speed of induction motor

$$N_r = (1-s)N_s = \frac{120 \times 60}{6}(1-0.05)$$

$$= 1140 \text{ rpm } \textbf{Ans.}$$

Example 4.26: A 3-ϕ, 4 pole induction motor is supplied from 3ϕ, 50 Hz AC supply. Calculate. **[UPTU 2005-06]**

(*i*) The synchronous speed.

(*ii*) The rotor speed when slip is 4%.

(*iii*) The rotor frequency when rotor runs at 600 rpm.

Solution: The synchronous speed

$$N_r = 1140 \text{ rpm } \textbf{Ans.}$$

$$N_s = \frac{120 \times f}{P} = \frac{120 \times 50}{4} = 1500 \text{ rpm}$$

(*ii*) $\qquad N_r = N_s(1-s) = 1500(1-0.04) = 1440 \text{ rpm}$

(*iii*) Slip at 600 rpm $= \dfrac{1500 - 600}{1500} = 0.06$

Rotor frequency $\quad f_r = s \times f = 0.06 \times 50 = 3 \text{ Hz}$

Example 4.27: A 12-pole, 3ϕ alternator is coupled to an engine running at 500 rpm it supplied a 3ϕ IM having a full-load speed of 1440 rpm. Find the percentage slip, frequency of rotor current and no. of poles of the motor. **[UPTU 2006-07]**

Solution: Frequency of supply from alternator

$$f = \frac{P_a \times N_a}{120} = \frac{12 \times 500}{120} = 50 \text{ Hz}$$

Full-load speed $\quad N_F = 1440 \text{ rpm}$

(Motor must have 4 pole)

$$P = \frac{120f}{N} = \frac{120 \times 50}{1440} \approx 4$$

$$N_s = \frac{120 \times 50}{4} = 1500 \text{ rpm}$$

$$s = \frac{N_s - N_r}{N_s} = \frac{1500 - 1440}{1500} \times 100 = 4\%$$

$$f_r = sf = 0.04 \times 50 = 2 \text{ Hz } \textbf{Ans.}$$

$$P = 4 \textbf{ Ans.}$$

Example 4.28: In a 3ϕ slip ring, four pole induction motor, the rotor frequency is found to be 2 Hz while connected to a 400 V, 3ϕ, 50 Hz supply. Determine the motor speed in rpm. **[UPTU 2003-04]**

Solution:
$$N_s = \frac{120 \times f}{P} = \frac{120 \times 50}{4} = 1500 \text{ rpm}$$

$$s = \frac{f_r}{f} = \frac{2}{50} = 0.04$$

$$N_r = 1500(1 - 0.04) = 1440 \text{ rpm } \textbf{Ans.}$$

Example 4.29: A 3ϕ, 50 Hz induction motor has 6 poles and operates with a slip of 5% at a certain load. Determine **[UPTU 2002-03]**

(i) The speed of the rotor with respect to the stator

(ii) The frequency of rotor current

(iii) The speed of the rotor magnetic field with respect to rotor

(iv) The speed of the rotor magnetic field with respect to stator and

(v) The speed of the stator magnetic field.

Solution: Given: $f = 50$, P = 6, $s = 0.05$, $N_S = \dfrac{120 \times 50}{6}$, $N_S = 1000$ rpm.

(i) $\quad\quad\quad\quad N_r = N_S(1 - s) = 1000(1 - 0.05) = 950 \text{ rpm } \textbf{Ans.}$

(ii) $\quad\quad\quad\quad f_r = s.f = 0.05 \times 50 = 2.5 \text{ Hz } \textbf{Ans.}$

(iii) $\quad\quad\quad N_r = \dfrac{120 \times f_r}{P} = \dfrac{120 \times 2.5}{6} = \textbf{50 rpm Ans.}$

(iv) $\quad\quad\quad N_S = 1000 \text{ rpm } \textbf{Ans.}$

(v) Rotor field and stator find are revolving at the same speed of 1000 rpm.

∴ Speed between stator at rotor is zero.

Example 4.30: A 3ϕ, 440 V, 50 hp, 50 Hz induction motor runs at 1450 rpm, when it delivers rated output power. Determine.

(i) No of poles in the machines.

(ii) Speed of rotating air gap.

(iii) Rotor induced voltage if stator to rotor turns ratio is 1:80. Assume the winding factors are the same.

(iv) Frequency of rotor current. **[UPTU 2007-08]**

Solution:

(i) $\quad\quad\quad P = \dfrac{120 \times f}{N} = \dfrac{120 \times 50}{1450} \approx 4 \textbf{ Ans.}$

(ii) Speed of rotating air gap field

$$N_s = \frac{120 \times 50}{4} = 1500 \text{ rpm } \textbf{Ans.}$$

(*iii*) Induced emf in rotor at standstill

$$kE_1 = kV_1 = 0.8 \times 440 = 352 \text{ V}$$

$$s = \frac{N_S - N_r}{N_S} = \frac{1500 - 1450}{1500} = 0.033$$

Rotor induced voltage

$$skV_1 = 0.033 \times 352 = 11.733 \text{ V Ans.}$$

(*iv*) $\qquad\qquad f_r = s \cdot f = 0.033 \times 50 = 1.66 \text{ Hz Ans.}$

Example 4.31: A 3φ delta connected 440 volts, 50 Hz, 4-pole induction motor has a rotor standstill emf per phase of 130 V. If the motor is running at 1440 rpm, calculate for this speed:

 (*i*) The slip

 (*ii*) The frequency of rotor induced emf

 (*iii*) The value of the rotor induced emf per phase and

 (*iv*) Stator to rotor turn ratio. **[UPTU 2009-10]**

Solution: $\qquad N_s = \dfrac{120 \times 50}{4} = 1500 \text{ rpm}$

(*i*) $\qquad\qquad s = \dfrac{1500 - 1400}{1500} = 0.04 = \textbf{4\% Ans.}$

(*ii*) $\qquad\qquad f_r = sf = 0.04 \times 50 = 2 \text{ Hz Ans.}$

(*iii*) Rotor induced emf per phase

$$= sE_2$$

$$= 0.04 \times 130$$

$$= 5.2 \text{ volts Ans.}$$

(*iv*) Stator for rotor turn ratio

$$= \frac{E_{P1}}{E_{P2}} = \frac{440}{130} = \frac{44}{13} \text{ Ans.}$$

Example 4.32: A 4-pole, 50 Hz, 3φ IM running at full load, develops a torque of 160 N·m. When the rotor makes 120 calculate cycle per minute. Calculate the shaft output power. **[MTU 2008-09]**

Solution: $\qquad f = 50$

$$f_r = \frac{120}{60} = 2 \text{ Hz}$$

$$s = \frac{f_r}{f} = \frac{2}{50} = 0.04$$

$$N_s = \frac{120 \times 50}{4} = 1500 \text{ rpm}$$

$$N_r = N_s(1 - s) = 1500(1 - 0.04) = 1440 \text{ rpm Ans.}$$

$$P_{out} = \omega\,T_{sh} = \frac{2\pi \times N_r}{60} \times T_{sh}$$

$$= \frac{2\pi \times 1440}{60} \times 160$$

$$= 24.127 \text{ kW} \quad \textbf{Ans.}$$

Example 4.33: A 3-pole, 440 V, 50 Hz, induction motor runs at 1450 rpm when it delivers rated output power. Determine: **[UPTU 2006-07]**

(a) No. of poles in the machine.

(b) Speed of rotating air gap field.

(c) Rotor induced emf of stator to rotor turn ratio is 1:0.8. Assume the winding factors are the same.

(d) Frequency of rotor current.

Solution: Given: $V = 440$ V, $f = 50$ Hz, $N_r = 1450$ rpm (as N_r and N_s are nearly equal).

(a) No. of poles is given by

$$P = \frac{120 f}{N_s} = \frac{120 f}{1450} = \frac{120 \times 50}{1450} = 4.13 \approx 4$$

∴ No. of poles will be 4. **Ans.**

(b) Speed of rotating air gap field.

$$N_S = \frac{120 f}{P} = \frac{120 \times 50}{4} = 1500 \text{ rpm}$$

(c) We know, $\quad k = \dfrac{0.8}{1} = 0.8$

Induced emf in rotor at stand still

$$= kE_1 = kV_1 = 0.8 \times 440 = 352 \text{ V}$$

slip, $\quad s = \dfrac{N_s - N_r}{N_s} = \dfrac{1500 - 1400}{1500} \times 100 = 3.33\% \textbf{ Ans.}$

∴ Rotor induced voltage $\quad = ksV_1 = 0.0333 \times 0.8 \times 440$

$$= 11.733 \text{ V} \quad \textbf{Ans.}$$

(d) Frequency of rotor current (f_r)

∴ $\quad\quad\quad f_r = s \cdot f = 0.0333 \times 50\, s\, f = 0.0333 \times 50$

$$= 1.66 \text{ Hz} \quad \textbf{Ans.}$$

Example 4.34: A 8-pole, three phase induction motor is supplied from 50 Hz, AC supply. On full load, the frequency of induced emf in rotor is 2 Hz. Find the full-load slip and the corresponding speed. **[UPTU 2006-07, MTU 2009-10]**

Solution: Given: $P = 8$, $f = 50$ Hz, $fr = 2$ Hz.

• The full load slip is calculated with the help of rotor frequency as

$$f_r = s \cdot f$$

$$\therefore \quad 2 = s \cdot 50$$

$$s = \frac{2}{50} = 0.04$$

$$\therefore \quad \%s = 0.04 \times 100 = 4\% \ \textbf{Ans.}$$

• The corresponding speed is given by

$$N_r = (1 - s) \, N_s$$

where

$$N_s = \frac{120f}{P} = \frac{120 \times 50}{8} = 750 \text{ rpm}$$

$$\therefore \quad N_r = (1 - 0.04) \, 750 = 720 \text{ rpm} \ \textbf{Ans.}$$

Example 4.35: A 4-pole, 50 Hz, 3φ induction motor has a rotor resistance of 0.024 Ω per phase and standstill reactance of 0.06 Ω per phase. Determine the speed at which maximum torque is developed. **[UPTU 2007-08, MTU 2009-10]**

Solution: Given: $r_2 = 0.024$ Ω/phase, $x_2 = 0.6$ Ω/phase.

• We know that the slip corresponding to maximum torque is find out by

$$s = \frac{r_2}{x_2} = \frac{0.024}{0.6} = 0.04$$

∴ Speed corresponding to maximum torque

$$N_r = (1 - s)N_s = (1 - 0.04)\frac{120 \times f}{P}$$

$$= (1 - 0.04) \times \frac{120 \times 50}{4}$$

$$= 1440 \text{ rpm} \ \textbf{Ans.}$$

Example 4.36: A 4-pole, 3φ induction motor is energized from a 60 Hz supply and is running at a load condition for which the slip is 0.03. Determine

(*i*) Rotor speed in rpm.

(*ii*) Rotor current frequency in Hz.

(*iii*) Speed of the rotors rotating magnetic fields with respect to the stator frame in rpm. **[UPTU 2008-09, MTU 2008-09]**

Solution: The given data

$$P = 4, f = 60 \text{ Hz}, s = 0.3$$

(*i*)

$$N_s = \frac{120f}{P} = \frac{120 \times 60}{4} = 1800 \text{ rpm}$$

$$= 1800 \text{ rpm}$$

$$\therefore \quad N_r = (1 - S) \, N_s = (1 - 0.03) \, 1800 = 1740 \text{ rpm} \ \textbf{Ans.}$$

(*ii*) The speed of rotor magnetic field with respect to stator frame is calculated as
$$f_r = s \cdot f = 0.03 \times 60 = 1.8 \text{ Hz}$$
(*iii*) Rotor magnetic field rotates at speed

$$= \frac{120 f_r}{P} = \frac{120 \times 1.8}{4} = 54 \text{ rpm} \text{ Ans.}$$

Example 4.37: A three phase, 50 Hz induction motor has a full load speed of 960 rpm calculate: **[UPTU 2009-10, MTU 2011-12]**

(*i*) Slip

(*ii*) Frequency of rotor induced emf.

(*iii*) No. of poles.

(*iv*) Speed of rotor field with respect to rotor structure.

(*v*) Speed of rotor field with respect to stator field.

Solution: Given: $N_r = 960$ rpm, $f = 50$ Hz.

(*i*) If $\qquad N_r = 960$ rpm, then we assume $N_s = 1000$ rpm

$$\%s = \frac{N_s - N_r}{N_s} \times 100 = \frac{1000 - 960}{1000} \times 100$$

∴ $\qquad = 4\% \text{ Ans.}$

(*ii*) The frequency of rotor induced emf is
$$f_r = s \cdot f$$
$$= 0.04 \times 50 = 2 \text{ Hz} \text{ Ans.}$$

(*iii*) The no. of poles is given by

$$N_S = \frac{120 f}{P}, \text{ i.e. } 1000 = \frac{120 \times 50}{P}$$

∴ $\qquad P = 6$

(*iv*) The speed of rotor field rotor structure is given by with the help of rotor frequency

∴ $\qquad \frac{120 f_r}{P} = \frac{120 \times 2}{6} = 40 \text{ rpm} \text{ Ans.}$

(*v*) Both stator and rotor fields are rotating at N_S with respect to stator structure hence speed of rotor field with respect to stator field is zero.

Example 4.38: A 3φ, 4-pole induction motor is running with 4% slip. The supply frequency is 50 Hz. Find out the speed of induction motor. **[GBTU 2010-11]**

Solution: Given: $P = 4$, $s = 0.04$, $f = 50$ Hz

$$N_s = \frac{120 f}{P} = \frac{120 \times 50}{4} = 1500 \text{ rpm}$$

$$N_r = (1 - s) N_s = (1 - 0.04) 1500$$
$$= 1440 \text{ rpm} \text{ Ans.}$$

EXERCISE

1. Define the following terms: (a) Slip (b) Slip speed (c) Rotor Current (d) Rotor power factor (e) Rotor frequency
2. Explain the principle operation of 3φ induction motor.
3. Draw slip/speed torque characteristics of 3φ induction motor.
4. What are the application of squirrel cage induction motors?
5. The rotor speed of a 6-pole, 50 Hz induction motor is 960 rpm. Find the percentage slip. **[UPTU 2008–09]**
6. Why single phase induction motor is not self starting?
7. Explain the types of single-phase induction motor and its applications.

Short Answer Questions

1. Explain the working principle of 3-phase induction motor. **[UPTU 2003-04]**
2. Draw and explain the torque-slip characteristics of 3-phase induction motor.
[UPTU 2004–06]
3. Draw the torque-slip characteristics of a 3-φ induction motor and explain its various region of operations.
4. Give the application of the following: (a) 3φ induction motor (b) 1φ induction motor.
5. Explain why single phase induction motor is not self starting and discuss briefly and two method used to produces starting torque is such motor?
[UPTU 2005–07, MTU 2009–10]

Long Answer Questions

1. Explain principle of operation of a single phase induction motor using two revolving field theory. Explain various method of starting.
[MTU 2008–10, UPTU 2008–10]
2. What are the different types of induction motor? Explain the principle of operation of 3φ induction motor.

   A 12 pole, 3φ induction alternator is coupled to an engine running at 1500 rpm. It supplied a 3φ induction motor having a full load speed of 1440 rpm. Find the percentage slip, frequency of rotor current and number of poles of motor.
[Ans. %5 = 4%, f_r = 2 Hz, P = 4]
3. Derive the torque equation of a 3φ induction motor. Draw the torque-slip characteristics of 3φ induction motor.
[UPTU 2003-008, MTU 2009–10, 2011–12]

5

Single Phase Induction Motors

5.1 INTRODUCTION

The most common type of electric motor is the single phase type, which finds domestic, commercial and industrial applications. Single phase motors are small size motors of fractional-kilowatt range. Domestic appliances like fans, hair driers, washing machines, vacuum cleaners, mixers, refrigerators, food processors and kitchen equipment employ these motors. These motors also find application in air-conditioning fans, blowers, office machinery, small power fools, dairy machinery, small framing equipment, etc. It is true that single phase motors are less efficient substitute for 3-phase motors but 3-phase power is normally not available except in large commercial and industrial establishment.

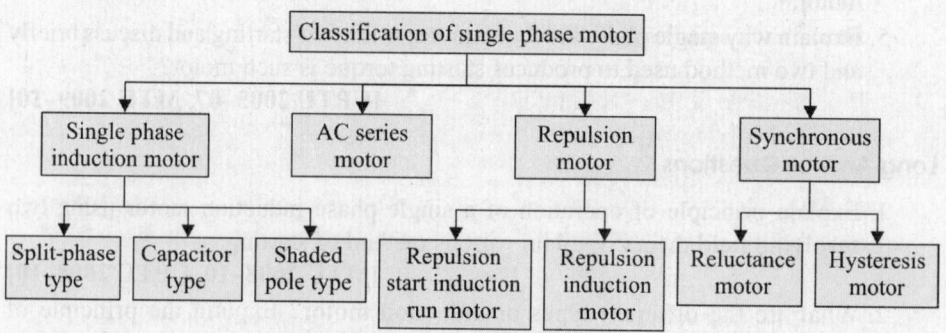

5.2 SINGLE PHASE INDUCTION MOTOR—PRINCIPLE

A single phase induction motor is very similar to a 3-phase squirrel cage induction motor.

It has (*i*) a squirrel-cage rotor identical to a 3-phase motor; (*ii*) a single-phase winding on the stator.

Unlike a 3-phase induction motor, a single phase induction motor is not self starting but requires some starting means. The single phase stator winding produces a magnetic field that pulsates in strength in a sinusoidal manner.

The field polarity reverses after each half cycle but the field does not rotate. Consequently, the alternating flux cannot produces rotation in a stationary squirrel-cage rotor.

However, if the rotor of a single phase motor is rotated in one direction by some mechanical means, it will continue to run in the direction of rotation (Fig. 5.1).

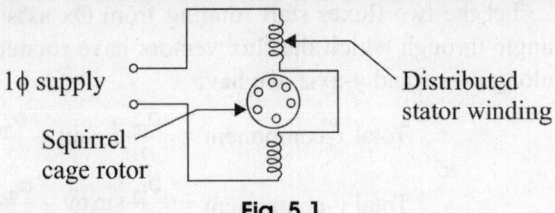

Fig. 5.1

As a matter of fact, the rotor quickly accelerated until it reaches a speed slightly below the synchronous speed.

Once the motor is running at this speed, it will continue to rotate even though single phase current is flowing through the stator winding.

This method of starting is generally not convenient for large motors.

Such a motor inhenently does not developed any starting torque and therefore, will not start to rotate if the stator winding is connected to single phase AC supply. However, if the rotor is started by auxiliary means, the motor will quickly attain the final speed. This starting behaviour can be explained by double-revolving field theory.

5.3 DOUBLE REVOLVING FIELD THEORY OF SINGLE PHASE INDUCTION MOTOR

- This theory makes use of the idea that an alternating uniaxial-quantity can be represented by two oppositely rotating vectors of half magnitude.
- Accordingly, an alternating sinusoidal flux can be represented by two revolving fluxes, each equal to half of the alternating flux and each rotating synchronously ($N_s = 120\ F/p$) in opposite direction with angular velocity $\pm\ \omega$.

  [For example, a flux given by $\phi = \phi_m \cos \omega t = \phi_m \cos 2\pi f t$ is equivalent to two fluxes revolving in opposite directions, each with a magnitude $\phi/2$, and an angular velocity $2\pi f$] (Fig. 5.2).

  It may be noted that Euler's expressions for $\cos\theta$ provides interesting justification for the decomposition of a pulsating flux. The expression is:

  $$\cos\theta = \frac{e^{j\theta} + e^{-j\theta}}{2}$$

  The term $e^{j\theta}$ represents a vector rotated clockwise through an angle θ, whereas $e^{-j\theta}$ represents rotation in anticlockwise direction. Now, the above given flux can be expressed as

  $$\phi_m \cos 2\pi f t = \frac{\phi_m}{2}\left(e^{j2\pi f t} + e^{-j2\pi f t}\right)$$

- The instantaneous value of flux due to the stator current of a single phase induction motor is given by

  $$\phi = \phi_m \cos \omega t$$

consider two rotating fluxes ϕ_1 and ϕ_2, and each of magnitude $\phi_m/2$, and rotating in opposite directions with angular velocity ω.

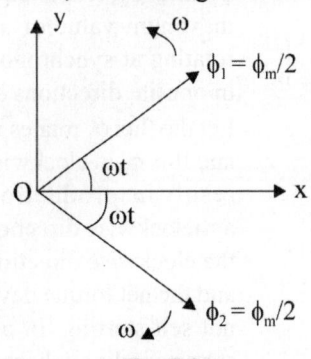

Fig. 5.2

Let the two fluxes start rotating from Ox axis at $t = 0$. After time t seconds, the angle through which the flux vectors have rotated is ωt. Resolving the flux vectors along x-axis and y-axis, we have

$$\text{Total } x\text{-component} = \frac{\phi_m}{2}\cos \omega t + \frac{\phi_m}{2}\cos \omega t = \phi_m \cos \omega t$$

$$\text{Total } y\text{-component} = \frac{\phi_m}{2}\sin \omega t - \frac{\phi_m}{2}\sin \omega t = 0$$

$$\text{Resultant flux, } \phi = \frac{\phi_m}{2}\sin \omega t - \frac{\phi_m}{2}\sin \omega t = 0$$

Thus the resultant flux vector is $\phi = \phi_m \cos \omega t$ along x-axis. Therefore an alternating field can be replaced by two rotating fields of half its amplitude rotating in opposite directions at synchronous speed.

• Resultant vector of two revolving flux vectors is a stationary vector that oscillates in length with time along x-axis. When the rotating flux vectors are in phase, the resultant vector is $\phi = \phi_m$.

When out of phase by $180°$ the resultant vector $\varphi = 0$ (Fig. 5.3).

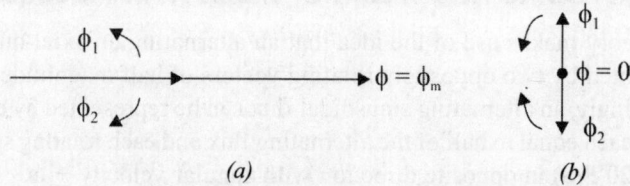

(a)   (b)

Fig. 5.3

5.3.1 Operation of Single Phase Motor by Double-Field-Revolving Theory

(*i*) **Rotor at standstill:** Consider the case that the rotor is stationary and the stator winding is connected to a single phase supply. The alternating flux produced by stator winding can be represented as the sum of two rotating fluxes ϕ_1 and ϕ_2, each equal to one half of the maximum value of alternating flux and each rotating at synchronous speed ($N_S = 120 \, f/P$) in oppsite directions shown in Fig. 5.4.

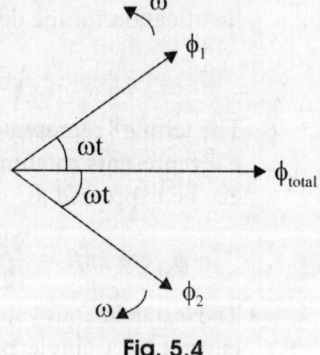

Fig. 5.4

Let the flux ϕ_1 rotates in anticlockwise direction and flux ϕ_2 in clockwise direction. The flux will result in production of torque T_1 in the anticlockwise direction and flux ϕ_2 will result in the production of torque T_2 in the clockwise direction. At standstill, these two torques are equal and opposite and the net torque developed is zero. Therefore single phase induction motor is not self starting. In other words, a single-phase induction motor with single-stator winding inherently has no starting torque.

(*ii*) **Rotor running or rotor slip with respect to two rotating fields:** If, the rotor is given an initial rotation by auxiliary means in either directions of the torque due to the rotating field acting in the direction of initial rotation will be more than the torque due to the other rotating field. Hence the motor will develop a net positive torque in the same direction as the initial rotation. The motor will, therefore, keep running in the direction of initial rotation.

If the rotor is started by auxiliary means, it will develop torque and continue to run in the same direction as one of the fields. By definitions the direction in which the rotor is started initially will be called the forward field (say clockwise direction).

Let N_s = sychronous speed, N = rotor speed.

The slip of the rotor with respect to the forward rotating field is,

$$s_f = s = \frac{N_s - N}{N_s} = 1 - \frac{N}{N_s} \qquad ...(5.1)$$

Since the backward rotating flux rotates (say anticlockwise) opposite to the stator, the sign of n must be changed in Eq. (5.1) to obtain the backward slip. Thus the slip of the rotor with respect to the backward rotating field is

$$s_b = \frac{N_s - (-N)}{N_s} = \frac{N_s + N}{N_s} = 1 - \frac{N}{N_s} \qquad ...(5.2)$$

Adding Eqs (5.1) and (5.2), we get

$$s + s_b = \left(1 - \frac{N}{N_s}\right) + \left(1 + \frac{N}{N_s}\right)$$

$$s_b = 2 - s \qquad ...(5.3)$$

Thus the rotor slips with respect to the different, and are given by Eqs (5.1) and (5.3).

[**Note:** Rotor current under running condition $I_r = sE_2 \big/ \sqrt{R_2^2 + (sX_2)^2}$.

Rotor forward and backward field slips are s_f and s_b.

$\therefore$ Rotor backward field slip increase and it will oppose the stator backward field. As a result stator backward field becomes weaker and the stator forward field becomes stronger which will cause the rotor to rotate in forward direction and motor gradually speedup]

In order to make clear the influence of the two rotating fluxes on the rotor, it will be assumed that $N < N_s$. Equation (5.1) corresponds to a motor operation and Eq. (5.3) denotes the braking region. Thus the two torques have an opposite influence on the rotor.

The equivalent circuits for forward and backward rotating fluxes are shown in Figs 5.5(*a*) and (*b*) respectively.

At standstill, the impedances are equal ($s = 1$) and therefore the current I_{2f} and I_{2b} are equal. These currents produce mmfs which oppose the stator mmfs equally. Therefore, the rotating forward and backward fluxes in the air gap are equal in magnitude, and no torque is developed.

However, when the rotor rotates, the impedances of the rotor circuits are unequal and the rotor current I_{2b} is greater than the rotor current I_{2f}.

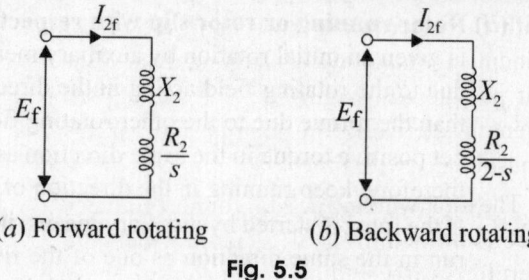

(a) Forward rotating   (b) Backward rotating

Fig. 5.5

Their mmfs, which oppose the stator mmfs, will result in the reduction of the backward rotating flux.

Consequently, the speed increases, the forward flux increases while the backward flux decreases. However, the resultant flux remains essentially constant.

This resultant flux induces voltage in the stator winding.

Both flux waves induce voltages in the rotor and produce torques in the rotor. These two torques are in opposite directions.

The net induced torque in the motor is equal to the difference between these torques. Figure 5.6 shows torques produced by the two revolving fields and also the resultant torque produced by the motor.

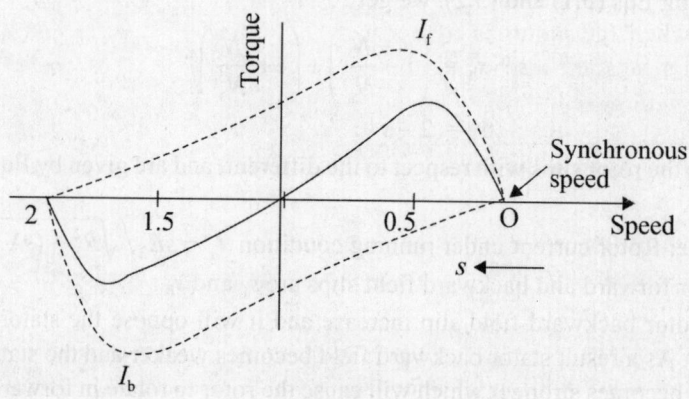

Fig. 5.6

If the voltage drops across the winding resistance and leakage reactance are neglected, the induced voltage is almost equal to the applied voltage. As the speed increases, forward torque increases and reverse torque decreases.

5.4 CROSS FIELD THEORY

During the positive half cycle of the supply, the stator current is flowing in the direction shown in Fig. 5.7. So N pole (looking from in side it is *CCW*) is formed at A and S pole at *C*. During the next half cycle their respective positions with be reversed. Thus with the alternating current is the stator winding, the stator wave is stationary in space but pulsates in magnitude, the stator field strength alternating in polarity and varying sinusoidially with time.

The rotor winding acts like a short circuited secondary winding of a transformer and therefore carry indicating currents. These currents are in reverse polarity in respect-stator current at any moment as shown as Fig. 5.7. Since the rotor poles (although reverse of stator poles) are always developed in either direction. Therefore, no starting torque in developed.

At standstill, only transformer emf generates. It lags from the stator current by 90°. Rotor current lags from the rotor voltage by 90°. So stator flux and rotor flux are at 180°. It is to be noted that the rotor winding has high inductance and negligible resistance.

However when the motor is rotated by some external means, the emf induced in the rotor winding will have a polarity as shown in Fig. 5.8 according to Fleming's right hand rule. This emf is known as speed emf.

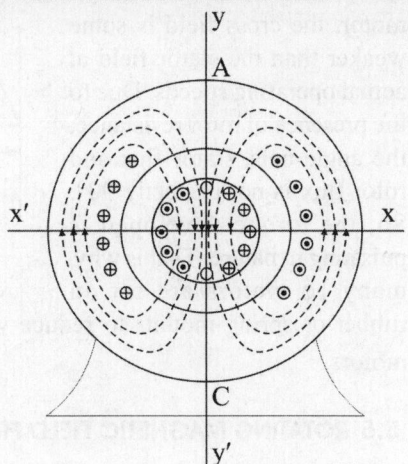

Fig. 5.7(a): Standstill condition

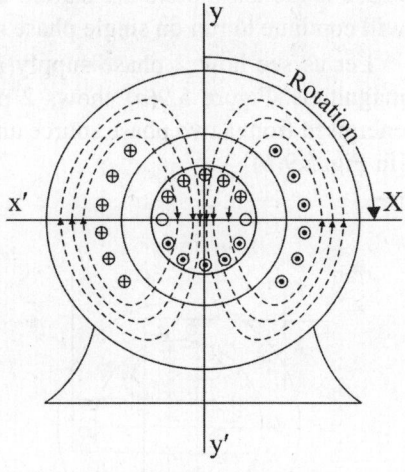

Fig. 5.7(b)

$$e = N\phi\,\omega_r\sin\omega_e t - N\frac{d\phi}{dt}\cos\omega_r t$$

Speed emf   Transformer emf

$\omega_r \rightarrow$ Space angle

It is seen in the equation of speed emf that if the flux is sinusoidal, i.e. $\phi = \phi_{max}\sin\omega t$ then emf is also sinusoidal, i.e. $e_{speed} = N\omega_r\sin\omega_r t\,\phi_{max}\sin\omega t$. So, the stator flux and the rotor induced emf are at same phase.

As the rotor winding consists of high inductance, rotor current lags rotor induced emf by almost 90°.

Since the field created by the rotor currents is at right angles to the field by the stator currents, it is known as cross-fields. These two fields combine to form a resulting rotating field that revolves at synchronous speed.

Since the cross-field is produced by a generator action, therefore, it is present only when the rotor is rotating and its strength is proportional to the speed of rotor. At N_s, the cross field is of same strength as the stator field. But as $N < N_s$ in case of induction motor, the cross field is some weaker than the stator field at actual operating speeds. Due to the presence of rotor resistance, the angle bets, stator flux and rotor flux is not perfectly 90°. So, the torque developed is pulsating in nature. That is why, many 1φ motors are set on

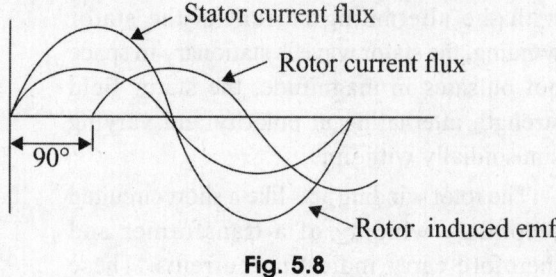

Fig. 5.8

rubber or spring mounts to reduce vibrations and noise which are inherent is such motors.

5.5 ROTATING MAGNETIC FIELD FROM 2-PHASE SUPPLY

As with a 3-phase supply, a 2-phase balanced supply also produces a rotating magnetic field of constant magnitude. With the exception of the shaded pole motor, all single-phase induction motors are started as 2-phase machine. Once so started, the motor will continue to run on single phase supply.

Let us see how 2 phase supply produces a rotating magnetic field of constant magnitude. Figure 5.9(*a*) shows 2 pole, 2 phase winding. The phases *x* and *y* are energised from a two phase source and current in this phases are indicated as I_x and I_y [in Fig. 5.9(*b*)].

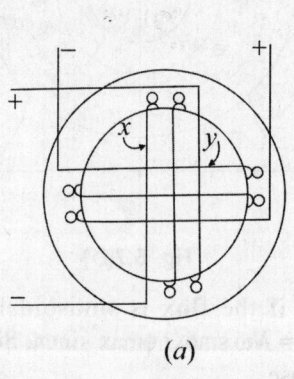
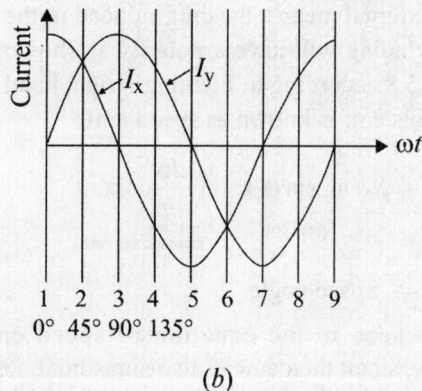

(*a*)　　　　　　　　　　　　(*b*)

Fig. 5.9

Referring to Fig. 5.9, the fluxes produced by these currents are given by

$$\phi_y = \phi_m \sin \omega t$$
$$\phi_x = \phi_m \sin(\omega t + 90) = \phi_m \cos \omega t$$

Here ϕ_m is the maximum flux due to either phase.

In two-phase system, the two windings (or phases) are 90° electrical apart.

We shall now prove that this 2-phase supply produces a rotating magnetic field of constant magnitude equal to ϕ_m.

(*i*) At instant 1, the current is zero in phase y and maximum in phase x. With the current in the direction shown in Figs 5.9(*b*) and (*c*), a resultant flux is established towards the right.

The magnitude of the resultant flux is constant and equal to ϕ_m as proved under:

At instant 1, $\qquad \omega t = 0°,\ \therefore\ \phi_y = 0$ and $\phi_x = \phi_m$

$\therefore$ Resultant flux, $\quad \phi_r = \sqrt{\phi_x^2 + \phi_y^2} = \sqrt{(\phi_m)^2 + (0)^2} = \phi_m$

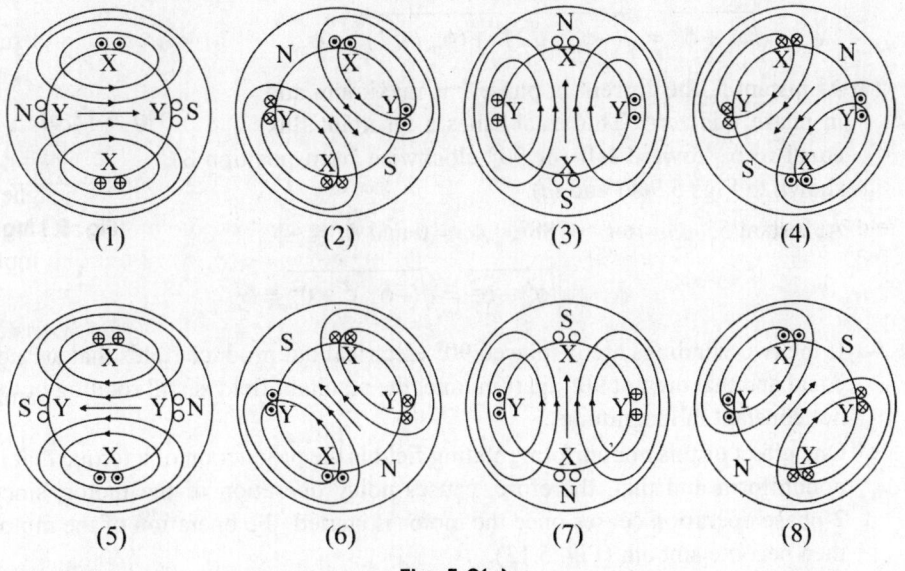

(1) (2) (3) (4)

(5) (6) (7) (8)

Fig. 5.9(c)

(*ii*) At instant 2, the current is still in the same direction in phase x and an equal current flowing in phase y. This establishes a resultant flux of the same value (i.e. $\phi_r = \phi_m$) as proved under (Figs 5.10 and 5.11):

At instant 2, $\qquad \omega t = 45°$

$\therefore \qquad\qquad \phi_y = \dfrac{\phi_m}{\sqrt 2}$ and $\phi_x = \dfrac{\phi_m}{\sqrt 2}$

$\therefore$ Resultant flux,

$$\phi_r = \sqrt{(\phi_x)^2 + (\phi_y)^2}$$

$$= \sqrt{(\phi_m/\sqrt 2)^2 + (\phi_m/\sqrt 2)^2} = \phi_m$$

Fig. 5.10

Fig. 5.11

(*iii*) At instant 3, the current in phase x has decreased to zero and current in phase y has increased to maximum.

$$\omega t = 90,\ \therefore\ \phi_y = \phi_m \text{ and } \phi_x = 0$$

$$\phi_r = \sqrt{\phi_x^2 + (\phi_y)^2} = \sqrt{(0)^2 + (\phi_m)^2}$$

Resultant flux has now turned 90° clockwise from position 1. This constant flux is shifting its position. In other words, the rotating flux is produced.

(*iv*) At instant 4, the current in phase *x* has reversed and has the same value as that of phase *y* (Fig. 5.12).

Fig. 5.12

$$\omega t = 135°, \therefore \phi_y = \frac{\phi}{\sqrt{2}} \text{ and } \phi_x = -\frac{\phi_m}{\sqrt{2}}$$

$$\phi_r = \sqrt{\phi_x^2 + \phi_y^2} = \sqrt{(-\phi_m/\sqrt{2})^2 + (\phi_m/\sqrt{2})^2} = \phi_m$$

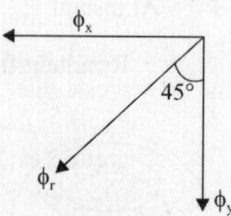

(*v*) At instant 5, the current in phase *x* is maximum and in phase *y* is zero. This establishes a resultant flux equal to ϕ_m toward left (or 90° clockwise from position 3), shown in Figs 5.9(*b*) and (*c*).

Fig. 5.13(a)

At instant 5, $\omega t = 188°, \therefore \phi_y = 0$ and $\phi_x = -\phi_m$

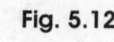

Fig. 5.13(b)

$$\phi_r = \sqrt{\phi_x^2 - \phi_y^2} = \sqrt{(-\phi_m)^2 + 0^2} = \phi_m$$

If the two windings are displaced 90° electrical but produce fields that are not equal and that are not 90° apart in time, the resultant field is still rotating but is not constant in magnitude.

One effect of this nonuniform rotating field is the production of a torque that is nonuniform and that, therefore, causes noisy operation of the motor, since 2-phase operation ceases once the motor is started, the operation of the motor then become smooth (Fig. 5.13).

5.6 EQUIVALENT CIRCUIT OF SINGLE PHASE INDUCTION MOTOR

When the stator of a single-phase induction motor is connected to single phase supply, the stator current produces a pulsating flux that is equivalent to two-constant amplitude fluxes revolving in opposite directions at the synchronous speed (double-field revolving theory).

Each of these fluxes induces currents in the rotor circuit and produces induction motor action similar to that in a 3-phase induction motor.

Therefore, a single-phase induction motor can be imagined to be consisting of two motors, having a common stator winding but with their respective rotors revolving in opposite directions.

Each rotor has resistance and reactance half the actual rotor values.

Let R_1 = resistance of stator winding

X_1 = Leakage reactance of stator winding

X_m = total magnetising reactance

R'_m = resistance of the rotor reffered to the stator

X'_m = Leakage reactance of the rotor reffered to the stator

We shall now develop the equivalent circuit of single phase induction motor using double-field revolving theory.

The equivalent circuit will be with motor operating with main winding only.

(*i*) **At standstill:** At standstill, the motor is simply a transformer with its secondary short-circuited. Therefore, the equivalent circuit of single phase motor at standstill will be as shown in Fig. 5.14(*a*). The double-field revolving theory suggests that characteristics associated with each revolving field will be just one half of the characteristics associated with the actual total flux.

Therefore each rotor has resistance and reactance equal to $R_2'/2$ and $X_2'/2$ respectively. Each rotor is associated with half of the total magnetising reactance. In the equivalent circuit, the core losses has been neglected. However coreloss can be represented by an equivalent resistance in parallel with the magnetising reactance.

$$E_f = 4.44 \ fN\phi_f; \ E_b = 4.44 \ fN\phi_b$$

At standstill, $\qquad \phi_f = \phi_b$, therefore $E_f = E_b$

$\therefore \qquad\qquad V_1 = E_f + E_b = I_1 Z_f + I_1 Z_b$

where Z_f = impedance of forward parallel branch.

Z_b = impedance of backward parallel branch.

(*ii*) **Rotor running:** Now consider that the motor is running at some speed in the direction of the forward revolving field, the slip being s. The rotor current produced by the forward field will have a frequency sf, where f is the stator frequency.

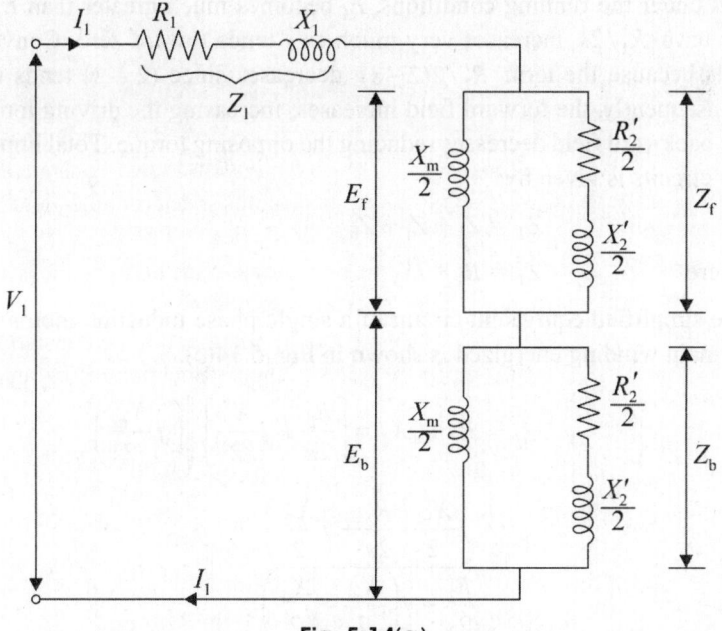

Fig. 5.14(a)

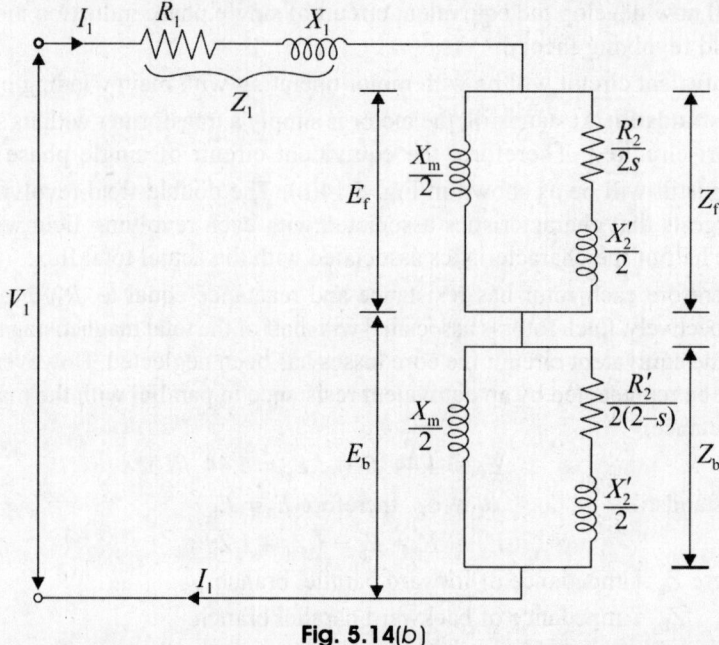

Fig. 5.14(b)

Also the rotor current produced by the backward field will have a frequency of $(2–s)f$. Figure 5.14(b) shows the equivalent circuit of a single phase induction motor when the rotor is rotating at slip s. It is clear from the equivalent circuit that under the running conditions, E_f becomes much greater than E_b because the term $R_2'/2s$ increases very much as s tends toward zero. Conversely, E_b falls because the term $R_2'/2(2-s)$ decreases since $(2 - s)$ tends toward 2. Consequently, the forward field increases, increasing the driving torque while the backward field decreases reducing the opposing torque. Total impedance of the circuits is given by:

$$Z_T = Z_1 + Z_f + Z_b$$

where $\qquad Z_1 = R_1 + jX_1$

The simplified equivalent circuit of a single phase induction motor with only its main winding energized is shown in Fig. 5.14(c).

$$Z_f = R_f + jX_f = \left[\frac{R_2'}{2s} + j\frac{X_2'}{2}\right] \left\| \left(j\frac{X_m}{2}\right)\right.$$

$$= \frac{j\dfrac{X_m}{2}\left(\dfrac{R_2'}{2s} + j\dfrac{X_2'}{2}\right)}{\dfrac{R_2'}{2s} + j\left(\dfrac{X_m}{2} + \dfrac{X_2'}{2}\right)}$$

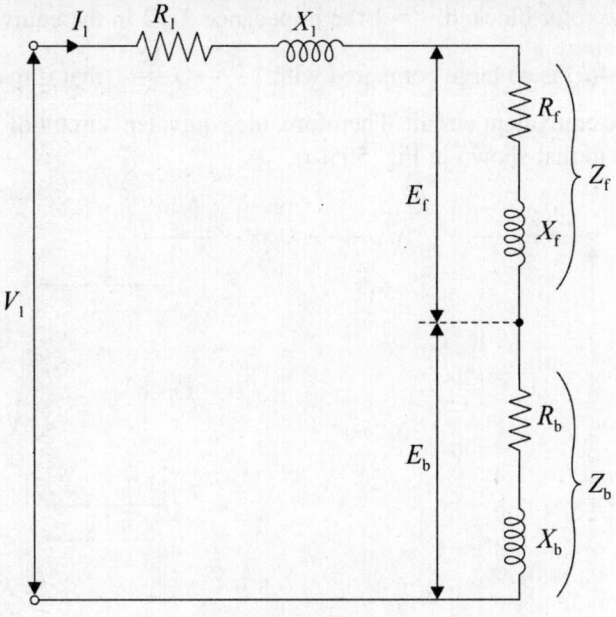

Fig. 5.14(c)

$$Z_b = R_b + jX_b = \left[\frac{R_2'}{2(2-s)} + j\frac{X_2'}{2} \right] \left\| \left(j\frac{X_m}{2} \right) \right.$$

$$= \frac{j\frac{X_m}{2} \left(\frac{R_2'}{2(2-s)} + j\frac{X_2'}{2} \right)}{\frac{R_2'}{2(2-s)} + j\left(\frac{X_m}{2} + \frac{X_2'}{2} \right)}$$

∴ $I_1 = V_1/Z_T$ [current in the stator winding].

5.7 DETERMINATION OF EQUIVALENT CIRCUIT PARAMETERS

The parameters of the equivalent circuit of a single phase induction motor can be determined from the blocked-rotor and no-load tests.

These tests are similar to those made on 3-phase induction motor. However, except for the capacitor-run motor, these tests are performed with auxiliary winding kept open.

1. **Blocked-rotor test:** In this test the rotor is at rest (blocked).

   A low voltage is applied to the stator so that rated current flows in the main winding. The voltage, current and power input are measured.

   Let V_{SC}, I_{SC} and P_{SC} denote the voltage, current and power respectively under the same conditions.

With the rotor blocked, $s = 1$ the impedance $X_m/2$ in the equivalent circuit of Fig. 5.14(b) is so large compared with $\left(\dfrac{R_2'}{2} + j\dfrac{X_2'}{2}\right)$ that it may be neglected from the equivalent circuit. Therefore, the equivalent circuit of 5.14(b) at $s = 1$ reduces to that shown in Fig. 5.15(a).

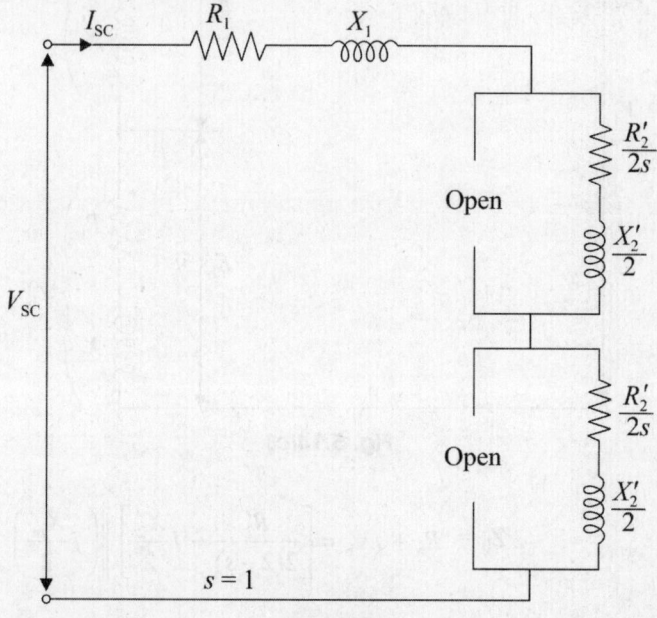

Fig. 5.15(a)

Note: Load current is too high so the magnetising current is very low in shunt path.

From Fig. 5.15(a), $Z_e = \dfrac{V_{SC}}{I_{SC}}$ the equivalent series resistance R_e of the motor is

$$R_e = R_1 + \frac{R_2'}{2} + \frac{R_2'}{2}$$

$$= R_1 + R_2' = \frac{P_{SC}}{I_{SC}^2}$$

Since the resistance of the main stator winding R_1 is already measured, the effective rotor resistance at line frequency is given by

$$R_2' = R_e - R_1 = \frac{P_{SC}}{I_{SC}^2} - R_1$$

Equivalent reactance X_e is given by

$$X_e = X_1 + \frac{X_2'}{2} + \frac{X_2'}{2}$$

$$= X_1 + X_2'$$

Since the leakage reactance X_1 and X_2' cannot be separated out we make a simplifying assumption that $X_1 = X_2'$.

$$\therefore \qquad X_1 = X_2' = \frac{1}{2}X_e = \frac{1}{2} \times \sqrt{Z_e^2 - R_e^2}$$

Thus, from blocked-rotor test, the parameters, R_2', X_1, X_2' can be found if R_1 is known.

2. **No-load test:** The motor is run without load at rated voltage and rated frequency. The voltage, current and input power are measured. At no load, the slip s is very small close to zero and $R_2'/2s$ is very large as compared to $X_m/2$.

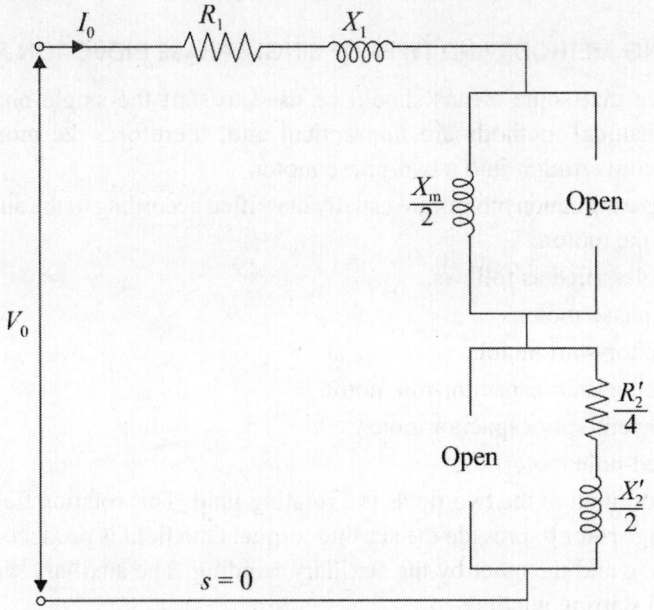

Fig. 5.15(b)

The resistance $\dfrac{R_2'}{2(2-s)}\left[\cong \dfrac{R_2'}{2(2-0)} = \dfrac{R_2'}{4}\right]$ associated with the backward rotating field is so small as compared to $X_m/2$, that the backward magnetizing current is negligible. Therefore under no-load conditions, the equivalent circuit simplified to Fig. 5.15(b).

The equivalent reactance at no load is given by

$$X_0 = X_1 + \frac{X_m}{2} + \frac{X_2'}{2}$$

Since X_1 and X_2' are already known from the block rotor test, the magnetizing reactance X_m can be calculated from the equation.

• Let V_0, I_0 and P_0 denote the voltage, current and power respectively in the no-load test.

Then the no-load power factor is

$$\cos \phi_0 = \frac{P_0}{V_0 I_0}$$

The no-load equivalent impedance is

$$Z_0 = \frac{V_0}{I_0}$$

The no-load equivalent reactance is

$$X_0 = Z_0 \sin \phi_0 = Z_0 \times \sqrt{1 - \cos^2 \phi_0}$$

5.8 STARTING METHODS AND TYPES OF SINGLE PHASE INDUCTION MOTOR

We have seen that some means should be used to start the single phase induction motor. Mechanical methods are impractical and, therefore, the motor is started temporarily converting it into a two-phase motor.

Single phase induction motors are usually classified according to the auxiliary means used to start the motor.

They are classified as follows:

1. Split-phase motor.
2. Capacitor-start motor.
3. Capacitor-start capacitor-run motor.
4. Permanent-split capacitor motor.
5. Shaded-pole motor.

The resultant of the two fields is a rotating field. This rotating field reacts with the cage rotor to provide the starting torque. One field is produced by the main winding and the other by the auxiliary winding. The auxiliary winding is also called starting winding.

5.9 SPLIT-PHASE INDUCTION MOTOR

Figure 5.16(a) shows a split-phase induction motor. It is also called a resistance-start motor.

It has a single-cage rotor and its stator has two windings:

(i) a main winding
(ii) a starting (auxiliary winding)

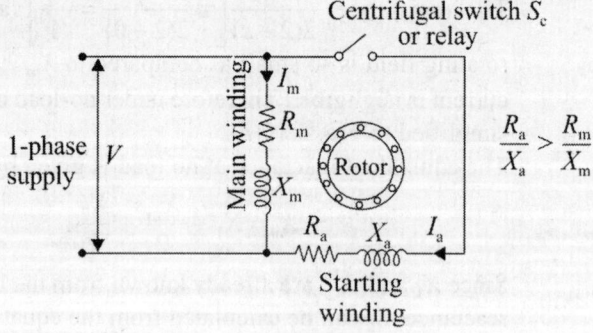

Fig. 5.16(a)

The main field winding and the starting winding are displaced 90° in space like the windings in a two-phase induction motor.

The main winding has very low resistance and high inductive reactance.

Thus the current I_m in the main winding lags behind the supply voltage V by nearly 90° shown in Fig. 5.16(*b*).

The auxiliary winding has a resistor connected in series with it. It has a high resistance and low inductive reactance so that the current I_a in the auxiliary winding is nearly in phase with the line voltage.

Thus there is time phase difference between the currents in the two windings. The time phase difference ϕ is not 90° but usually of the order of 25° to 30°.

This phase difference is enough to produce a rotating magnetic field. Since the currents in the two windings are not equal, the rotating field is not uniform, and the starting torque is small of the order of 1.5 to 2 times the rated running torque.

The main and auxiliary windings are connected in parallel during starting. The starting winding is disconnected from the supply automatically when the motor reaches speed about 70% to 80% of synchronous speed.

For motors rated about 100 W or more, a centrifugally operated switch is used to disconnect the starting winding. For smaller motors a relay is often used. The relay is connected in series with the main winding. At the time of starting, a heavy current flows in the relay coil causing its contacts to close. This brings the starting winding into the circuit.

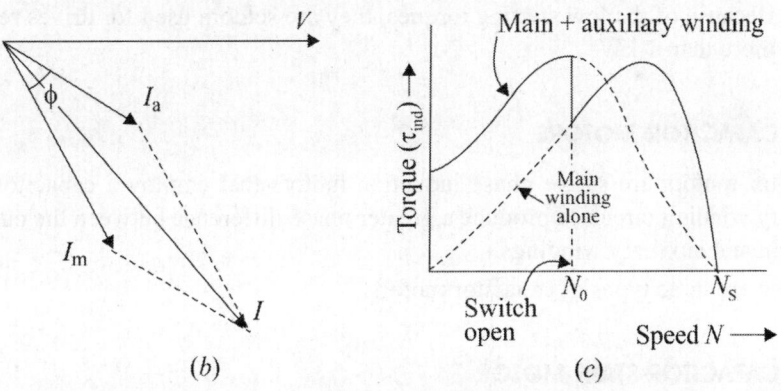

Fig. 5.16(*b*) and (*c*)

As the motor reaches its predetermined speed of the order of 70% to 80% of synchronous speed, the current through the relay coil decreases. Consequently the relay opens and disconnects the auxiliary winding from the main supply and the motor then runs only on the main windings.

The torque speed characteristic of this motor is shown in Fig. 5.16(*c*) below, which also shows the speed N_0 at which the centrifugal switch operates.

5.9.1 Reversal of Direction of Rotation

This motor continues to run in the direction in which it is started. The direction of rotation of the resistance-start induction motor may be reversed by reversing the line connections of either the main winding or the starting winding. The motor must be brought to rest for this purpose. This is the reversal of rotation can be made only when the motor is standstill but not while running.

5.9.2 Motor Characteristics

(i) The starting torque of a resistance start induction motor is about 1.5 times full load torque.

(ii) The maximum or pull-out torque is about 2–5 times full load torque at about 75% of synchronous speed.

(iii) The split phase motor has a high starting current which is usually 7 to 8 times the full load value.

5.9.3 Applications

(i) Split phase motors are cheap and they are mot suitable for easily started loads where frequency of starting is limited.

(ii) The common applications are washing machines, air-conditioning fans, food mixer, grinder, floor polisher, blowers, centrifugal pumps, small drills, lathes, office machinery, dairy machinery etc.

(iii) Because of the low starting torques, they are seldom used for drives requiring more than 1 kW.

5.10 CAPACITOR MOTORS

Capacitor motors are single phase induction motors that employ a capacitor in the auxiliary winding circuit to produce a greater phase difference between the current in the main and auxiliary windings.

These are three types of capacitor motors.

5.11 CAPACITOR-START MOTOR

Figure 5.17(a) shows the connections of a capacitor-start motor. It has a cage rotor and its stator has two windings namely, the main winding and the auxiliary winding (starting winding). The two windings are displaced 90° in space. A capacitor C_S is connected in series with the starting windings. A centrifugal switch S_C is also connected as shown in Fig. 5.17(a). By choosing the capacitor of the proper rating the current I_m in the main winding may be made to lag the current I_a in the auxiliary winding by 90°, shown in Fig. 5.17(b).

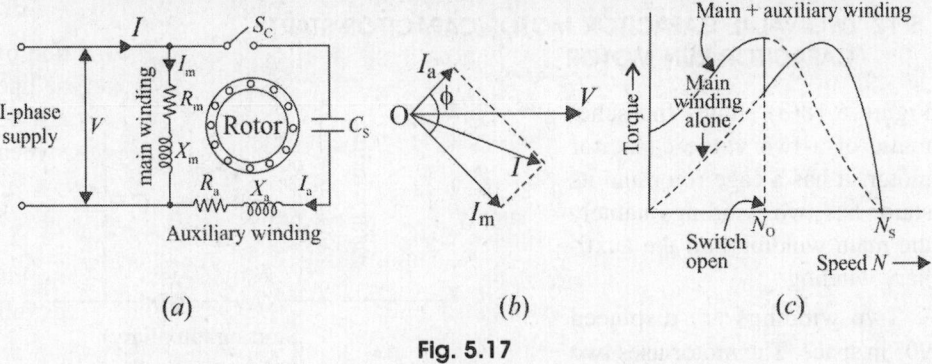

Fig. 5.17

Thus a single phase supply current is split into two phases to be applied to the stator windings. Thus the windings are displaced 90° electrical and their mmfs are equal in magnitude but 90° apart in time phase.

Therefore the motor acts like a balanced two-phase motor. As the motor approaches its rated speed, the auxiliary winding and the starting capacitor C_S are disconnected automatically by the centrifugal switch S_C mounted on the shaft. The motor is so named because it used the capacitor only for the purpose of starting.

5.11.1 Motor Characteristics

The capacitor start motor develops a much higher starting torque (3.0 to 4.5 times the full-load torque) than does an equally rated resistance start motor.

The value of the starting capacitor must be large and the starting winding resistance low to obtain a high starting torque.

Because of the high V_{Ar} rating of the capacitor required, electrolytic capacitors of the order of 250 μF are used. The capacitor C_S is short-time rated. The torque speed characteristics of the motor is shown in Fig. 5.17(c).

Capacitor start motors are more costly than split-phase motors because of the additional cost of the capacitor.

5.11.2 Reversal of Direction of Rotation

The capacitor-start motor may be reversed by reversing the connections of one of the winding. The motor is first brought to rest for this purpose.

5.11.3 Applications

(i) Capacitor-start motors are used for loads of high inertia where frequent starts are required.

(ii) These motors are most suitable for pumps and compressor, and therefore they are widely used in refrigerators and compressors of air conditioner.

(iii) They are also used for conveyors and some machine tools.

5.12 TWO-VALUE CAPACITOR MOTOR/CAPACITOR-START CAPACITOR-RUN MOTOR

Figure 5.18(a) shows the schematic of a two value capacitor motor. It has a cage rotor and its stator has two windings namely the main winding and the auxiliary winding.

Two windings are displaced 90° in space. The motor uses two capacitors C_S and C_R and connected in parallel.

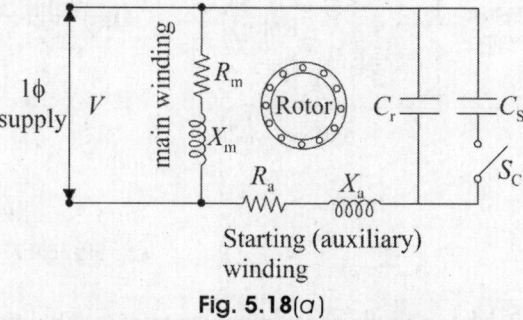

Fig. 5.18(a)

Capacitor C_S is called the starting capacitor. In order to obtain a high starting torque, a large current is required. For this purpose, the capacitive reactance X in the starting winding should be low. Since $X_a = 1/2\pi f C_a$, the value of C_S should be large. The capacitor C_S is short time rated and is almost always electrolytic.

During normal operation, the rated line current is smaller than the starting current. Hence the capacitive reactance should be large. Since $X_r = 1/(2\pi f C_a)$, the value of C_r should be small.

As the motor approaches synchronous speed, the capacitor C_S is disconnected by a centrifugal switch S_C. The capacitor C_r is permanently connected in circuit. It is called run-capcitor. It is long time rated for continuous running. It is usually oil-filled paper construction. Since one capacitor C_S is used only at starting and the other C_r for continuous running, this motor is also called capacitor-start capacitor run motor.

At starting both the capacitors are in the circuit and $\phi > 90°$ [Fig. 5.18(b)] when the capacitor C_S is disconnected ϕ becomes 90° (electric) [shown in Fig. 5.18(c)].

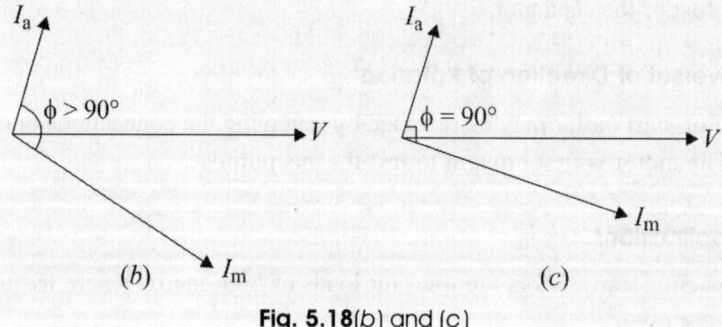

Fig. 5.18(b) and (c)

• Two-value capacitor motors are quite and smooth running. They have a higher efficiency than motors run on the main windings alone (Fig. 5.18(d)).

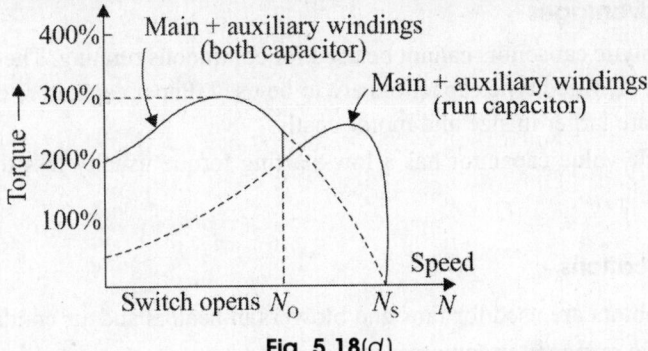

Fig. 5.18(d)

5.12.1 Application

Two-value capacitor motors are used for loads of higher intertia requiring starts where the maximum pull out torque and efficiency required are higher. They are used in pumping equipment, refrigeration, air compressors, etc.

5.13 PERMANENT-SPLIT CAPACITOR (PSC) MOTOR/SINGLE VALUE CAPACITOR MOTOR

It has cage rotor and its stator has two winding, namely the main winding and the auxillary winding. This motor has only one capacitor C which is connected in series with the starting winding is shown in Fig. 5.19. The capacitor is permanently connected in the circuit both at starting and running conditions. A PSC motor is called the single value capacitor motor. Since the capacitor C is always in the circuit, this type of

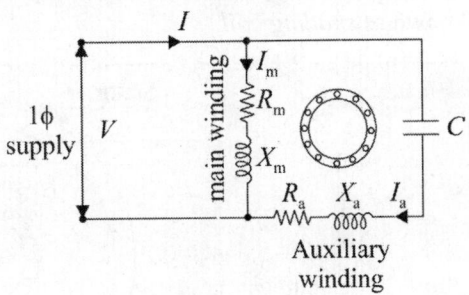

Fig. 5.19

motor has no starting switch. The auxiliary winding is always in the circuit, and therefore this motor operates in the same way as a balanced two-phase motor. Consequently, it produces a uniform torque. The motor is therefore less noisy during operations.

5.13.1 Advantages

A single value capacitor motor possesses the following advantages:

    (*i*) No centrifugal switch is required.

    (*ii*) It has higher efficiency.

    (*iii*) It has higher power factor because of permanently–connected capacitor.

    (*iv*) It has a higher pull out torque.

5.13.2 Disadvantages

(*i*) Electrolytic capacitors cannot be used for continuous running. Therefore paper-spaced oil-filled type capacitors are to be used. Paper capacitors of equipment rating are larger in size and motor costly.

(*ii*) A single-value capacitor has a low starting torque usually less than full-load torque.

5.13.3 Applications

(*i*) PSC motors are used for fans and blowers in heaters and air conditioners.

(*ii*) To drive refrigerator compressor.

(*iii*) Also used to drive office machinery.

5.14 SHADED POLE MOTORS

A shaded-pole motor is a simple type of self-starting single-phase induction motor. It consists of a stator and a cage type rotor.

The stator is made up of salient pole.

Each pole is slotted on side and a copper ring is fitted on the smaller part a, shown in Fig. 5.20. This part is called shaded pole. The ring usually a single-turn coil and is known as *shading coil*.

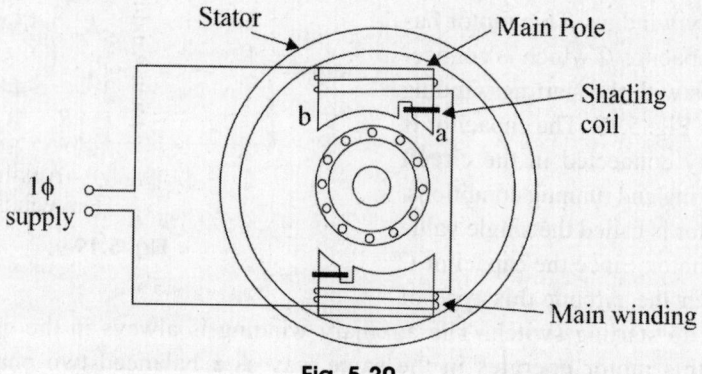

Fig. 5.20

When alternating current flows in the field winding, an alternating flux is produced in the field core. A portion of this flux with the shading coil, which behaves as a short-circuited secondary of a transformer. A voltage is induced in the shading coil, and this voltage circulates a current in it. The induced current produces a flux, called *induced flux* which opposes the main core flux. The shading coil, thus, causes the flux in the shaded portion a to lag behind the flux in the unshaded portion b of the pole. At the same time, the main flux and the shaded pole flux are displaced in space. This space displacement is less than 90°. Since there is time and space displacement between the two fluxes, the conditions for setting up a rotating magnetic field are produced. Under the action of the rotating flux a starting torque is developed on the cage rotor. The direction of this rotating field (flux) is form the unshaded to the shaded portion of the pole. In this motor, the reversal of direction of rotation is not possible.

5.14.1 Applications

(*i*) Shaded pole motors are very cheap.

(*ii*) The starting torque developed by a shaded pole motor is very low.

(*iii*) The losses are high and the power factor is low.

(*iv*) The efficiency is also very low.

(*v*) For this reason, this motors are built only in small sizes of the power rating in the order of 40 W (below 0.05 HP) or less.

(*vi*) They are used to drive devices which require low starting torque.

(*vii*) The most common applications are table fans, exhaust fans, hair drivers, fans for refrigeration and air conditioning equipments, electronic equipment, cooling fans etc.

(*ix*) They are also used in record players, tape recorders, slide projection photo-copying machines, in starting electric clockes and other synchronous (single phase) timing motors.

5.15 COMPARISON BETWEEN 1φ AND 3φ INDUCTION MOTORS

1. Single phase motors develop about 50% of the output of that of three phase motors for the same size and temperature rise.

2. Single phase motors have low power factor.

3. The starting torque is low in 1φ motor.

4. Single phase motors have lower efficiency.

5. Single phase motors are costlier than 3φ motors of the same rating. However 1φ induction motors are simple robust, reliable and less expensive for small ratings.

   They are generally available upto 1 kW rating.

5.16 AC SERIES MOTOR OR UNIVERSAL MOTOR

A dc series motor will rotate in the same direction regardless of the polarity of the supply.

In series motor, both armature field windings are in series. Since the armature current and flux reverse simultaneously, the torque always acts in the same same direction regardless of the polarity of the supply.

One can expect that a DC motor would also operate on a single phase supply. It is then called an AC series motor. However some changes must be made in a DC motor that it is to operate satisfactorily on AC supply.

The changes effected are:

(*i*) The entire magnetic circuit is laminated in order to reduce eddy current loss. Hence an ac series motor requires a more expensive construction than dc motor.

(*ii*) A high field is obtained by using a low reluctance magnetic circuit.

(*iii*) There is considerable sparking between the brushes and the commutor when the motor is used on AC supply.

It is because the alternating flux establishes high current in the coil short circuited by the brushes.

When the short circuited coils break contact from the commutator, excessive sparking is produced.

This can be eliminated by using high resistance leads to connect the coils to the commutator segments.

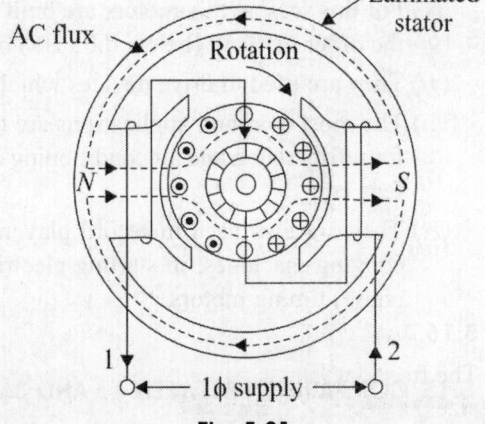

Fig. 5.21

(*iv*) In order to reduce the effect of armature reaction, thereby improving commutation and reducing armature reactance, a compensating winding is used. This winding is put in the stator slots in shown in Fig. 5.21. The axis of the compensating winding is 90° (electrical) with the main field axis. It may be connected in series with both the armature and field as in Fig 5.21. In such a case the motor is conductivity compensated. The compensating winding may be short circuited on itself, in which case the motor is said to be inductively compensated [Figs 5.22(*a*) and (*b*)].

5.16.1 Construction

Construction of an AC series motor is very similar to a DC series motor except the above modification are incorporated. Such a motor can be operated either on AC or DC supply and the resulting torque-speed curve is about the same in each case. For this reason it is sometimes called a *universal motor*.

5.16.2 Operation

When the motor is connected to an AC supply, the same alternator current flows through the field and armature windings. The field windings produces an alternating flux ϕ that reacts with the current flowing in the armature to produce a torque. Since both

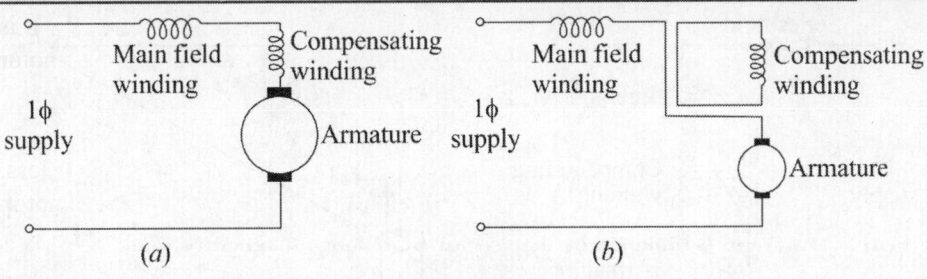

Fig. 5.22

armature current and flux reverse simultaneously, the torque always acts in the same direction. It may be noted that no rotating flux is produced in this type of machine.

5.16.3 Characteristics

(*i*) The speed increases to a high value with a decrease in load. In very small series motors, the losses are usually large enough at no load that limit the speed to a definite value (1500–15000 rpm).

(*ii*) The motor torque is high for large armature currents, thus giving a high starting torque.

(*iii*) At full-load, the power factor is about 90%, However, at starting or when carrying an overload, the power factor is lower.

5.16.4 Applications

The fractional horsepower AC series motors have high speed (corresponding to small sizes) and large starting torque.

They can therefore, be used to drive:

(*a*) High-speed vacuum cleaners

(*b*) Sewing shavers

(*c*) Electric shavers

(*d*) Drills

(*e*) Machine tools, etc.

5.16.5 Phasor Diagram

The circuit diagram and phasor diagram for the conductively coupled single-phase AC series motor are shown in Fig. 5.23(*a*) and (*b*) respectively.

$$Y_P = E_g + I_a Z_{se} + I_a Z_1 + I_a Z_C + I_a Z_a$$
$$E_g = \text{The generated armature counter emf}$$

5.17 SINGLE PHASE REPULSION MOTOR

A repulsion motor is similar to an AC series motor except:

(*i*) brushes are not connected to supply but are short-circuited [see Fig. 5.24]. Consequently, currents are induced in the armature conductors by transformer action.

(*ii*) The field structure has non-salient pole construction.

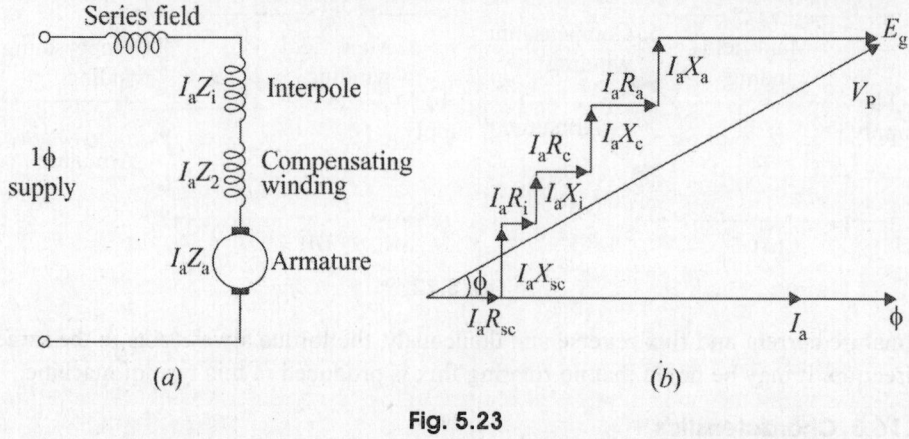

Fig. 5.23

By adjusting the position of short-circuited brushes on the commutator, the starting torque can be developed in the motor.

5.17.1 Construction

The field of stator winding is wound like the main winding of a split-phase motor and is connected directly to a single-phase source. The armature or rotor is similar to a DC motor armature with drum type winding connected to a commutator (not shown in the Fig. 5.24). However, the brushes are not connected to supply but are connected to each other or short-circuited. Short-circuiting the brushes effectively makes the rotor into a type of squirrel cage. The major difficulty with an ordinary single-phase induction motor is the low starting torque. By using a commutator motor with brushes short-circuited, it is possible to vary the starting torque by changing the brush axis. It has also better power factor than the conventional single-phase motor.

5.17.2 Principle of Operation

The principle of operation is illustrated in Fig. 5.24 which shows a two-pole repulsion motor with its two short-circuited brushes. The two drawings of Fig. 5.24 represent a time at which the field current is increasing in the direction shown so that the left-hand pole is N-pole and the right-hand pole is S-pole at the instant shown.

(i) In Fig. 5.24(a), the brush axis is parallel to the stator field. When the stator winding is energised from single-phase supply, emf is induced in the armature conductors (rotor) by induction. By Lenz's law, the direction of the emf is such that the magnetic effect of the resulting armature currents will oppose the increase in flux. The direction of current in armature conductors will be as shown in Fig. 5.24(a). With the brush axis in the position shown in Fig. 5.24(a), current will flow trom brush B to brush A where it enters the armature and flows back to brush B through the two paths ACB and ADB. With brushes set in this position, half of the armature conductors under the N-pole carry current

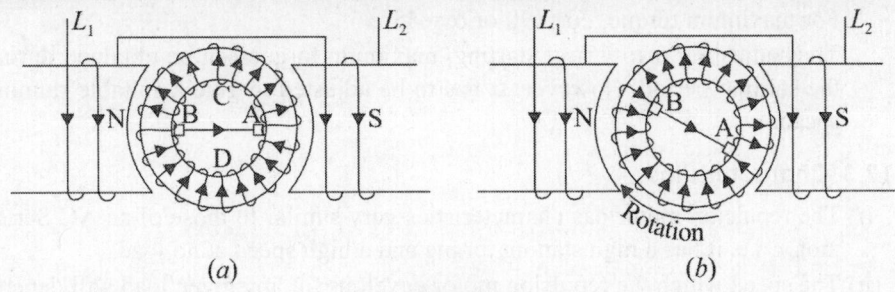

(a)   (b)

Fig. 5.24

inward and half carry current outward. The same is true under S-pole. Therefore, as much torque is developed in one direction as in the other and the armature remains stationary. The armature will also remain stationary if the brush axis is perpendicular to the stator field axis. It is because even then net torque is zero.

(ii) If the brush axis is at some angle other than 0° or 90° to the axis of the stator field, a net torque is developed on the rotor and the rotor accelerates to its final speed. Figure 5.24(b) represents the motor at the same instant as that in Fig. 5.24(a), but the brushes have been shifted clockwise through some angle from the stator field axis. Now emf is still induced in the direction indicated in Fig. 5.24(a), and current flows through the two paths of the armature winding from brush A to brush B. However, because of the new brush positions, the greater pan of the conductors under the N-pole carry current in one direction while the greater pan of conductors under S-pole carry current in the opposite direction. With brushes in the position shown in Fig. 5.24(b), torque is developed in the clockwise direction and the rotor quickly attains the speed.

(iii) The direction of rotation of the rotor depends upon the direction in which the brushes are shifted. If the brushes are shifted in clockwise direction from the stator field axis, the net torque acts in the clockwise direction and the rotor accelerates in the clockwise direction. If the brushes are shifted in anticlockwise direction as in Fig. 5.24(c), the armature current under the pole faces is

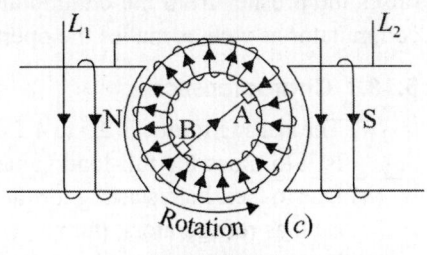

Fig. 5.24(C)

reversed and the net torque is developed in the anticlockwise direction. Thus a repulsion motor may be made to rotate in either direction depending upon the direction in which the brushes are shifted.

(iv) The total armature torque in a repulsion motor can be shown to be

$$T_a \propto \sin 2\alpha$$

where α = angle between brush axis and stator field axis

For maximum torque, $2\alpha = 90$ or $\alpha = 45°$

Thus adjusting α to 45° at starting, maximum torque can be obtained during the starting period. However, α has to be adjusted to give a suitable running speed.

5.17.3 Characteristics

(*i*) The repulsion motor has characteristics very similar to those of an AC Series motor, i.e. it has a high staning torque and a high speed at no load.

(*ii*) The speed which the repulsion motor develops for any given load will depend upon the position of the brushes.

(*iii*) In comparison with other single-phase motors, the repulsion motor has a high starting torque and relatively low starting current.

5.18 REPULSION-START INDUCTION-RUN MOTOR

Sometimes the action of a repulsion motor is combined with that of a single-phase induction motor to produce repulsion-start induction-run motor (also called repulsion-start motor). The machine is started as a repulsion motor with a corresponding high starting torque. At some predetermined speed, a centrifugal device short-circuits the commutator so that the machine then operates as a single-phase induction motor.

The repulsion-start induction-run motor has the same general construction of a repulsion motor. The only difference is that in addition to the basic repulsion-motor construction, it is equipped with a centrifugal device titted on the armature shaft. When the motorreaches 75% of its full running speed, the centrifugal device forces a short-circuiting ring to come in contact with the inner surface of the commutator. This short-circuits all the commutator bars. The rotor then resembles squirrel-cage type and the motor runs as a single-phase induction motor. At the same time, the centrifugal device raises the brushes trom the commutator which reduces the wear of the brushes and commutator as well as makes the operation quiet.

5.18.1 Characteristics

(*i*) The starting torque is 2.5 to 4.5 times the full-load torque and the starting current is 3.75 times the full-load value.

(*ii*) Due to their high starting torque, repulsion-motors were used to operate devices such as refrigerators, pumps, compressors etc.

However, they possed a serious problem of maintainance of brushes, commutator and the centrifugal device. Consequently, manufacturers have stopped making them in view of the development of capacitor motors which are small in size, reliable and low-priced.

5.19 REPULSION INDUCTION MOTOR

The repulsion induction motor produces a high starting torque entirely due to repulsion motor action. When running, it functions through a combination of induction-motor and repulsion motor action.

5.19.1 Construction

Figure 5.25 shows the connections of a 4-pole repulsion-induction motor for 230 V operation. It consists of a stator and a rotor (or armature).

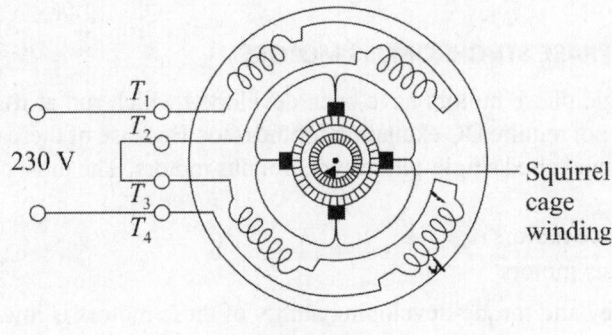

Fig. 5.25

(*i*) The stator carries a single distributed winding fed from single-phase supply.

(*ii*) The rotor is provided with two independent windings placed one inside the other. The inner winding is a squirrel-cage winding with rotor bars permanently short-circuited. Placed over the squirrel cage winding is a repulsion commutator armature winding. The repulsion winding is connected to a commutator on which ride short-circuited brushes. There is no centrifugal device and the repulsion winding functions at all times.

5.19.2 Operation

(*i*) When single-phase supply is given to the stator winding, the repulsion winding (i.e. outer winding) is active. Consequently, the motor starts as a repulsion motor with a corresponding high starting torque.

(*ii*) As the motor speed increases, the current shifts fiom the outer to inner winding due to the decreasing impedance of the inner winding with increasing speed. Consequently, at running speed, the squirrel cage winding carries the greater part of rotor current This shifting of repulsion–motor action to induction–motor action is thus achieved without any switching arrangement.

(*iii*) It may be seen that the motor starts as a repulsion motor, When running, it functions through a combination of principle of induction and repulsion; the former being predominant.

5.19.3 Characteristics

(*i*) The no-load speed of a repulsion-induction motor is somewhat above the synchronous speed because of the effect of repulsion winding. However, the speed at full-load is slightly less than the synchronous speed as in an induction motor.

(*ii*) The speed regulation of the motor is about 6%.

(*iii*) The starting torque is 2.25 to 3 times the full-load torque; the lower value being for large motors. The starting current is 3 to 4 times the full-load current.

This type of motor is used for applications requiring a high starting torque with essentially a constant running speed. The common sizes are 0.25 to 5 hp.

5.20 SINGLE PHASE SYNCHRONOUS MOTORS

Very small single phase motors have been developed which run at true synchronous speed. They do not require DC excitation for the rotor. Because of these characteristics, they are called unexcited single-phase synchronous motors. The most commonly used types are:

(*i*) Reluctance motors

(*ii*) Hysteresis motors

The efficiency and torque-developing ability of these motors is low. The output of most of the commercial motors is only a few watts.

5.21 RELUCTANCE MOTOR

It is a single phase synchronous motor which does not require DC excitation to the rotor. Its operation is based upon the following principle:

Whenever a piece of ferromagnetic material is located in a magnetic field, a force is exerted on the material, tending to align the material so that reluctance of the magnetic path that passes through the material is minimum.

5.21.1 Construction

A reluctance motor (also called synchronous reluctance motor) consists of: (*i*) a stator carrying a single-phase winding along with an auxiliary winding to produce a synchronous–revolving magnetic field.

(*i*) A squirrel-cage rotor having unsymmetrical magnetic construction. This is achieved by symmetrically removing some of the teeth from the squirrel-cage rotor to produce salient poles on the rolor. As shown in Fig. 5.26(*a*), 4 salient

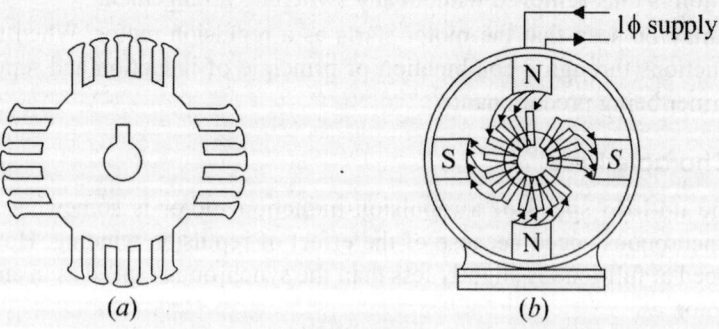

(*a*)   (*b*)

Fig. 5.26

poles have been produced on the rotor. The salient poles created on the rotor must be equal to the poles on the stator. Note that rotor salient poles offer low reluctance to the stator flux and, therefore, become strongly magnetised.

5.21.2 Operation

(*i*) When single-phase stator having an auxillary winding is energised, a synchronously-revolving field is produced. The motor starts as a standard squirrel-cage induction motor and will accelerate to near its synchronous speed.

(*ii*) As the rotor approaches synchronous speed, the rotating stator flux will exert reluctance torque on the rotor poles tending to align the salient-pole axis with the axis is of the rotating field. The rotor assumes a position where its salient poles lock with the poles of the revolving field [see Fig. 5.26(*b*)]. Consequently, the motor will continue to run at the speed of revolving flux, i.e. at the synchronous speed.

(*iii*) When we apply a mechanical load, the rotor poles fall slightly behind the stator poles, while continuing to turn at synchronous speed. As the load on the motor is increased, the mechanical angle between the poles increases progressively. Nevertheless, magnetic attraction keeps the rotor locked to the rotating flux. If the load is increased beyond the amount under which the reluctance torque can maintain synchronous speed, the rotor drops out of step with the revolving field. The speed, then, drops to some value at which the slip is sufficient to develop the necessary torque to drive the load by induction–motor action.

5.21.3 Characteristics

(*i*) These motors have poor torque, power factor and efficiency.

(*ii*) These motors cannot accelerate high-inertia loads to synchronous speed.

(*iii*) The pull-in and pull-out torques of such motors are weak.

Despite the above drawbacks, the reluctance motor is cheaper than any other type of synchronous motor. They are widely used for constant-speed applications such as timing devices, signalling devices etc.

5.22 HYSTERESIS MOTOR

It is a single-phase motor whose operation depends upon the hysteresis effect, i.e. magnetisation produced in a ferromagnetic material lags behind the magnetising force.

5.22.1 Construction

It consists of:

(*i*) a stator designed to produce a synchronously-revolving field from a single-phase supply. This is accomplished by using permanent-split capacitor type construction. Consequently, both the windings (i.e. starting as main winding) remain connected in the circuit during running operation as well as at starting.

The value of capacitance is so adjusted as to result in a flux revolving at synchronous speed.

(*ii*) a rotor consisting of a smooth cylinder of magnetically hard steel, without winding or teeth.

5.22.2 Operation

(*i*) When the stator is energised from a single-phase supply, a synchronously–revolving field (assumed in anticlockwise direction) is produced due to split–phase operation.

(*ii*) The revolving stator flux magnetises the rotor. Due to hysteresis effect, the axis of magnetisation of rotor will lag behind the axis of stator field by hysteresis lag angle a as shown in Fig. 5.27. Thus the rotor and stator poles are locked. If the rotor is stationary, the starting torque produced is given by:

$$T_S \propto \phi_S \phi_r \sin \alpha$$

where ϕ_S = stator flux

ϕ_r = rotor flux

From now onwards, the rotor accelerates to synchronous speed with a uniform torque.

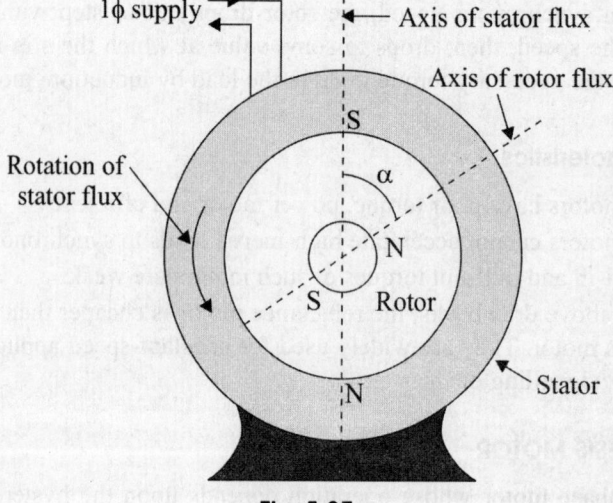

1φ supply

Axis of stator flux

Axis of rotor flux

Rotation of stator flux

S

α

N

S   Rotor

N

Stator

Fig. 5.27

(*iii*) After reaching synchronism, the motor continues to run at synchronous speed and adjusts its torque angle so as to develop the torque required by the load.

5.22.3 Characteristics

(*i*) A hysteresis motor can synchronise any load which it can accelerate, no matter how great the inertia. It is because the torque is uniform from standstill to synchronous speed.

(*ii*) Since the rotor has no teeth or salient poles or winding, a hysteresis motor is inherently quiet and produces smooth rotation of the load.

(*iii*) The rotor takes on the same number of poles as the stator field. Thus by changing the number of stator poles through pole-changing connections, we can get a set of synchronous speeds for the motor.

5.22.4 Applications

Due to their quiet operation and ability to drive high-inertia loads, hysteresis motors are particularly well suited for driving (*i*) electric clocks (*ii*) timing devices (*iii*) tape-decks (*iv*) turn-tables and other precision auto-equipment.

SOLVED EXAMPLES

Example 5.1: A resistance split-phase induction motor is rated at 187 W, 1725 rpm 115 V, 50 Hz. When the rotor is locked, a test at reduced voltage on the main and starting windings yields the following results:

	Main winding	Starting winding
Applied voltage	23 volts	23 volts
Current	4 A	1.5 A
Active power (*P*)	60 W	30 W

Calculate: (*i*) the phase angle between I_m and I_s (*ii*) the locked-rotor current drawn from the line at 115 V.

Solution: (*i*) We shall first calculate the phaee angle between I_m and the applied voltage *V* of the main winding.

$$\text{Apparent power, } S = VI_m = 23 \times 4 = 92 \text{ VA}$$

∴ Power factor, $\cos \phi_m = P/S = 60/92 = 0.65$ ∴ $\phi_m = 49.2°$

Therefore, ϕ_m lags 49.2° behind *V*.

We now calculate the phase angle-between I_s and *V* of the starting winding.

$$\text{Apparent power, } S = VI_s = 23 \times 1.5 = 34.5 \text{ VA}$$

∴ Power factor, $\cos \phi_s = P/S = 30/34.5 = 0.87$ ∴ $\phi_s = 29.6°$

Therefore, I_s lags *V* by 29.6°.

The phase angle between I_m and I_s, $\alpha = \phi_m - \phi_s = 49.2° - 29.6° = 19.6°$

(*ii*)   Total active power, $P = P_m + P_s = 60 + 30 = 90$ W

Total reactive power, $Q = Q_m + Q_s$

$$= \sqrt{90^2 - 60^2} + \sqrt{34.5^2 - 30^2} = 86.8 \text{ VAR}$$

Total apparent power, $S = \sqrt{P^2 + Q^2} = \sqrt{90^2 + 86.6^2} = 125 \text{ VA}$

Line current, $I = S/V = 125/23 = 5.44$ A (at 23 V)

∴ Current drawn at 115 V, $I_L = 5.44 \times (115/23) = 27.2$ **Ans.**

Example 5.2: A 0.5 h.p, 230 V, single-phase induction motor (split-phase) takes a current of 4.2 A lagging the voltage by 10° for the starting winding and a current of 6.2 A lagging the voltage by 40° for its main winding. Find (*i*) total current and power factor at the time of starting (*ii*) total current and power factor during running (*iii*) phase angle between the main winding current and starting winding current (*iv*) power drawn by starting winding (*v*) power drawn by main winding (*vi*) total power drawn during starting (*vii*) total power drawn during running.

Solution: (*i*) Total current drawn by the motor is

$$I = I_m + I_s = 6.2\angle - 40° + 4.2\angle -10°$$
$$= (4.75 - j\,3.986) + (4.136 - j\,30.73)$$
$$= (8.886 - j\,4.716)\ A = 10.06\angle - 27.96°A$$

∴ $I = 10.6$ A; Power factor = cos (−27.06°) = 0.883 log

(*ii*) During running, the starting winding is not connected. Therefore, the total current drawn is equal to that drawn by the main winding.

∴ $I = I_m = 6.2$ A; power factor = cos (−40°) = 0.766 log

(*iii*) Phase angle between I_m and $I_s = 40° - 10° = 30°$.

(*iv*) Power drawn by the starting winding is

$$P_s = VI_s \cos 10° = 230 \times 4.2 \times \cos 10° = 951.32\ W$$

(*v*) Power drawn by the main winding is

$$P_m = VI_m \cos 40° = 230 \times 6.2 \times \cos 40° = 1092.38\ W$$

(*vi*) Total power drawn during the starting of motor is

$$P = P_s + P_m = 951.32 + 1092.38 = 2043.7\ W$$

(*vii*) During running, only the main winding is in the motor circuit.

∴ Power drawn by the motor during running is

$$P = P_m = 1092.38\ W$$

Example 5.3: A single-phase split-phase motor is rated at 220 V, 175 W, 180 rad/s, 0.5 pf and 45% efficiency at rated load. Calculate its rated load (*i*) line current (*ii*) torque.

Solution: (*i*) Input power, $P_i = \dfrac{\text{Output power}}{\eta} = \dfrac{175}{0.45} = 388.89\,W$

Line current, $I_L = \dfrac{P_i}{V\cos\phi} = \dfrac{388.89}{220\times 0.5} = 3.535\ A$

(*ii*) Output power = $T\omega$

$$175 = T \times 180$$

∴ $T = 0.972$ Nm

Example 5.4: A 4-pole, 250 W, 115 V, 60 Hz capacitor-start induction motor takes a full-load line current of 5.3 A while running at 1760 rpm. If the full-load efficiency of the motor is 64%, find (*i*) motor slip (*ii*) power factor and (*iii*) full-load torque.

Solution: (*i*) $\qquad N_S = 120 \, f/P = 120 \times 60/4 = 1800$ rpm

$$s = \frac{N_S - N}{N_S} = \frac{1800 - 1760}{1800} = 0.022$$

(*ii*) $\qquad$ Input power $= 250/0.64 = 390.6$ W

$$\text{Power factor} = \frac{390.6}{115 \times 5.3} = 0.64 \text{ log}$$

(*ii*) $\qquad$ Full load torque, $T = 9.55 \dfrac{P_{\text{out}}}{N} = 9.55 \dfrac{250}{1760} = 1.35$ N-m

Example 5.5: A 0.25 HP, 110 V, single-phase inductor motor (split-phase) takes a current of 4 A lagging the supply voltage by 15° for the starting winding and a current of 5 A lagging the voltage by 40° for its main winding. Calculate (*i*) the total starting current and the power factor (*ii*) the phase angle between the main winding current and starting winding current.

Solution: (*i*) Starting winding current, $I_s = 4 \angle -15°\text{A} = (3.86 - j1.035)\,\text{A}$

$\qquad$ Main winding current, $I_m = 6 \angle -40°\text{A} = (4.60 - j3.86)\,\text{A}$

$\qquad$ Total starting current, $I = I_s + I_m = (3.86 - j1.035) + (4.60 - j3.86)$

$$= (8.46 - j4.895)\,\text{A} = 9.77 \angle -30°\text{A}$$

∴ $\qquad$ Total starting current $= 9.77$ A; Power factor $= \cos(-30°) = 0.866$ lagging

$\qquad$ (*ii*) Phase angle between I_m and I_s, $\alpha = 40° - 15° = 25°$

Example 5.6: A 2-pole, 240 V, 50 Hz, single-phase inductor motor has the following constants referred to the stator.

$$R_1 = 2.2 \, \Omega, \; X_1 = 3.0 \, \Omega; \; R_2' = 3.8\Omega; \; X_2' = 2.1 \, \Omega; \; X_m = 86\Omega$$

Find the stator current and the input power when the motor is operating at a full-load speed of 2820 rpm.

Solution: Figure 5.28 shows the equivalent circuit of single-phase induction motor with circuit values.

$\qquad$ Synchronous speed, $N_s = 120 \, f/P = 120 \times 50/2 = 3000$ rpm

$$\text{Slip, } s = \frac{N_s - N}{N_s} = \frac{3000 - 2820}{3000} = 0.06$$

∴ $\dfrac{3.8}{2s} = \dfrac{3.8}{2 \times 0.06} = 31.67\Omega; \; \dfrac{3.8}{2(2 - s)} = \dfrac{3.8}{2(2 - 0.06)} = 0.979\Omega; \; \dfrac{X_m}{2} = \dfrac{86}{2} = 43\Omega$

∴ $Z_f = \dfrac{j43(31.67 + j2.1/2)}{31.67 + j(43 + 2.1/2)} = 25.22 \angle 37.61°\Omega = (19.98 + j15.39)\Omega$

∴ $Z_b = \dfrac{j43(0.979 + j2.1/2)}{0.979 + j(43 + 2.1/2)} = 1.40 \angle 48.3°\Omega = (0.933 + j1.05)\Omega$

Total impedance, $Z_T = Z_1 + Z_f + Z_b$

$$= (2.2 + j3.0) + (19.98 + j15.39) + (0.933 + j1.05)$$

$\therefore$ Stator current, $I_1 = \dfrac{V_1}{Z_T} = \dfrac{240\angle0°}{30.2\angle40.1°} = 7.95\angle-40.1°\,A$

Magnitude of stator current, $I_1 = 7.95$ A at 40.1° lagging

Input power $= V_1 I_1 \cos\phi$

$$= 240 \times 7.95 \times \cos 40.1°$$

$$= 1459.5 \text{ W}$$

EXERCISE

1. (a) Discuss why single phase induction motors do not have a starting torque.

   (b) Describe with the help of diagrams of connections and phasor diagrams two methods of producing starting torque in a single phase induction motor.

2. Draw the construction of a capacitor-start capacitor-run single phase induction and explain and working. Where this type of motor is commonly used?

3. Describe the construction and working of a capacitor-start single-phase induction motor.

4. Describe the construction and working of a shaded-pole motor.

5. Explain the working principle of (a) split phase, (b) capacitor-start single-phase induction motor with the help of neat sketches. How can you reverse the direction of rotation of such motor? What are the industrial and domestic applications of such motors?

6. Discuss the modifications necessary to operate a DC series motor satisfactorily on single phase AC supply.

7. Write short notes on the following:

   (a) Starting of single phase induction motors

   (b) Capacitor motors

   (c) Shaded-pole motor

   (d) Universal motor

8. What type of motor would you use in the following applications: washing machine, sewing machine, dishwasher, portable electric drill, food mixer? State your reasons.

9. Explain simply why a universal motor can operate from DC as well as AC supplies.

10. Give details of four methods of starting small single-phase induction motors, and mention typical applications for which these would be suitable.

11. Using double-revolving field theory, explain why a single-phase induction motor is not self starting.

12. Explain the double-revolving field theory for single-phase induction motors.

13. Draw a torque-speed curve of a single-phase induction motor on the basis of double-revolving-field theory.

14. Draw and explain the equivalent circuit of a single-phase induction motor. How can the porformance of the motor be analysed?

15. Discuss the procedure for determining the parameters of equivalent circuit of a single-phase induction motor.

16. What are the disadvantages of a single-phase induction motor when compared with a 3-phase induction motor?

17. Draw and explain the phasor diagram of an AC series motor.

6

Alternator

6.1 INTRODUCTION

AC generators are usually called alternators. Rotating machines that rotate a speed fixed by the supply frequency and the number of poles are called *synchronous machines*. Alternators is a machine for converting mechanical power from a prime mover to AC electric power at a specific voltage and frequency. Synchronous generators are usually of 3-phase type because of the several advantages of 3-phase generation, transmission and distribution. Large synchronous generators are used to generate bulk power at thermal, hydro- and nuclear-power stations. Synchronous generators with power ratings of several hundred MVA are very commonly used in generating stations. The biggest size used in India has a rating 500 MVA used in super-power thermal stations. For bulk generation, stator windings of synchronous generator are designed for voltages ranging from 6.6 kV to 33 kV.

6.2 BASIC PRINCIPLE

Alternators operate on the same fundamental principles of electromagnetic inductions as DC generators. They also consists of an armature winding and a magnetic field. But there is one important difference between the two whereas in DC generators, the armature rotates and the field system is stationary, the arrangement of alternators is just the reverse of it. In their case, standard construction consists of armature winding mounted or a stationary element called stator and field winding or a rotating element called rotor. In other words a synchronous machine works as a generator when its rotor carying the field system is rotated by a prime mover. According to faraday's law of electromagnetic induction, when there is cutting of magnetic flux by a conductor or when their is a change of flux linkage by a coil, emf is induced in the conductor or the coil.

The field winding is energised from the DC exciter and alternate N and S poles are developed on the rotor. When the rotor is rotated in anticlockwise direction by a prime mover, the stator or armature conductors are cut by the magnetic flux of rotor poles.

Consequently, emf is induced in the armature conductors due to electromagnetic induction.Because the magnetic poles are alternately N and S, they induce an emf and hence current in armature conductors, which first flows in one direction and their in the other. Hence an alternating emf is produced in the stator conductors (*i*) whose frequency depends on the number of N and S poles moving past a conductor in one sec. and (*ii*) whose direction is given by fleming's right hand rule.

6.3 ADVANTAGES OF ROTATING FIELD ALTERNATOR

Most alternators have the rotating field and stationary armature. The rotating-field type alternator has several advantages over the rotating–armature type alternator:

(*i*) A stationary armature is more easily insulted for the high voltage for which the alternator is designed. This generated voltage may be as high as 33 kV.

(*ii*) The armature windings can be braced better mechanically against high electromagnetic forces due to large short circuit currents when the armature windings are in the stator.

(*iii*) The armature windings, being stationary are not subjected to vibration and centrifugal forces.

(*iv*) The output current can be taken directly from fixed terminals on the stationary armature without using sliprings, brushes, etc.

(*v*) The rotating field is supplied with direct current. Usually the field voltage is between 100 to 500 volts. Only two slip rings are required to provide direct current for the rotating field, while at least three sliprings would be required for a rotating armature. The insulation of the two relatively low voltage sliprings form the shaft can be provided easily.

(*vi*) The bulk and weight of the armature windings are substationally greater than the windings of the field poles. The size of the machine is therefore reduced.

(*vii*) Rotating field is comparatively light and can be constructed for high speed rotation. The armatures of large alternators are forced cooled with circulating gas or liquids.

(*viii*) The stationary armature may be cooled more easily because the armature can be made large to provide a number of cooling ducts.

6.4 SPEED AND FREQUENCY

The frequency of the generated voltage depends upon the number of field poles and on the speed at which the field poles are rotated. One complete cycle of voltage is generated in an armature coil when a pair of field poles (one north and one south pole) passes over the coil.

Let P = Total no. field poles

p = Pair of field poles

N = Speed of the field poles in rpm

n = Speed of the field poles in rps

f = Frequency of the generated voltage in Hz

Obviously $$\frac{N}{60} = n \qquad \qquad ...(6.1)$$

and $$\frac{P}{2} = p \qquad \qquad ...(6.2)$$

In one revoltion of the rotor, an armature coil is cut by $P/2$ north poles and $P/2$ south poles. Since one cycle is generated in an armature coil when a pair of field poles passes over the coil, the number of cycles generated in one revolution of the rotor will be equal to the number of pairs of poles. That is

number of cycles per revolution = p

Also revolutions per second = n

Now   frequency = Number of cycles per second

$$= \frac{\text{Number of cycles}}{\text{revolutions}} \times \frac{\text{revolutions}}{\text{seconds}}$$

$$f = p \times n$$

Since $n = N/60$ and $p = P/2$

$\therefore$

$$f = \frac{PN}{120}$$

6.5 CONSTRUCTION OF ALTERNATORS

An alternator has 3-phase winding on the stator and DC field winding on the rotor.

1. **Stator:** It is the stationary part of the machine and is built up of sheet–steel laminations having slots on its inner periphery. A 3-phase winding is placed in these slots and serves as the armature winding of the alternator. The armature winding is always connected in star and the neutral is connected to ground.

2. **Rotor:** The rotor carries a field winding which is supplied with direct current through slip rings by a seperate DC source. This DC source (called exciter) is generally a small DC shunt or compound generator mounted on the shaft of the alternator. Rotor construction is of two types, namely, (*i*) salient (or projecting) pole type (*ii*) non-salient (or cyclinderical) pole type.

(*i*) **Salient pole type:** In this type, salient or projecting poles are mounted on a large circular steel frame which is fixed to the shaft of the alternator as shown in Fig. 6.1(*a*). This individual field pole windings are connected in series in such a way that when the field winding is energised by the DC exciter, adjacent poles have opposite polarities.

Low or medium speed alternators (120–400 rpm) such as those driven by diesel engine or water turbines have salient pole type rotors due to the following reasons.

(*a*) The salient field poles would cause an excessive windage loss if driven at high speed and wound tend to produce noise.

(*b*) Salient pole construction cannot be made strong enough to withstand the mechanical stresses to which they may be subjected at higher speeds.

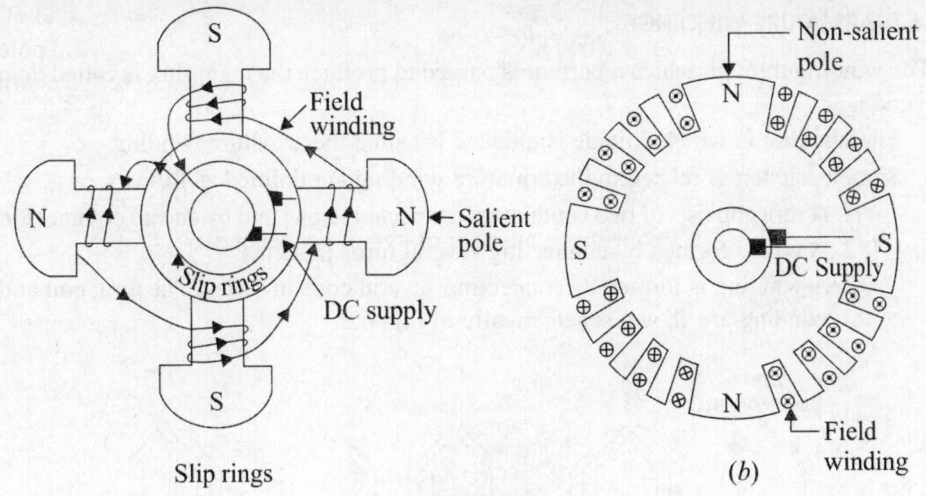

Slip rings   (b)

Fig. 6.1

Since a frequency of 50 Hz is required, we must use a large no. of poles on the rotor of slow–speed alternators.

Low speed rotors always possess a large diameter to provide the necessary space for the the poles. Consequently, salient-pole type rotors have large diameters and short axial lengths.

(*ii*) **Non-salient pole type:** In this type, the rotor is made of smooth solid forged–steel radial cylinder having a number of slots along the outer periphery. The field winds are embedded in these slots and are connected in series to the slip rings through which they are energiced by the DC exiter.

The regions forming the poles are usually left unslotted as shown in Fig. 6.1(*b*). It is clear that the poles formed are non-salient, i.e. they do not project out from the rotor surface.

High speed alternators (1500 or 3000 rpm) are driven by steam turbines and use non-salient type rotor due to the following reasons:

(*a*) This type of construction has mechanical robustness and gives noiseless operations at high speed.

(*b*) The flux distribution around the periphery is nearby a sine wave and hence a better emf waveform is obtained than in the case of salient pole type.

Since steam turbines run at high speed and a frequency of 50 Hz is required, we need a small number of poles on the rotor of high-speed alternators (also called *turbo alternators*). We can use not less than 2 poles and this fixes the highest possible speed. For a frequency of 50 Hz, it is 3000 rpm.

The next lower speed is 1500 rpm for a 4 pole machine. Consequently, turbo alternators possess 2 or 4 poles and have small diameters and very long axial length.

6.6 ARMATURE WINDINGS

The winding through which a current is passed to produce the main flux is called *field winding*.

The winding in which voltage is induced is called the armature winding.

Some basic terms related to the armature winding are defined as follows:

1. A turn consists of two conductors connected to one end by an end connnector.
2. A coil is formed by connecting several turns in series.
3. A winding is formed by connecting several coils in series. The turn, coil and winding are shown schematically in Fig. 6.2.

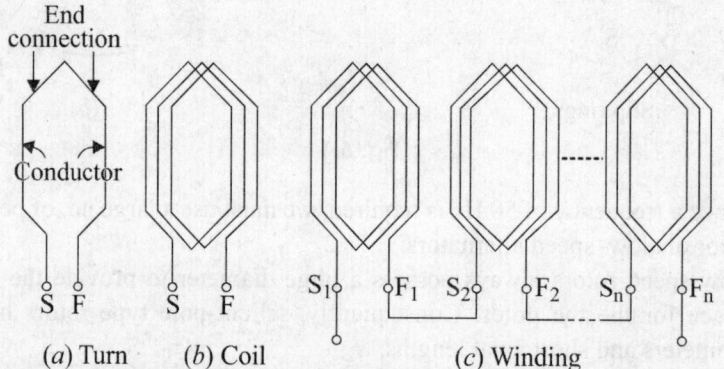

(*a*) Turn   (*b*) Coil   (*c*) Winding

Fig. 6.2

The begining of the turn, or coil, is defined by the symbol S (start) and the end of the turn or coil by the symbol F (finish).

The conept of electrical degrees is very useful in the study of machine.

$$\theta_{md} = \text{Mechanical degrees or angular measure in space}$$

$$\theta_{ed} = \text{Electrical degrees or angular measure in cycles.}$$

For *P*-pole machine, electrical degree is defined as follows:

$$\theta_{ed} = \frac{P}{2} \cdot \theta_{md}$$

The advantage of this rotation is that expressions written in terms of electrical angles apply to machines having any number of poles. The angular distance between the centres of two adjacent poles or a machine is known as pole pitch or pole span.

$$\text{One pole pitch} = 180^\circ_{ed} = \frac{360^\circ_{md}}{P}$$

Regardless of the no. of poles in a machine, a pole-pitch is always 180° electrical degrees.The two sides of the coil are placed in two slots on the stators surface.

The distance between the two sides of a coil is called the *coil-pitch*. If the coil pitch is one pole-pitch, it is called the full-pitch coil or a coil having a span equal to 180° electrical is called a full-pitch coil (Fig. 6.3). If the coil pitch is less than one pole pitch, the coil is called short pitch–coil or traditional pitch coil. In other word A coil having a span less than 180° electrical is called a short–pitch coil or fractional-pitch coil. It is also called a chorded coil. A stator winding using fractional pitch coils is called a chorded winding. If the span of the coil is reduced by an angle and electrical degrees, the coil span will be (180 − α) electrical degrees as shown in Fig. 6.3(*b*).

In case of a full-pitch coil, the two coil sides span a distance exactly equal to the pole pitch of 180° electrical. As a result, the voltage generated in a full pitch coil is such that the coil side voltages are in phase as shown in Fig. 6.3(*a*).

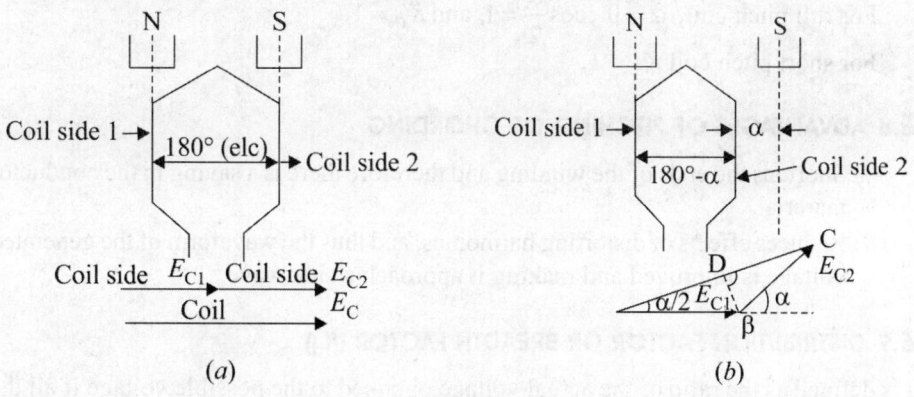

Fig. 6.3

Let E_{C1} and E_{C2} be the voltages generated in the coil sides and B_C the resultant coil voltage.

Then
$$E_C = E_{C1} + EC_2$$
$$|E_{C1}| = |E_{C2}| = E_1 \text{(say)}$$

Since E_{C1} and E_{C2} are in phase, the resultant coil voltage E_C is equal their arithmetic sum.

∴
$$E_C = E_{C1} + E_{C2} = 2E_1$$

If the coil span of a single coil is less than the pole pitch of 180° (electrical), the voltages generated in each coil side are not in phase. The resultant coil voltage E_C is equal to the phasor sum of E_{C1} and E_{C2}.

If the coil span is reduced by an angle α electrical degrees, the coil span is (180 − α) electrical degrees. The voltages generated B_{C1} and E_{C2} in the two coil sides will be out of phase with respect to each other by an angle a electrical degrees as shown in Fig. 6.4. The phasor sum of E_{C1} and E_{C2} is E_C (= AC).

6.7 PITCH FACTOR/COIL SPAN FACTOR (K_C)

It is defined as the ratio of the voltage generated in the short pitch coil to the voltage generated in the full pitch coil. It is also called chording factor.

$$K_C = \frac{\text{actual voltage generated in the coil}}{\text{voltage generated in the coil of span } 180°\text{electrical}}$$

$$= \frac{\text{Phasor sum of the voltages of two coil sides}}{\text{arithmetic sum of the voltages of two coil sides}}$$

$$= \frac{AC}{2AB} = \frac{2AD}{2AB} = \cos\frac{\alpha}{2}$$

$$\therefore \qquad K_C = \cos\frac{\alpha}{2}$$

For full pitch coil, $\alpha = 0$, $\cos\frac{\alpha}{2} = 1$, and $K_C = 1$

For short pitch coil $K_C < 1$.

6.8 ADVANTAGES OF PITCHING OR CHORDING

(*i*) Shortens the ends of the winding and therefore there is a saving in the conductor material.

(*ii*) Reduces effects of distorting harmonics, and thus the waveform of the generated voltage is improved and making it approach a sine wave.

6.9 DISTRIBUTION FACTOR OR BREADTH FACTOR (K_d)

It is defined as the ratio of the actual voltage obtained to the possible voltage if all the coils of a polar group were concentrated in a single slot.

$$K_d = \frac{\text{Phasor sum of coil voltages per phase}}{\text{Arithmetic sum of coil voltages per phase}}$$

Let m = Slots per pole per phase

B = Angular displacement between adjacent slots in electrical degree.

$$\beta = \frac{180}{\text{slots/pole}} = \frac{180 \times \text{poles}}{\text{slots}}$$

Thus, one phase of the winding consists of coils arranged in m consecutive slots. Voltages E_{C1}, E_{C2}, E_{C3}, E_{C4} are the individual coil voltages.

Each coil voltage E_C will be out of phase with the next coil voltage by the slot pitch B. Fig. 6.4 shows the voltage polygon of the induced voltages in the four coils group ($m = 4$).

The voltages E_{C1}, E_{C2}, E_{C3} and E_{C4} are represented by phasors AB, BC, CD and DF respectively in Fig. 6.4.

Each of the phasors is chord of a circle with centre O and subtends an angle B at 0.

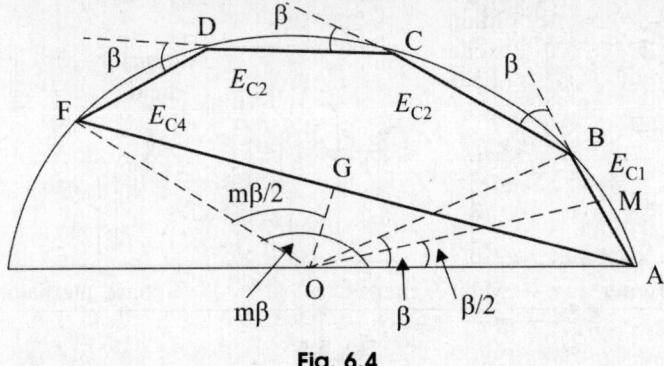

Fig. 6.4

The phasor sum *Af*, representing the resultant winding voltage, subtends an angle $m\beta$ at the centre.

Arithmetic sum of individual coil voltages

$$= mE_C = mAB = m(2\ AM)$$

$$= 2m\ OA\ \sin\ A\hat{O}M = 2m\ OA\ \sin\ \beta/2$$

Phasor sum of individual coil voltages

$$= AF = 2AG = 2OA \sin A\hat{O}G = 2OA \sin \frac{m\beta}{2}$$

$\therefore$ $$k_\alpha = \frac{\text{Phasor sum of coil voltage per phase}}{\text{Arithmetic sum of coil voltage per phase}} = \frac{2OA\sin m\beta/2}{2OA\ m\ \sin\beta/2}$$

$$\boxed{k_d^* = \frac{\sin m\beta/2}{m\sin\beta/2}}$$

6.10 EXCITATION SYSTEMS FOR SYNCHRONOUS MACHINES

Some of the important excitation systems are given below.

    (*i*) **DC exciters:** This is an old conventional method of exciting of the alternators. In the this; three machines, namely pilot exciter, main exciter and the main three phase alternator are mechanically coupled and therefore driven by the same shaft. The pilot exciter is a DC shunt generator feeding the field winding of a main exciter. The main exciter is a separately excited DC generator. The DC output from the main exciter is given to the field winding of the main alternator through brushes on slip rings as shown in Fig. 6.5.

    The conventional method of excitation suffers from cooling and maintenance problem associated with slip rings, brushes and commutators as the alternator ratings rise.

    The trend toward modern excitation system has been to decrease those problem by minimising the number of sliding contacts and brushes. This trend has led to the development of static excitation and brushless-excitation systems.

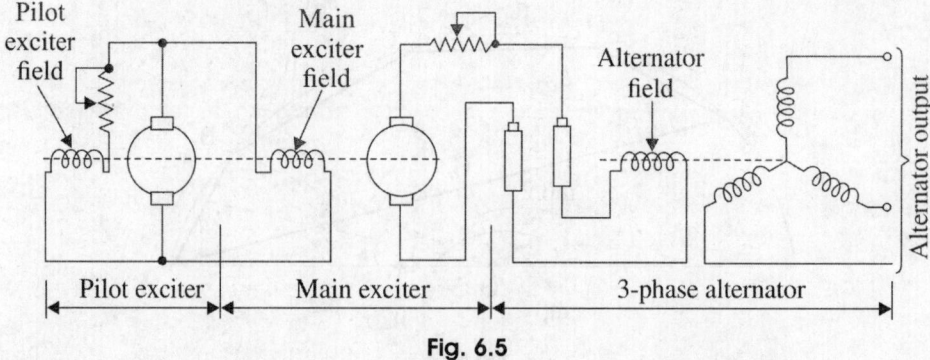

Fig. 6.5

(*ii*) **Static excitation:** In this method, the excitation power for the main alternator field is drawn form output terminals of the main 3-phase alternator. For this purpose, a 3-phase transformer 'TR' steps down the alternator voltage to the desired value. This 3-phase voltage is fed to the 3-phase full-converter bridge using thyristors. The firing angle of these thyristors is controlled by mean of a regulator which picks up the signal from alternator terminals through potential transformer PT and current transformer CT, as shown in Fig. 6.6. The controlled power output from thyristor is delivered to the field winding of main alternator through brushes and ship rings as shown in Fig. 6.6.

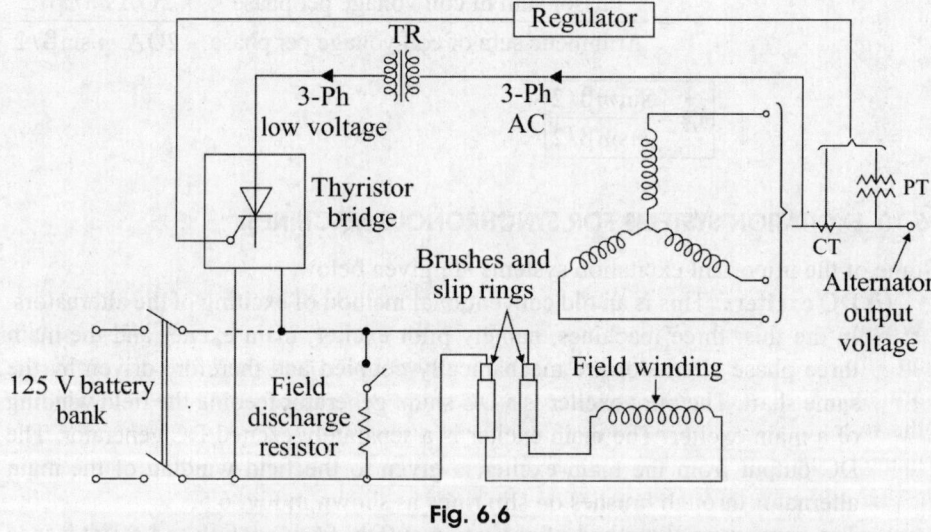

Fig. 6.6

For initiating the process of static excitation, first of all, field winding is switched on the station bank to establish the field current in alternator.

The alternator speed is adjusted to rated speed. After the output voltage from alternator has built up sufficiently, the alternator field winding is disconnected from battery bank and is switched on to the thyristor bridge output.

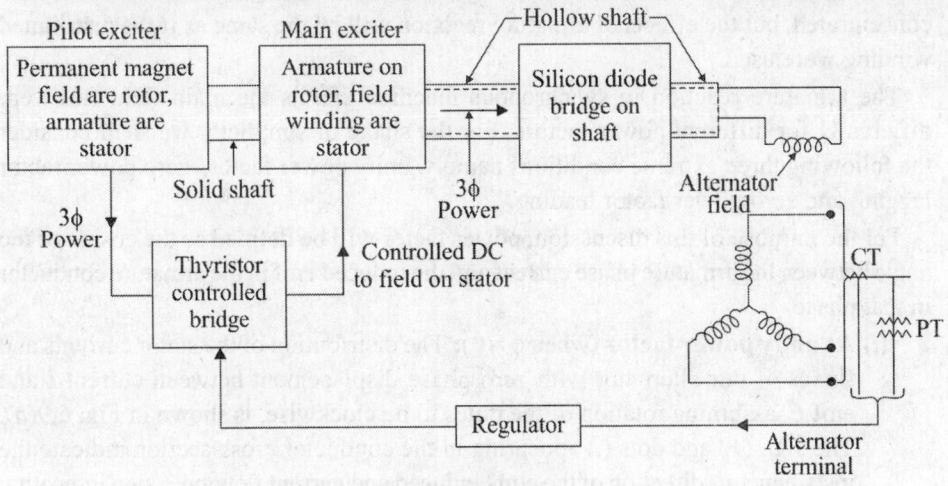

Fig. 6.7

(*iii*) **Brushless excitation:** In this scheme, main shaft of prime mover drives pilot exciter, main excitor and the main alternator: Silicon diode reactifiers are also mounted on the main shaff as shown in Fig. 6.7.

Pilot excitor is a permanent magnet alternator with permanent magnet poles on the rotor and three phase armature winding on the stator. Three phase power, from pilot excitor is fed to thyristor controlled bridge placed on the floor. After rectification the controlled DC output is supplied to stationary field winding of main exciter. The three phase power, developed in the rotor of main exciter is fed through hollow shaft to the rotating silicon diode reactifier mounted on the same shaft. The DC power from the rectifier bridge is delivered, along the main hollow shaft, to the main alternator field without brushes and sliprings.

A signal, picked from alternator terminals through CT and PT, controls and firing angle of thyristor bridge.

6.11 ARMATURE REACTION IN SYNCHRONOUS MACHINE

When current flows through the armature winding of an alternator, the resulting mmf produces flux. The armature flux reacts with the main pole flux, causing the resultant flux to become either less than or more than original main field flux.

The effect of armature (stator) flux on the flux produced by the rotor field poles is called armature reaction.

For simplicity, we consider a 3-phase, 2-pole alternator having a single layer winding as shown in Fig. 6.8. But this treatement is valid for any number of poles. Also the winding of each phase is assumed to be

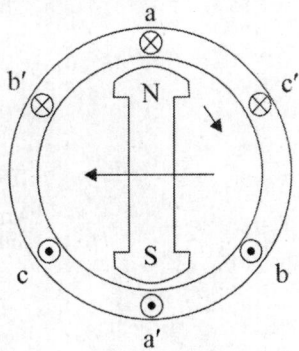

Fig. 6.8

concentrated, but the effects of armature reaction will be the same as if the distributed winding were used.

The armature reaction in synchronous machine affects the main field flux very differently for different power factors. For the shake of simplicity we shall consider the following three extreme conditions namely, unity power factor, zero power factor lagging and zero power factor leading.

For the purpose of this discussion, power factor will be defined as the cosine of the angle between the armature phase current and the induced emf in the armature conductor in that phase.

(*i*) **At unity power factor** (when $\phi = 0$): The distribution of the stator currents and fluxes of our alternator with zero phase displacement between current I and emf E, assuming rotation of the poles to be clockwise, is shown in Fig. 6.9(*a*). The sign (+) and dots (.) appearing in the conductor cross-section indicate the instantaneous direction of the emfs induced and current flowing + sign indicating the inward direction and dot indicating the outward direction.

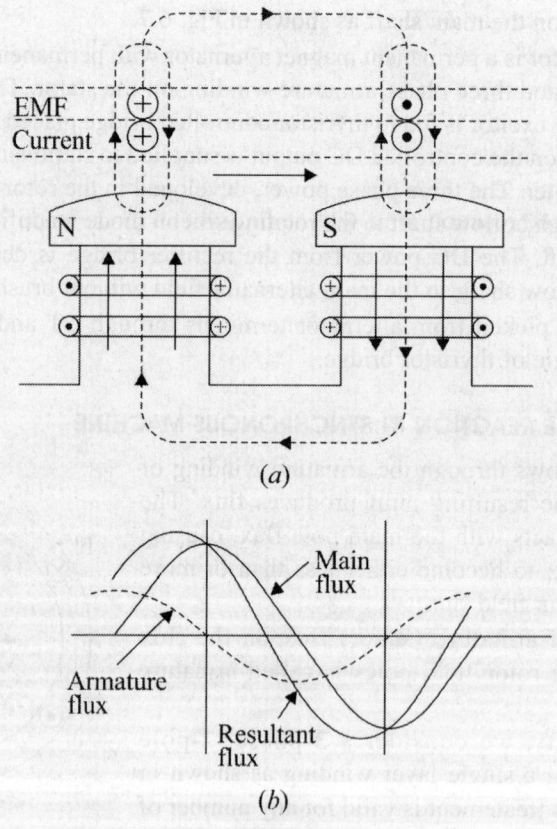

(*a*)

(*b*)

Fig. 6.9

The maximum of the fundamental wave of the field will be opposite to the pole centres, and at the some points the conductors have their maximum induced emf with $\phi = 0$ the conductors carrying the maximum current will also be at the same points as shown in Fig. 6.9(b).

The armature reaction mmf is directed perpendicular to the main field mmf, as in case of a DC machine with brushes on the neutral axis. This causes the distortion of the flux due to the main field and asymmetrical distribution of the flux density under the pole shoe.

The flux density under the trailing pole tips increases some what, while under the leading pole tips it increases. The axis of the resultant field is displaced under the action of armature reaction mmf in a direction opposite to that of rotation of the rotor.

Hence armature reaction at unity power factor has got **distorting effect.**

(ii) **At lagging zero power factor (when** $\phi = \dfrac{+\pi}{2}$**):** The distribution of currents

and fluxes with phase angle radius $\phi = \dfrac{+\pi}{2}$, i.e. for a purely inductive load is shown in Fig 6.10. The current maximum will be shifted in space by an angle $\pi/2$ from the emf maximum, which coinsides with the centres of the poles. This shift will be opposite to the direction of rotation, because the fundamental

armature reaction waves rotates in step with the field poles, while when $\phi = \dfrac{+\pi}{2}$,

current wave lags behind the emf wave by an angle $\pi/2$. The field created by the armature reaction mmf will be in opposite to the main field flux and will therfore have a wholly **demagnetising effect.**

Zero lagging mean p.f. = 0. So power factor angle is 90° and currents lags voltage by 90°.

(iii) **At leading zero power factor:** The distribution of currents and fluxes with phase angle $\phi = -\pi/2$, i.e. for a purely capacitive load is shown in Fig. 6.11. The current maximum will be shifted to the right from the emf maximum, which remains as before under the pole centres, and the armature reaction will, therefore have a wholly **magnetizing effect** on the main field.

For any powerfactor $\cos\phi$ of load, the armature reaction has cross magnetizing component proportional to $I\cos\phi$ and demagnetizing component proportional to $I\sin\phi$ and is −ve for leading power factor.

The phasor sum of the fundamental armature winding and field winding mmf waves constitutes, in a synchronous machine, the mmf creating the resultant magnetic flux.

6.12 ALTERNATOR ON LOAD

Figure 6.10 shows Y-connected alternator supplying inductive load (lagging pf). When The load on the alternator is increased (i.e. annature current I_a is increased), the field

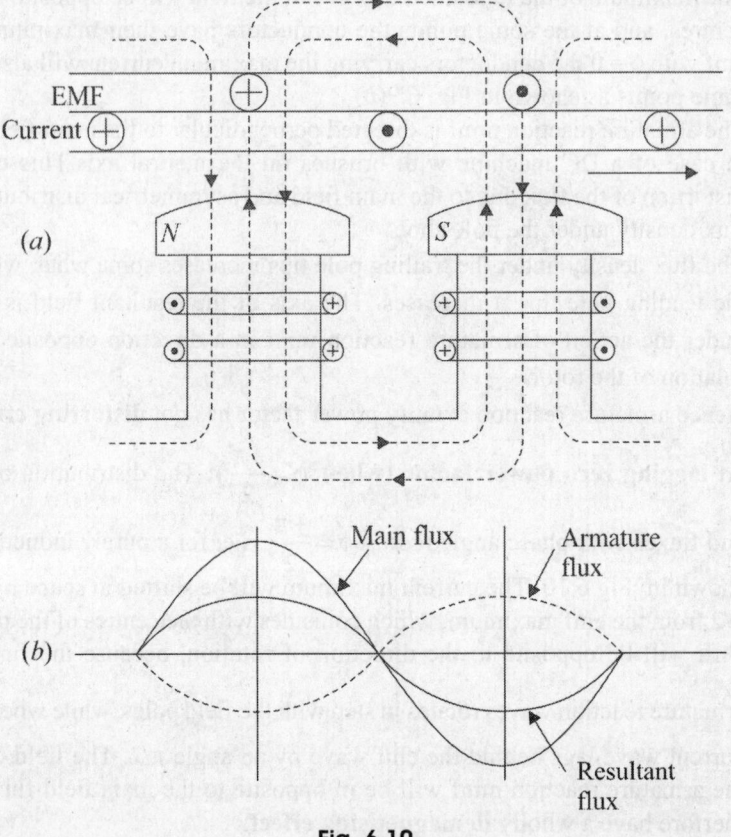

EMF

Current

(a)

N

S

(b)

Main flux

Armature flux

Resultant flux

Fig. 6.10

excitation and speed being kept constant, the terminal voltage V(phase value) of the alternator decreases. This is due to:

(i) Voltage drop I_aR_a where R_a is the armature resistance per phase.

(ii) Voltage drop I_aX_L where X_L is the armature leakage reactance per phase.

(iii) Voltage drop because of armature reaction.

(i) **Armature resistance (R_a):** Since the armature or stator winding has some resistance, there will be an I_aR_a drop when current (I_a) flows through it. The armature resistance per phase is generally small so that I_aR_a drop is negligible for all practical purposes.

(ii) **Armature leakage resistance (X_L):** When armature current flows through the winding, alternating flux produced which is the self flux of those coils. This flux link with the armature coil itself and due to it leakage reactance is developed (Fig. 6.12(b).

(iii) **Armature reaction:** The load is generally inductive and the effect of armature reaction is to reduce the generated voltage. Since armature reaction results in a voltage effect in a circuit caused by the change in flux produced by current in

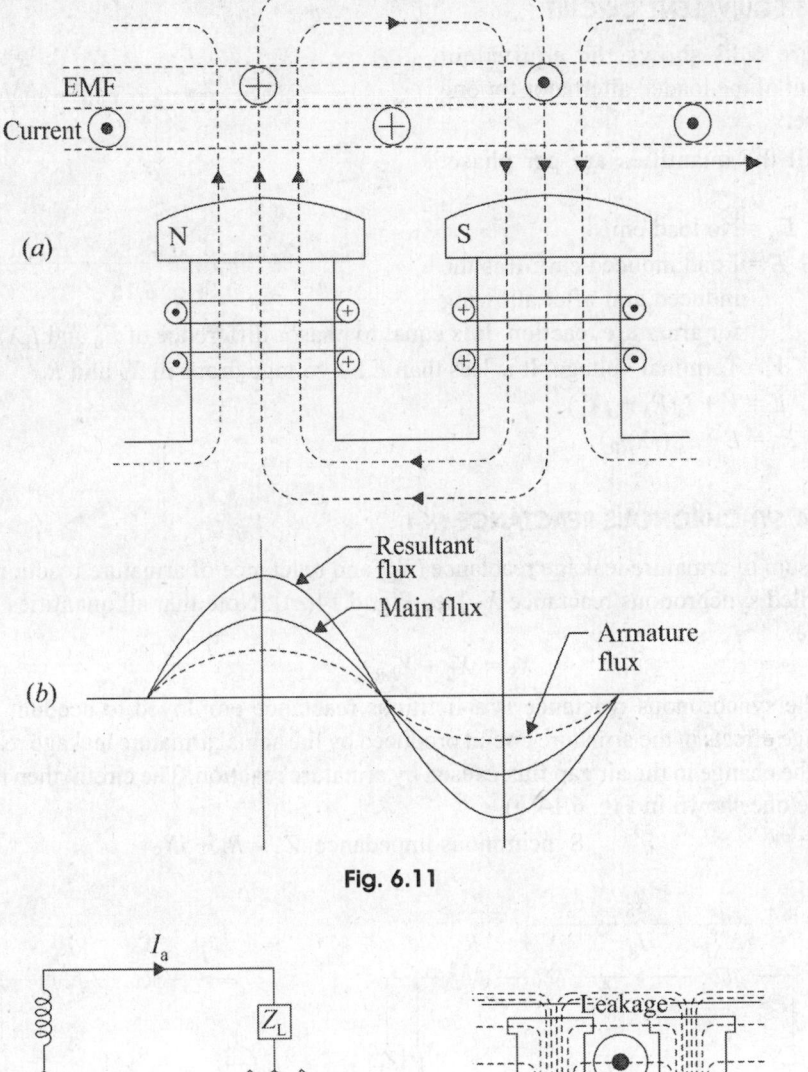

Fig. 6.11

Fig. 6.12(a)

Fig. 6.12(b)

the same circuit, its effect is of the nature of an inductive reactance. Therefore, armature reaction effect is accounted for by assuming the presence of a fictitious reactance X_{AR} in the armature winding. The quantity X_{AR} is called reactance of armature reaction. The value of X_{AR} is such that $I_a X_{AR}$ represents the voltage drop due to armature reaction.

6.13 EQUIVALENT CIRCUIT

Figure 6.13 shows the equivalent circuit of the loaded alternator for one phase.

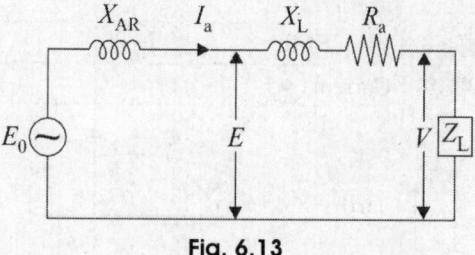

Fig. 6.13

All the quantities are per phase. Here

E_0 = No load emf

E = Load induced emf. It is the induced emf after allowing for armature reaction. It is equal to phasor difference of E_0 and $I_a X_{AR}$.

V = Terminal voltage. It is less than E by voltage drops in X_L and R_a.

$$E = V + I_a(R_a + jX_L)$$

and $E_0 = E + I_a(jX_{AR})$

6.14 SYNCHRONOUS REACTANCE (X_S)

The sum of armature leakage reactance (X_L) and reactance of armature reaction (X_{AR}) is called synchronous reactance X_S [see Fig. 6.14(a)]. Note that all quantities are per phase.

$$X_S = X_L + X_{AR}$$

The synchronous reactance is a fictitious reactance employed to account for the voltage effects in the armature circuit produced by the actual armature leakage reactance and the change in the air gap flux caused by armature reaction. The circuit then reduces to the one shown in Fig. 6.14(b).

Synchronous impedance, $Z_S = R_a + jX_S$

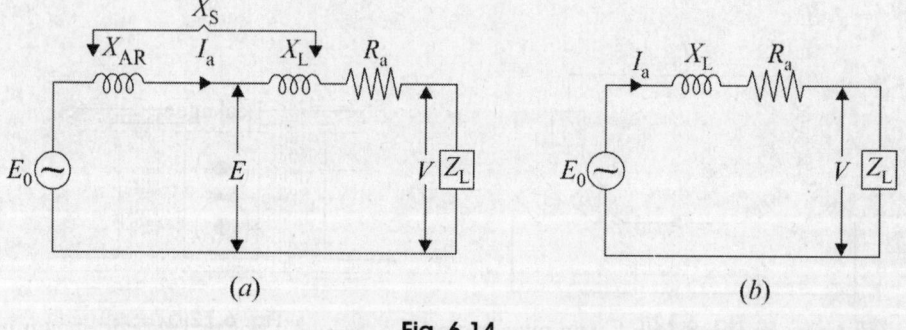

(a)   (b)

Fig. 6.14

The synchronous impedance is the fictitious impedance employed to account for the voltage effects in the armature circuit produced by the actual armature resistance, The actual armature leakage reactance and the change in the air-gap flux produced by armature reaction.

$$E_0 = V + I_a Z_S = V + I_a(R_a + jX_S)$$

6.15 EQUIVALENT CIRCUIT AND PHASOR DIAGRAMS OF A SYNCHRONOUS GENERATOR

The equivalent ckt of a synchronous generator is shown in Fig. 6.15(a). It is redrawn in Fig. 6.15(b) by taking $X_S = X_{AR} + X_L$.

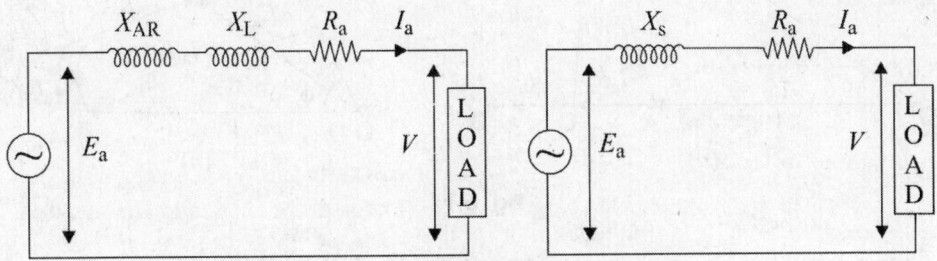

Fig. 6.15

(*i*) **For lagging power factor cos φ:** Figure 6.16 shows the phasor diagram for lagging load. The power factor is cosφ lagging. In this diagram the terminal voltage V is taken as reference phasor along OA such that OA = V.

For lagging power factor cos φ, the direction of the armature current I_a lags behind V by an angle φ along OB, where OB = I_a. The voltage drop in the armature resistance is $I_a R_a$. It is represented by CD. It leads the current I_a by 90° and therefore, CD is drawn in a direction perpendicular to OB. The total voltage drop in the synchronous impedance

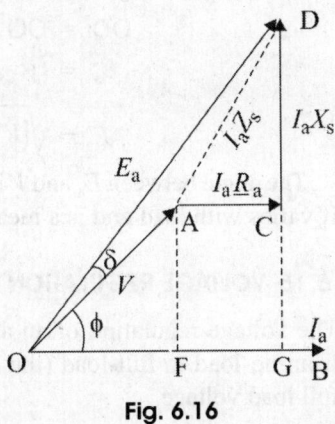

Fig. 6.16

is the phasor sum of $I_a R_a$ and $I_a X_S$. It is represented by AD. The phasor OD represents E_a.

The magnitude of E_a can be found from the right–angled triangle OGD.

$$OD^2 = OG^2 + GD^2 (OF + FG)^2 + (GC + CD)^2$$

$$E_a^2 = \left(V\cos\phi + I_a R_a\right)^2 + \left(V\sin\phi + I_a \times s\right)^2$$

$$E_a = \sqrt{\left(V\cos\phi + I_a R_a\right)^2 + \left(V\sin\phi + I_a \times s\right)^2}$$

(*ii*) **Unity power factor:** The phasor diagram for unity power factor is shown in Fig. 6.17(a) from right angled triangle OCD.

$$(OD)^2 = (OC)^2 + (CD)^2 = (OA + AC)^2 + (CD)^2$$

$$E_a^2 = (V + I_a R_a)^2 + (I_a X_S)^2$$

$$E_a = \sqrt{\left(V + I_a R_a\right)^2 + \left(I_a X_s\right)^2}$$

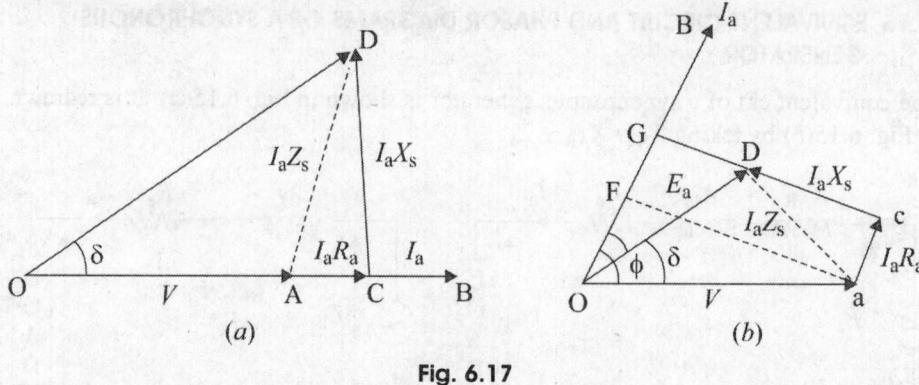

Fig. 6.17

(*iii*) **Leading power factor cosφ:** The phasor diagram for leading power factor is shown in Fig. 6.17(*b*) from right-angled triangle OGD

$$OD^2 = OG^2 + GD^2 = (OF + FG)^2 + (GC - CD)^2$$

$$E_a^2 = (V\cos\varphi + I_a R_a)^2 + (V\sin\phi + I_a X_s)^2$$

$$E_a = \sqrt{(V\cos\varphi + I_a R_a)^2 + (V\sin\phi + I_a \times s)^2}$$

The angle between E_a and V is called the power angle or torque angle of the machine. If varies with load and is a measure of air gap power developed in the machine.

6.16 VOLTAGE REGULATION

The voltage regulation of an alternator is defined as the change in terminal voltage from no-load to full-load (the speed and field excitation being constant) divided by full load voltage.

$$\text{Voltage regulation} = \frac{|E_a| - |V|}{|V|} \times 100$$

where $|E_a|$ = Magnitude of generated voltage per phase.

$|V|$ = Magnitude of rated terminal voltage per phase.

The voltage regulation depends upon the power factor of the load. For unity and lagging power factors, there is always a voltage drop with the increase of load, but for a certain leading power, the full-load voltage regulation is zero. In this case, the terminal voltage is the same for both full-load and no-load conditions. At lower leading power factors the voltage rises with the increase of load, and the regulation is negative.

6.17 DETERMINATION OF VOLTAGE REGULATION

The following methods are used to determine the voltage regulation of smooth cylindrical rotor type alternators: (*a*) Direct load test (*b*) Indirect methods.

(*a*) **Direct load test:** The alternator is run at synchronous speed and its terminal voltage is adjusted to its rated value V. The load is varied until the ammeter and wattmeter indicate the rated values at the given power factor. Then the load is removed and the speed and field excitation are kept constant. The open-circuit or no-load voltage E_a is recorded. The voltage regulation is found from percentage $V \cdot R = \dfrac{E_a - V}{V} \times 100$. The method of direct loading is suitable only from small alternators of power rating less than 5 kVA.

(b) **Indirect methods:** For large alternators, the three indirect methods which are used to predetermine the VR of smooth cylinderical-rotor type alternators are as follow:

    1. Synchronous Impedance method or EMF method.

    2. Ampere-turn method or MMF method.

    3. Zero power factor method.

6.18 SYNCHRONOUS IMPEDANCE METHOD OR EMF METHOD

The synchronous impedance method is based on the concept of replacing the effect of armature reaction by ficticious reactance.

For a synchronous generator

$$V = E_a - Z_S I_a$$

where $Z_S = R_a + jX_S$.

In order to determine the synchronous impedance Z_S is measured and then the value of E_a is calculated.

6.18.1 Measurement of Synchronous Impedance

The following tests are performed on an alternator to know its performance: (a) DC resistance test (b) Open-circuit test (c) Short-circuit test.

(a) **DC resistance test:** Assume that the alternator is star connected with DC field winding open (Fig. 6.18), measure the DC resistance between each pair of terminals either by using ammeter–voltmeter method or by using wheat stone's bridge. The average of three sets of resistance value R_t is taken. This value of R_t is divided by 2 to obtain the DC resistance (ohmic resistance) per phase. The alternator should be at rest. Since the effective AC resistance is larger than DC resistance due to skin effect, therefore the effective AC resistance per phase is obtained by multiplying the DC resistance by a factor 1.20 to 1.75 depending on the size of the machine. A typical value to use in the calculation would be 1.25.

(b) **Open-circuit test:** The alternator is run at rated synchronous speed and the load terminals are kept open (Fig. 6.19). That is all the loads are disconnected. The field current is set to zero. Thus the field current is gradually increased in steps, and the terminal voltage E_t is measured at each step. The excitation current may be increased to get 25% more than rated voltage of the alternator. A graph is plotted between the open-circuit phase voltage $E_P \left(= \dfrac{E_t}{\sqrt{3}} \right)$ and field current I_f.

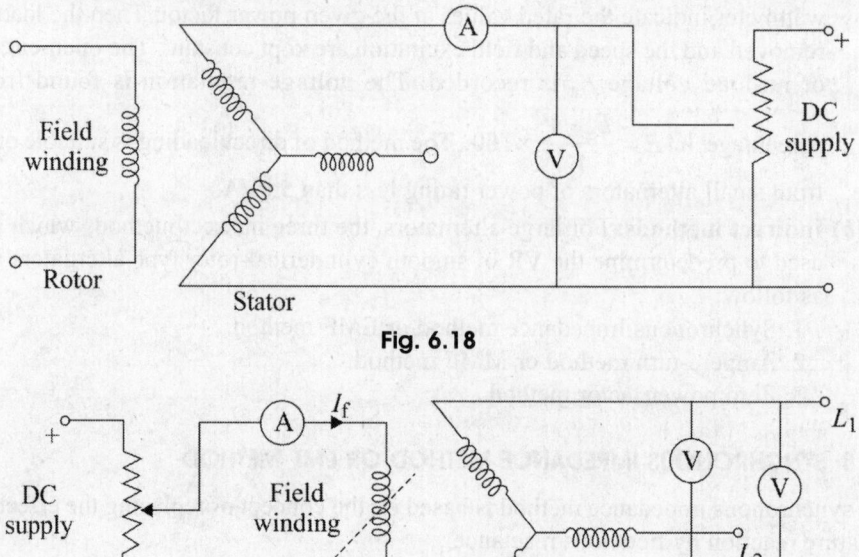

Fig. 6.18

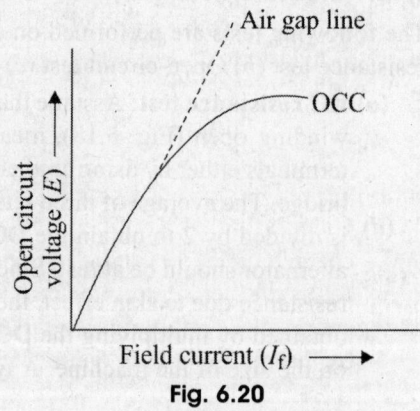

Fig. 6.19

The characteristic curve so obtained is called open circuit characteristic (OCC). It takes the shape of a normal magnetisation curve. The extension of the linear portion of an OCC is called *air gap* line of the characteristic. The OCC and the air gap line are shown in Fig. 6.20.

(c) **Short-circuit test:** The armature terminals are shorted through three ammeters (Fig. 6.21). Care should be taken in performing this test, and the field current should first be decreased to zero before starting the alternator. Each ammeter should have a range greater than the rated full-load value. The alternator is then run at synchronous speed. Then the field current is gradually increased in step, and the armature current is measured at each steps. The field current may be increased to get armature current upto 15% of the rated value. The field current I_f and the average of three ammeter readings at each step is taken. A graph is plotted between the armature current I_a and the field current I_f. The characteristic so obtained is

Fig. 6.20

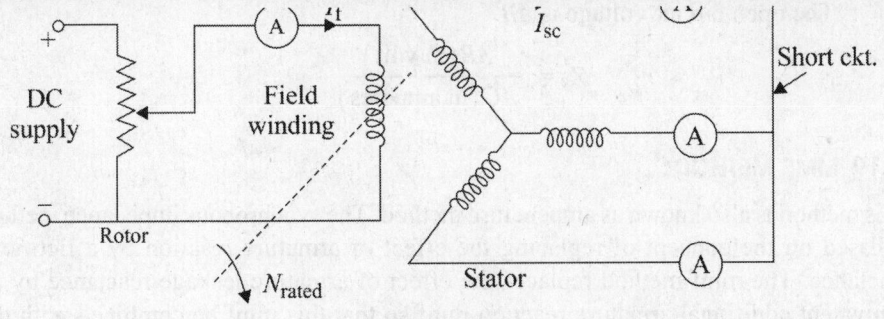

Fig. 6.21

called short-circuit characteristic (SSC). This characteristic is a straight line as shown in Fig. 6.22(a).

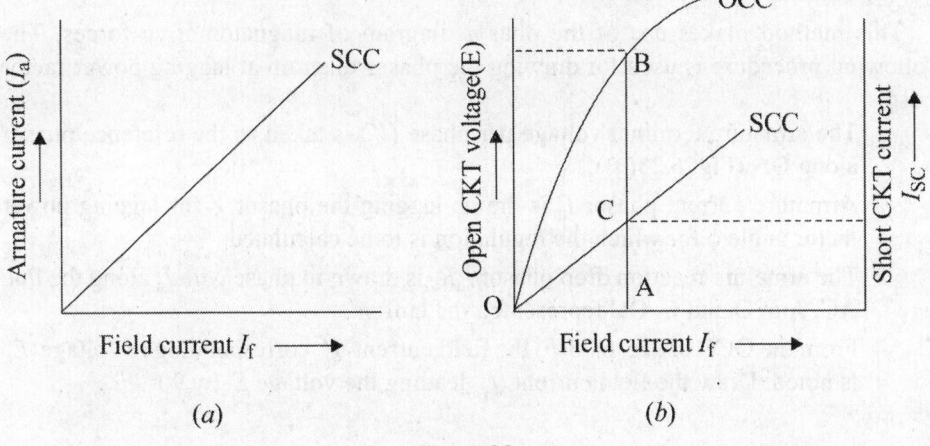

(a)   (b)

Fig. 6.22

(**d**) **Calculation of Z_S:** The open-circuit characteristic (OCC) and short-circuit characteristic (SSC) are drawn on the same curve sheet. Determine the value of I_{SC} at the field current that gives the rated alternator voltage per phase. The synchronous impedance Z_S will than be equal to the open-circuit voltage divided by the short-circuit at that field current which gives the rated emf per phase.

$$Z_S = \frac{\text{Open cut voltage per phase}}{\text{shortcut armature current}}$$

For the same value of field current.

The synchronous reactance is found as follows: $X_S = \sqrt{Z_s^2 - R_a^2}$

In Fig. 6.22(b) consider the field current $I_f = OA$ that produces rated alternator voltage per phase, corresponding to this field current the open circuit voltage is AB.

The open circuit voltage is AB.

$$\therefore \qquad Z_S = \frac{AB \text{(in volt)}}{AC \text{(in amperes)}}$$

6.19 MMF METHOD

This method is also known as ampere turn method. The synchronous impedance method is based on the concept of replacing the effect of armature reaction by a fictitious reactance. The mmf method replaces the effect of armature leakage reactance by an equivalent additional armature reaction mmf so that this mmf be combined with the armature reaction mmf f_{ar}.

The following information is required is predict the regulation by the mmf method.
(a) The resistance of the stator winding per phase.
(b) Open circuit characteristic at synchronous speed.
(c) Short circuit characteristic.

This method makes use of the phasor diagram of magnetomotive forces. The following procedure is used for drawing the phasor diagram at lagging power factor $\cos\phi$.

1. The armature terminal voltage per phase (V) is taken as the reference phasor along OA (Fig. 6.23(a)).

2. Armature current phasor I_a is drawn lagging the phasor V for lagging power factor angle ϕ for which the regulation is to be calculated.

3. The armature reaction drop phasor $I_a R_a$ is drawn in phase with I_a along the line AC. Join O and C. OC represented the emf E'.

4. From the OCC of Fig. 6.23(b) the field current I'_f corresponding to voltage E' is noted. Draw the field current I'_f leading the voltage E' by 90°.

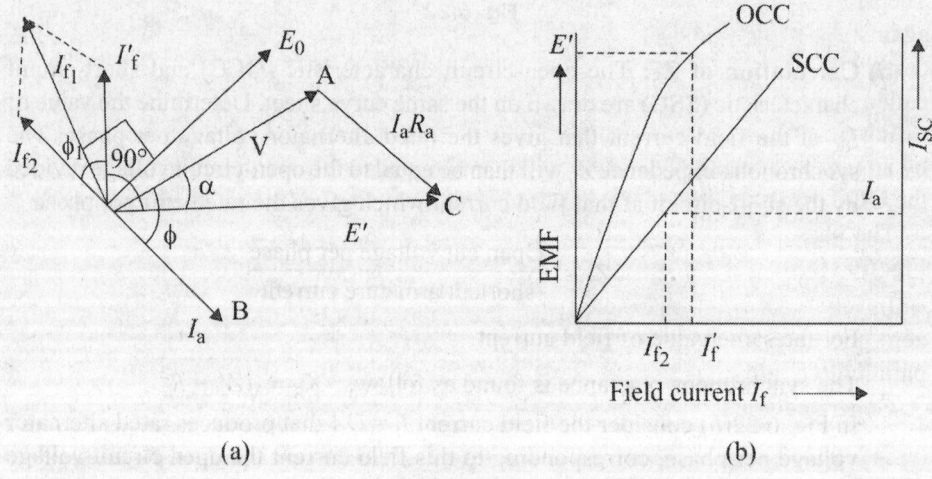

(a)   (b)

Fig. 6.23

It is assumed that an short circuit all the excitation is opposed by the mmf of armature reaction and armature reactance. Thus $(I'_f = I'_f \angle 90 - \alpha)$.

5. From the SSC of Fig. 6.23(b) determine the field current I_{f2} required to circulate the rated current as short circuit. This is the field current required to overcome the synchronous reactance drop $I_a X_S$. Draw field current I_{f2} in phase opposition to current I_a. Thus $I_{f_2} = I_{f_2} \angle 180° - \phi$.

6. Determine the phasor sum of field currents I'_f and I_{f_2}. This gives resultant field current I_f which would generate a voltage E_0 under no-load conditions of the alternator. The open-circuit emf E_0 corresponding to field current I_f is found from the open-circuit characterisitic.

7. The regulation of the alternator is found from the relation.

$$\text{Regulation} = \frac{E_0 - V}{V} \times 100\%$$

6.19.1 Ampere-Turn Method with R_a Neglected

Here $I_{f2} = I_{f_2} \angle 180 - \phi$

$\quad I'_f = I'_f \angle 90°$

$\quad I_f = I_{f_2} + I'_f$

Alternatively, from Fig. 6.24.

$I_f^2 = I'^2_f + I^2_{f_2} + 2I'_f I_{f_2} \cos(90 - \phi)$

$\quad = I'^2_f + I^2_{f_2} + 2I'_f I_{f_2} \sin\phi$

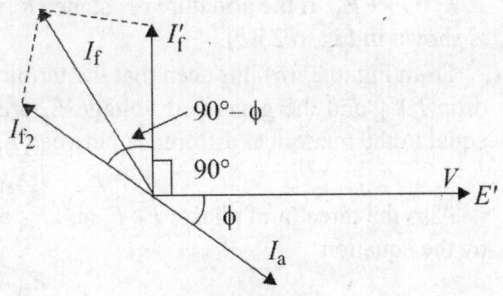

Fig. 6.24

6.20 Zero-Power Factor Characteristic (ZPFC)

The ZPFC of an alternator is a curve of the armature terminal voltage per phase plotted against the field current obtain by operating the machine with constant rated armature current at synchronous speed and zero lagging power factor. The ZPFC is some times called *potier characteristic* after its originator.

For maintaining very low pf, alternator is loaded by means of reactors or alternatively by an under excited synchronous motor. The shape of ZPFC is very much like that of the OCC displaced downwards and to the right.

The phasor diagram corresponding to zero-power factor lagging load is shown in Fig. 6.25(a).

In Fig. 6.25(a), the terminal phase voltage V is taken as the reference phasor. At zero power factor lagging, the armature current I_a lags behind V by 90°. Draw $I_a R_a$ parallel to I_a and $I_a X_{aL}$ perpendicular to I_a.

∴ Generated voltage per phase $E_g = V + I_a R_a + I_a X_{aL}$

$\quad\quad\quad F_{ar}$ = armature reaction mmf. It is in phase with I_a.

$\quad\quad\quad F_f$ = mmf of the main field winding (field mmf)

$\quad\quad\quad F_r$ = resultant mmf.

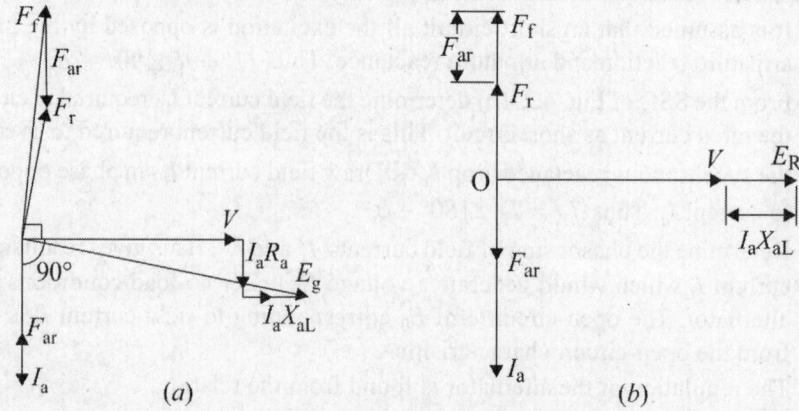

Fig. 6.25

$\therefore$ $F_r = F_f + F_{ar}$. If the armature resistance R_a is neglected, the resulting phasor diagram is shown in Fig. 6.25(b).

From Fig. 6.25(b) it is seen that the terminal phase voltage V, the reactance voltage drop $I_a X_{aL}$ and the generated voltage E_g are all in phase. Therefore V is practically equal to the numerical difference between E_g and $I_a X_{aL}$.

$$V = E_g - I_a X_{aL} \qquad ...(6.1)$$

Also the three mmf phasor F_f, F_r and F_{ar} are in phase. Their magnitudes are related by the equation

$$F_f = F_r + F_{ar} \qquad ...(6.2)$$

The arithmatical relations given in Eqs (6.1) and (6.2) from the basis for the potier triangle.

Equation can be converted into its equivalent field current from by dividing its both sides by I_f, the effective no. of turns per pole on the rotor field.

$$\therefore \qquad \frac{F_f}{T_f} = \frac{F_r}{T_f} + \frac{F_{ar}}{T_f}$$

or $\qquad I_f = I_r + I_{ar}$

6.20.1 Potier Triangle

In Fig. 6.26 , consider a point b on the ZPFC corresponding to rated terminal voltage V and a field current $OM = I_f = \dfrac{F_f}{I_f}$.

If, for this condition of operation, the armature reaction mmf has a value expressed in equivalent field current of $LM\left(= I_{ar} = \dfrac{F_{ar}}{I_f} \right)$, then the equivalent field current of

the resultant mmf would be $OL\left(= I_r = \dfrac{F_{ar}}{I_f} \right)$.

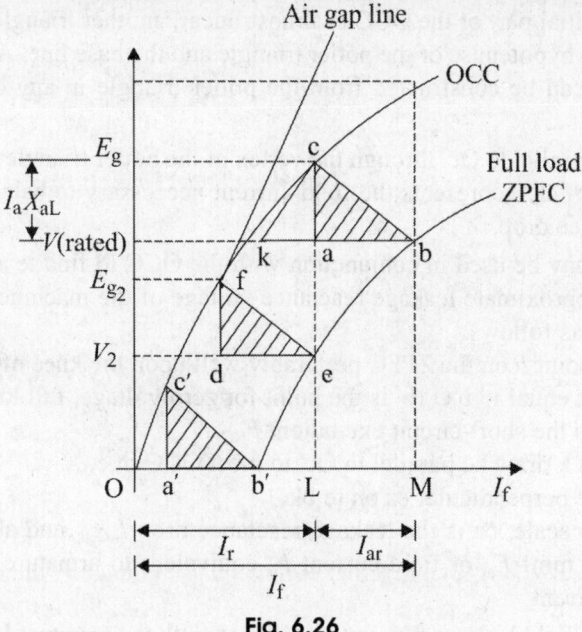

Fig. 6.26

This field current OL would result in a generated voltage $E_g \, (= LC)$ from the no-load saturation curve. Since for lagging zero-power factor operation.

$$E_g = V + I_a + X_{aL}$$

Vertical distance must be equal to the leakage reactance voltage drop $I_a X_{aL}$ where I_a is the rated armature current.

$$\therefore \qquad X_{aL} = \frac{\text{Voltage AC per phase}}{\text{Rated armature current}}$$

The triangle formed by the vertices a, b, c is called the potier triangle.

For zero power factor operation rated current at any other terminal voltage. Such as V_2, Since the armature current is of the same value, both the $I_a X_{aL}$ voltage and the armature mmf must be of the same respective values as they were for operation with rated terminal voltage V.

Therefore, for all condition of operation with rated armature current at zero lagging pf, the same potier triangle must be located between the terminal voltage V point on the ZPPC and the corresponding E_g point on the OCC. Thus, if the potier triangle 'cab' is moved downward so that the side 'ab' is kept horizontal and b is kept on the ZPFC, the point c will move on the OCC. When the point b is reaches the point e, the potier triangle cab will be in the position fde and the location of pff on the OCC will determine the voltage E_{g2} which will be generated for zero power factor operation with terminal voltage V_2. When the point b reaches point b', the potier triangle will be in the position $c'a'b'$. This is the limiting position which corresponds to short–circuit conditions, because at b', the terminal voltage V is zero.

Since the intitial part of the OCC is almost linear, another triangle Oc'b' is formed by the OCC the hypotenuse of the potier triangle and the base line. A similar triangle, such as 'ckb', can be constructed from the potier triangle in any other location by drawing.

A line 'kc' parallel to Oc' through the vertex of the potier triangle which lies on the OCC the length 'ak' represents the field current neccessary to balance the armature leakage reactance drop.

The ZPFC may be used in conjunction with the OCC to find te armature reaction mmf and the approximate leakage reactance voltage of the machine (Fig. 6.26). The construction is as follows:

(i) Take a point b on the ZPFC preferably well upon the knee of the curve.

(ii) Draw bk equal to b'O (b' is the point for zero voltage, full load current). That is, Ob' is the short-circuit excitation, F_{SC}.

(iii) Through k draw kc parallel to Oc' to meet OCC in c.

(iv) Drop the perpendicular ca on to bk.

(v) Then, to scale, ca is the leakage reactance drop $I_a X_{aL}$ and ab is the armature reaction mmf F_{ar} or field current I_{ar} equivalent to armature reaction mmf at rated current.

The effect of field leakage flux in combination with the armature leakage flux gives rise to an equivalent leakage reactance X_P, known as the potier reactance. It is greater than the armature leakage reactance.

$$\text{Also potier reactance } X_P = \frac{\text{Voltage drop per phase}(= AC)}{\left(\text{ZPF rated armature current per phase } I_a\right)}$$

For equal rotor machines, potier reactance X_P is approximately equal to leakage reactance X_{aL}. In salient pole m/c X_P may be as large as 3 times X_{aL}.

6.20.2 Assumptions

The following assumptions are made in the potier method.

1. The armature resistance R_a is neglected.
2. The OCC taken on no-load accurately represents the relation between mmf and voltage on load.
3. The leakage reactance voltage $I_a X_{aL}$ is indendent of excitation.
4. The armature–reaction mmf is constant.

It is not necessary to plot the entire ZPFC for determining X_{aL} and F_a experiementally. Only two points b and b' in Fig. 6.26 are sufficient.

6.20.3 Procedure to Obtain the Regulation by Zero-Power Factor Method

The following procedure is used to obtain regulation by the zero-power factor method.

The phasor diagram for lagging power factor is drawn as shown in Fig. 6.27. In the phasor diagram:

OA = V = terminal phase voltage at full load. It is taken as reference phasor and drawn horizontally.

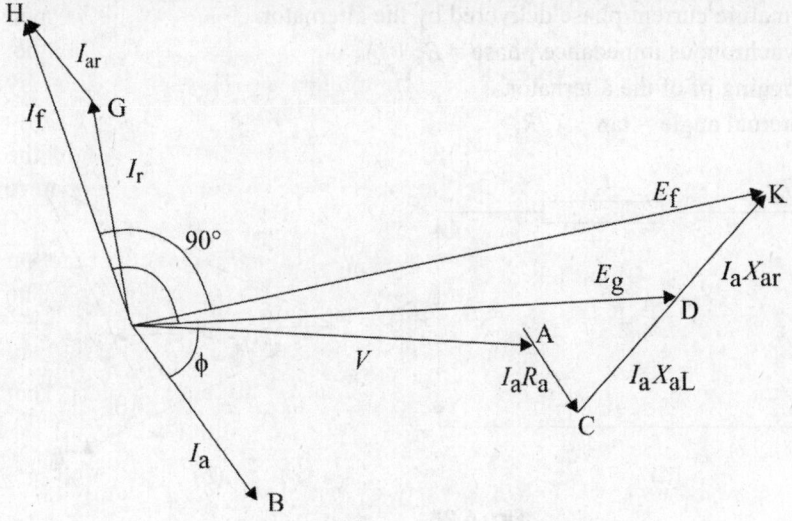

Fig. 6.27

OB = I_a = full load current lagging behind V by an angle ϕ, $\cos\phi$ is the power factor os the load.

AC = voltage drop I_aR_a in the armature resistance (If R_a is given). It is drawn parallel to $I_a(OB)$.

CD = I_aX_{aL} = leakage reactance voltage drop. It is perpendicular AC. Join OD. It represents the generated emf E_g.

Find the field excitation current I_R corresponding to this generated emf, e.g. from the OCC.

Draw OG (equal to in perpendicular to OD).

Draw GH parallel to load current OB (= I_a) to represent excitation (field current) equivalent to full-load armature reaction I_{ar}. OH gives the total field current I_f.

If the load is thrown off, then terminal voltage will be equal to generated emf, corresponding to field excitation OH.

Determine the E_f (= OK) corresponding to field excitation OH from the OCC phasor OK will log behind phasor OH by 90°. DK represents the voltage drop due to armature reaction.

Now voltage regulation is obtained from the relation:

$$\text{Voltage regulation} = \frac{E_f - V}{V} \times 100\%$$

6.21 POWER OPERATING IN CYLINDERICAL ROTOR

Consider a star-connected cylinderical rotor alternator operating on infinite busbar.

Let, V = Bus bar voltage/phase

E_0 = Generated emf/phase

I_a = armature current/phase delivered by the alternator.

Z_s = Synchronous impedance/phase = $R_a + jX_s$

$\cos \phi$ = Lagging pf of the alternator.

θ = Internal angle = $\tan^{-1} X_s/R_s$

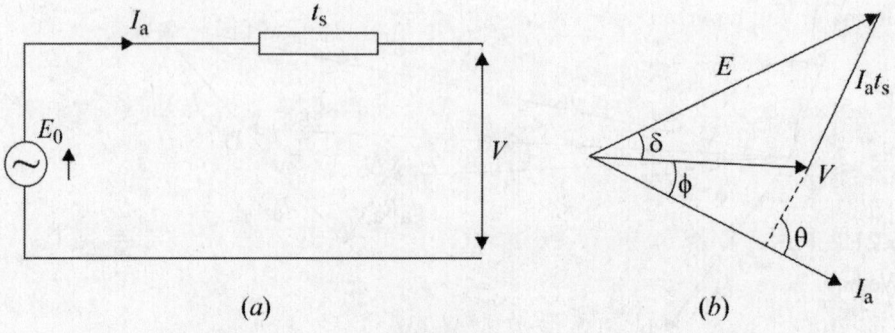

(a)   (b)

Fig. 6.28

Fig. 6.28(a), shows the equivalent circuit for one phase of the alternator whereas Fig. 6.28(b), shows the phasor diagram. Note that in drawing the phasor diagram, V has been taken as the reference phasor. We shall now drive an expression for the power delivered by the alternator (Fig. 6.28(b)).

$$I_a = \frac{E_0\angle\delta - V\angle 0}{Z_s\angle\theta} = \frac{E_0}{Z_s}\angle(\delta-\theta) - \frac{V}{Z_s}\angle-\theta$$

power output/phase, $P = V \times$ real part of I_a.

$$= V \times \left[\frac{E_0}{Z_s}\cos(\delta-\theta) - \frac{V}{Z_s}\cos(-\theta)\right]$$

$$= \frac{V}{Z_s}\left[E_0\cos(\delta-\theta) - V\cos\theta\right]$$

$$= \frac{V^2}{Z_s}\left[\frac{E_0}{V}\cos(\delta-\theta) - \cos\theta\right]$$

$$\therefore \qquad P = \frac{E_0 V}{Z_s}\cos(\theta-\delta) - \frac{V^2}{Z_s}\cos\theta$$

The above relation gives the electrical output of the alternator in terms of E_0, V, Z_S, θ and load angle δ.

6.21.1 Maximum Power Output

For given E_0, V and frequency, the condition for maximum power output can be obtained by differntiating equation with respect to δ and equating the resulting to zero, i.e.

$$\frac{dP}{d\delta} = 0$$

or $$\frac{E_0 V}{Z_s} \sin(\theta - \delta) = 0$$

or $$\sin(\theta - \delta) = 0$$

or $$\theta = 0$$

For constant busbar voltage V and fixed excitation (i.e. E_0), the power output of the alternator will be maximum when $\theta = \delta$.

$$P_{max}/\text{phase} = \frac{B_0 V}{Z_s} \cos 0^\circ - \frac{V^2}{Z_s} \cos\theta$$

$$= \frac{E_0 V}{Z_s} - \frac{V^2}{Z_s} \cos\theta.$$

6.21.2 Power Output/Phase Expression

We have seen above that

Power output/phase, $$P = \frac{V^2}{Z_s} \left[\frac{E_0}{V} \cos(\theta - \delta) - \cos\theta \right]$$

$$= \frac{V^2}{Z_s} \left[\frac{E_0}{V} \cos(\theta - \delta) - \frac{R_a}{Z_S} \right] \qquad \left(\because \cos\theta = \frac{R_a}{Z_S} \right)$$

If $R_a \ll Z_s$, then $\theta = 90^\circ$ and $Z_S = X_S$.

$$P = \frac{V^2}{X_s} \left[\frac{E_0}{V} \cos(90^\circ - \delta) - 0 \right]$$

$$P = \frac{E_0}{X_s} \sin\delta$$

6.22 POWER ANGLE/POWER ANGLE CHARACTERISTIC

The power operating system of an alternator is given by:

Power output/phase, $$P = \frac{E_0 V}{X_S} \sin\delta$$

Total power output $$= \frac{3 E_0 V}{X_S} \sin\delta$$

Note that power output varies sinusoidally with power angle δ. Fig. 6.29, shows the power angle characteristic of the alternator. The alternator delivers maximum power when $\delta = 90^\circ$. It δ becomes greater than 90°, the machine will less synchronism.

Note that stability of the alternator is determined by the power angle characteristic. Suppose the operating position of the alternator is represented by point P on the curve. If unsteadiness occurs due to a transient spike of mechanical input, then load angle δ increases by a small amount. The additional electrical output caused by an increase in δ produce a torque which is not balanced by the driving torque once the spike has passed.

This torque causes retardation of the rotor and the alternator returns to the operating point P. The torque causing the return of the alternator to the steady–state position is called the *synchronising torque* and the power associated with it is known as *synchronising power*.

Power output of the alternator will be maximum when $\delta = 90°$.

$$P_{max}/\text{phase} = \frac{E_0 V}{X_S}$$

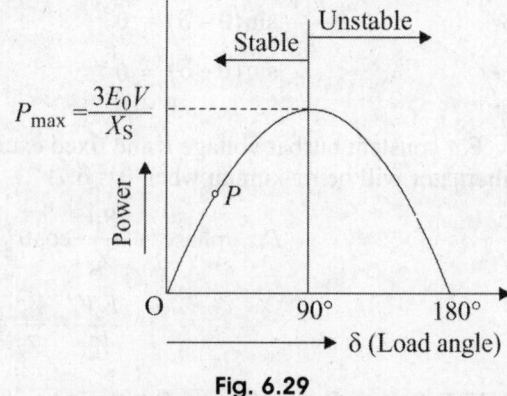

Fig. 6.29

Comments: We have seen above that if armature resistance R_a is very small (as is usually the case), then the total power for three phases is given by

$$\text{Total power} = \frac{E_0 V}{X_S} \sin \delta$$

When the alternator operates at a constant speed with constant field current, X_S and E_0 are both constant. Also applied volt/ph (i.e. V) is usually held constants. Therefore, power output of the alternator is directly proportional is $\sin\delta$, where δ is the angle between V and E_0 and is called power angle.

Equation is also known as the power-angle relation. Note that in deriving equation we have assumed cylinderical rotor. For a salient-pole rotor, the curve is somewhat modified.

The torque developed by a cylinderical rotor machine is given by

$$T_d = \frac{\text{output power}}{W_S} = \frac{3 E_0 V}{W_S X_S} \sin \delta$$

where W_S = angular velocity of the rotor.

Since torque developed is also directly proportional to $\sin \delta$ is also referred to as the torque angle. The torque developed by an alternator opposes the torque applied by the prime movers.

6.23 EFFECT OF SALIENT POLES

The treatement developed so far is applicable only to cylinderical rotor machines. In these machines, the air gap is uniform so that the reluctance of the magnetic path is the same in all directions. Therefore, the effect of armature reaction can be accounted for one reactance, i.e. the synchronous reactance X_S (Fig. 6.30).

It is because the value of X_S is constant for all directions of armature flux relative to the rotor. However, in a salient pole machine, the radial length of the air gap varies [Fig. 6.30(a)], so that reluctance of the magnetic circuit along the polar axis (called direct axis or d-axis) is much less than the reluctance along the interpolar axis (called quadrature axis or q-axis).

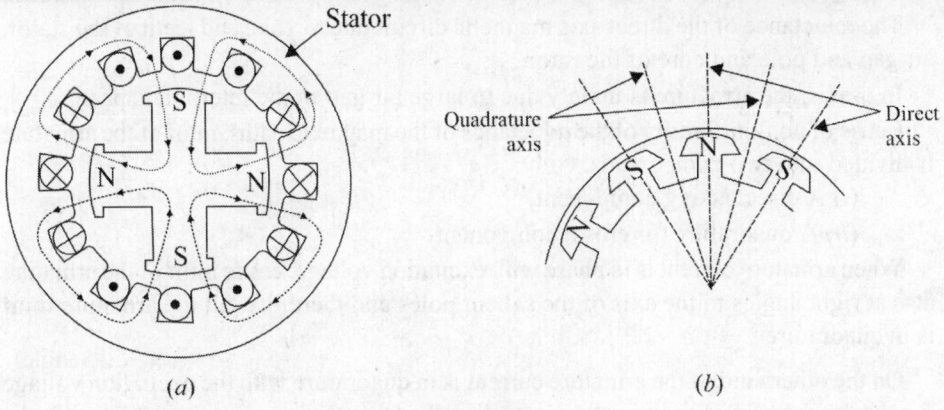

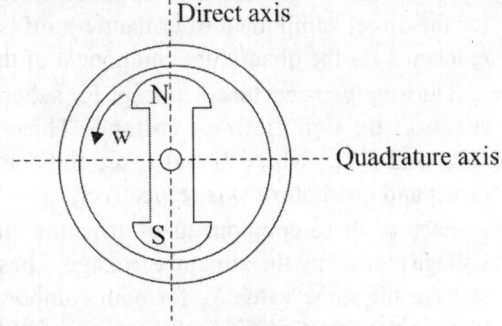

(a)   (b)

Fig. 6.30

Because of the lower reluctance along the polar axis (i.e. *d*-axis), more flux is produced along d-axis than along q-axis. Therefore reactance due to armature reaction will be different along d-axis and q-axis. These are

$$X_{ad} = \text{Direct axis reactance due to AR}$$
$$X_{aq} = \text{Quadrature axis reactance due to AR}$$

6.24 TWO–REACTANCE CONCEPT FOR SALIENT POLE SYNCHRONOUS MACHINE

A multipolar machine with cylinderical rotor has a uniform air gap and therefore, its reactance remains the same, irrespective of the spatial position of rotor. The effect of armature reaction, fluxes, and voltage induced can, therefore, be treated in a simple way with concept of a synchronous reactance and taking it as constant for all potions of field poles with respect to the armature.

However, a salient pole synchro-nous machine has non-uniform air gap due to which its reactance varies with the rotor position.

Thus the salient pole machine possesses two axes of geometric symmetry:

(*i*) Field pole axis, called direct axis or d-axis.

(*ii*) Axis passing through the centres of the interpolar space, called the quadrature axis or q-axis as shown in Fig. 6.31, whereas in case of a cylinderical rotor m/c there is only one axis of symmetry (pole axis or direct axis).

Fig. 6.31

In case of salient pole m/c, the reluctance of the magnetic paths on which the emf acts are different along the direct axis and quadrature axis.

The reluctance of the direct axis magnetic circuit due to yoke and teeth of the stator, air gap and pole and core of the rotor.

In q-axis, the armature is mainly due to large air gap in the interpolar space.

Cause of non-uniformity of the reluctance of the magnetic paths, mmf of the armature is divided into two components. Only:

(i) A direct acting component,

(ii) A quadrature (or cross) component.

When armature current is in phase with excitation voltage, entire mmf of the armature acts at right angles to the axis of the salient poles and therefore, all the armature mmf is in quadrature.

On the otherhand, if the armature current is in quadrature with the excitation voltage E_0, the entire mmf of the armature acts directly upon the magnetic paths through the direct poles and all of the armature mmf is direct acting, either directly opposing or directly aiding the mmf of the salient pole field windings.

When the phase difference beween armature current and excitation voltage is of some angle in between 0 and 90°, the armature mmf will have both a direct acting and a quadrature component.

The direct acting component is proportional to the sine of the those angle between the armature current and the excitation voltage whereas the quadrature (or cross) component is proportional of the cosine of the phase angle between the armature current and the excitation voltage.

The two reaction concept is similar to the synchronous impedance concept in the effect of armature reaction is taken into account by means equivalent armature reaction voltage.

However, owing to the difference in the reluctance of the magnetic effect upon which the two components of the armature mmf the value of the equivalent reactance for the direct component of armature mmf is greater than the value of the equivalent reactance for the quadrature component of the armature mmf.

Thus the two-reactance concept for salient-pole machine is the effect of armature reaction by two fictious voltage. These reactance voltages are respectively $I_d X_{ad}$ and $I_q X_{aq}$, where I_d and I_q are the components of the armature currents along direct and quadrature axis respectively.

Each of these components of armature currents also produce a leakage reactance voltage caused by the armature leakage. These armature leakage reactance is assumed to have the same value X_L for both components of the armature current. Therefore, there is synchronous for each component of the armature mmf as follows.

Direct axis synchronous, $X_d = X_{ad} + X_L$

Quadrature axis synchronous reactance, $X_q = X_{aq} + X_L$

The voltage equation for each phase of the armature based on the two reaction concept is

$$V = E_0 - I_a R_a - I_d X_d - I_q X_q$$

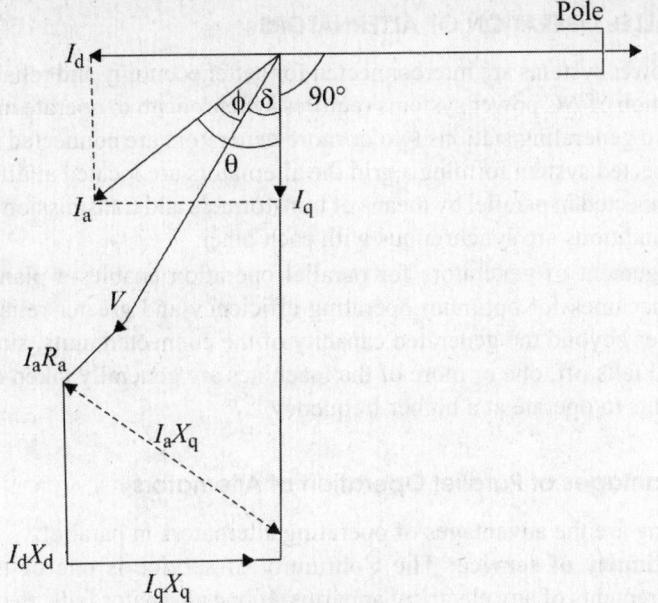

Fig. 6.32

In salient pole synchronous machine $X_q = 0.6$ to 0.7 times X_a, whereas in cylindrical rotor machine $X_q = X_a$.

Note that in drawing the phasor diagram, the armature resistance R_a is neglected since it is quite small. Further all values are phase values.

Hence V is terminal voltage/phase and E_0 is the emf per phase to which the generator is excited.

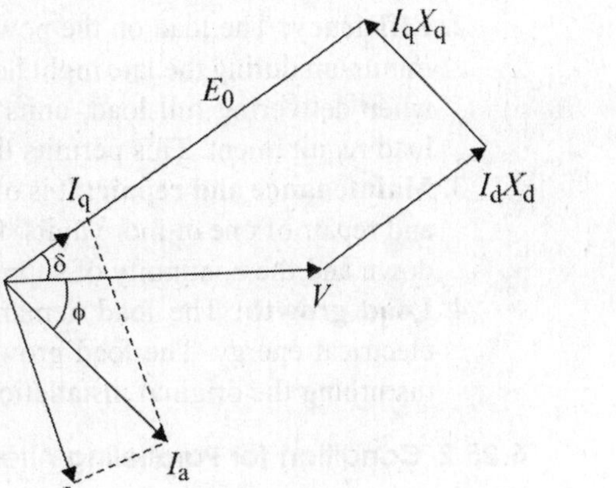

$$\therefore \quad I_q = I_a \cos(\delta + \phi)$$
$$\text{and} \quad I_d = I_a \sin(\delta + \phi)$$

The angle δ between E_0 and V is called the power angle or torque angle.

Fig. 6.33

$$E_0 = V \cos \delta + I_d X_d$$
$$= V \cos \delta + I_a X_d \sin(\delta + \phi) \quad [\because I_d = I_a \sin(\delta + \phi)]$$
$$E_0 = V \cos \delta + I_a X_d (\sin \delta \cos \phi + \cos \delta \sin \phi)$$

Also $\quad V \sin \delta = I_a X_q$
$$= I_a X_q \cos(\delta + \phi) \quad [\; I_q = I_a \cos(\delta + \phi)]$$
$$\therefore \quad V \sin \delta = I_a X_q (\cos \delta \cos - \sin \delta \sin \phi)$$

6.25 PARALLEL OPERATION OF ALTERNATORS

Electrical power systems are interconnected for better economy and reliable operation. Interconnection of AC power systems requires synchronism to operate in parallel with each other. In generating stations two or more generators are connected in parallel. In an interconnected system forming a grid the alternators are located at different places. They are connected in parallel by means of transformers and transmission under normal operating conditions are synchronous with each other.

An arrangement of generators for parallel operation enables a plant engineer to adjust the machines for optimum operating efficiency and greater reliability. As the load increases beyond the generated capacity of the connected units, similarly, as the load demand falls off, one or more of the machines are generally taken off the line to allow the units to operate at a higher frequency.

6.25.1 Advantages of Parallel Operation of Alternators

The following are the advantages of operating alternators in parallel:
1. **Continuity of service:** The Continuity of service is one of the important requirements of any electrical appartus. If one alternator fails, the continuity of supply can be maintained through the other healthy units. This will ensure uninterrupted supply to the consumers.
2. **Efficiency:** The load on the power system varies during the whole day; being minimum during the late night hours. Since alternators operate most efficiently when delivering full-load, units can be added or put off depending upon the load requirement. This permits the efficient operation of the power system.
3. **Maintenance and repair:** It is often desirable to carry out routine maintenance and repair of one or more units, for this purpose, desired unit/units can be shut down and the continuity of supply is maintained through the other units.
4. **Load growth:** The load demand is increased due to the increasing use of electrical energy. The load growth can be met by adding more units without disturbing the original installation.

6.25.2 Condition for Paralleling Alternator with Infinite Busbars

Most synchronous machines will operate in parallel with other and the process of connecting one machine in parallel with another with an infinite busbar is known as synchronizing.

Those machines already carrying load are known running machines while the alternator which is to be connected in parallel with the system is known as the incoming machine. Before the incoming machine is to be connected to the system, the following condition should be satisfied:
1. The phase sequence of the bus bar voltages and the incoming machine voltage must be the same.
2. The bus bar voltages and the incoming machine terminal voltage must be in phase.

3. The terminal voltage of the incoming machine should be equal to that of the alternator with which it is to be run in parallel or with the busbar voltage.

4. The frequency of the generated voltage of the incoming machine must be equal to the frequency of the voltage of the supply busbar.

6.26 METHOD OF SYNCHRONIZING

A stationary alternator must not be connected to supply busbar because the induced emf is zero at standstill and a short circuit will occur.

The synchronizing procedure and the equipment for checking it are needed, in order to check wheather one alternator is to be connected in parallel with another alternator or other alternator is connected to the infinite busbar or not.

The following methods are used for synchronization.

    1. Synchronizing lamps or three lamp method.

    2. Synchroscope

6.26.1 Three Lamp Method

In this method of synchronizing, three lamps L_1, L_2 and L_3 are connected as shown in Fig. 6.34. The lamp L_1 is straight connected between the corresponding phases (R_1 and R_2) and the other two are cross–connected between the other two phases.

Thus lamp L_2 is connected beween Y_1 and B_2 and lamp L_3 between B_1 and Y_2. When the frequency and phase voltage of the incoming alternator is same as the burbars, the straight connected Lamp L_1 will be dark while cross connected lamps L_2 and L_3 will be equally bright.

At this instant, the synchronisation is perfect and the switch of the incoming alternator can be closed to connect it to the burbars.

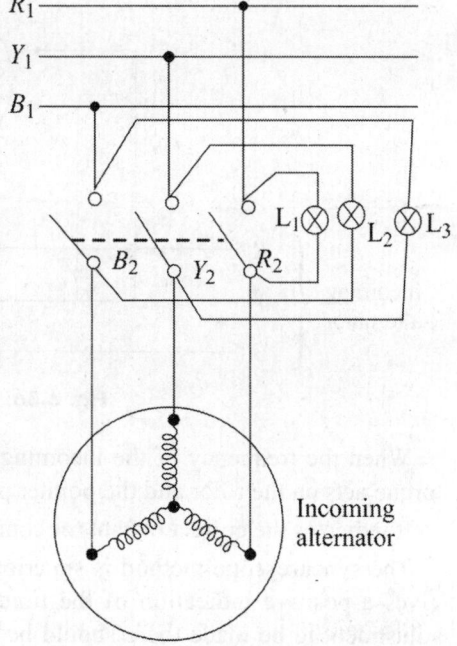

Fig. 6.34

In Fig. 6.35, phasors R_1, Y_1 and B_1 represent the busbar voltages and phasors R_2, Y_2 and B_2 represent the voltages of the incoming alternator. At the instant when R_1 is in phase with R_2, voltage across lamp L, is zero and voltages across lamps L_2 and L_3 are equal. Therefore, the Lamp L_1 is dark while Lamp L_2 and L_3 will be equally bright.

At this instant, the switch of the incoming alternator can be closed. Thus incoming alternator gets connected in parallel with the busbar.

6.26.2 Synchroscope

A synchroscope is an instrument that is indicated by means of a revolving pointer, the phase difference and frequency difference between the voltage of the incoming alternator and the busbars.

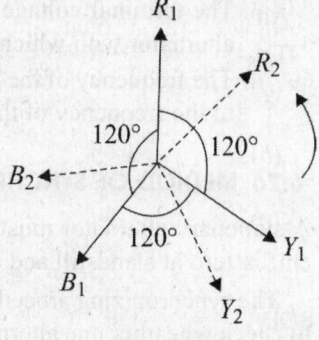

Fig. 6.35

It is essentially a small motor, the field being supplied from the busbar through a potential transformer and the rotor from the incoming alternator.

A pointer is attached to the rotor. When the incoming alternator is running fast (i.e. frequency of the incoming alternator is higher than that of the busbar), the rotor and hence the pointer moves in the clockwise direction.

When the incoming alternator is running slow (i.e. frequency of the incoming alternator is lower than that of the busbar), the pointer moves in anticlockwise direction.

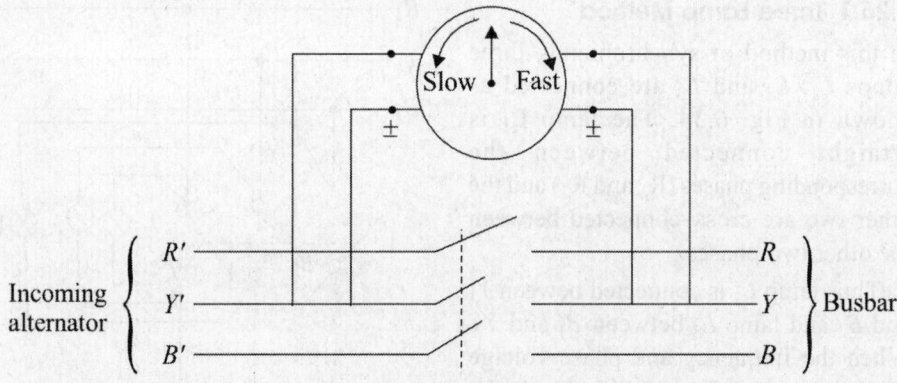

Fig. 6.36: Synchroscope

When the frequency of the incoming alternator is equal to that of the busbar, no torque acts on the rotor and the pointer points vertically upwards. (12 O' clock).

It indicates the correct instant for connecting the incoming alternator to the busbar.

The synchroscope method is superior, than the lamp method because it not only gives a positive indication of the time to close the switch but also indicates the adjustment to be made there should be a difference between the frequencies of the incoming alternator and the busbar (Figs 6.35 and 6.36).

6.27 ALTERNATOR ON INFINITE BUSBAR

Upto this point, we have considered only a single alternator working on an isolated load or two such machine operating in parallel. In practice, generating stations do not operate as isolated units but are interconnected by the national grid. The result is that a very large number of alternators operate in parallel.

An alternator connected to such a network is said to be operating on infinite busbar.

The characteristics of an infite busbar as follows:

(a) The terminal voltage remains constant, because the incoming machine is too small to increase or decrease it.

(b) The frequency remains constant because the rotational inertia is too large to enable the incoming machine to alter the speed of the system, and

(c) The synchronous impedance is very small since the system has a large number of alternators in parallel.

SOLVED UNIVERSITY PROBLEMS FOR PRACTICE

Example 6.1: A 3-phase, 50 Hz, 2 pole star connected turbo alternator has 54 slots with 4 conductors per slot. The pitch of the coil is 2 slots less than the pole pitch the machine gives 3300 V between the lines on open circuit with sinusoidal distribution, determine the useful flux per pole. **[RTU 2007–08]**

Solution: Given: $f = 50$ Hz, $P = 2$, 54 slots, 4 conductors/slot, $E_{line} = 3300$ V.

$$n = \text{slots/pole} = \frac{54}{2} = 27$$

$$\beta = \frac{180}{n} = \frac{180}{27} = 6.6666°$$

$$\alpha = 2 \text{ slot less} = 2 \times \beta = 2 \times 6.6666 = 13.333°$$

$$m = \text{slot/phase} = \frac{n}{3} = \frac{27}{3} = 9$$

$$K_d = \frac{\sin\dfrac{m\beta}{2}}{m\sin\dfrac{\beta}{2}} = \frac{\sin\left(\dfrac{9 \times 6.6666°}{2}\right)}{9\sin\left(\dfrac{6.6666°}{2}\right)} = 0.95547$$

$$K_c = \cos\frac{\alpha}{2} = \cos\left(\frac{13.333°}{2}\right) = 0.9932$$

$$E_{ph} = \frac{E_{line}}{\sqrt{3}} = \frac{3300}{\sqrt{3}} = 1905.2558 \text{ V}$$

$$Z = \text{Slots} \times \text{conductor/slot} = 54 \times 4 = 216$$

$$Z_{ph} = \frac{Z}{3} = \frac{216}{3} = 72$$

$$T_{ph} = \frac{Z_{ph}}{2} = \frac{72}{2} = 36$$

$$E_{ph} = 4.44\, K_C \cdot Kd \cdot f\phi \cdot T_{ph}$$

$$1905.2558 = 4.44 K_C \times K_d \times f \times T_{ph}$$

$$= 4.44 \times 0.9932 \times 0.95547 \times \phi \times 50 \times 36$$

$$\boxed{\phi = 0.2512 \ w_b} \text{ flux/pole.}$$

Example 6.2: A 3-phase star connected aternator has the following data: voltage required to be generated on open circuit = 4000 V at 50 Hz speed = 500 rpm, stator has slots/pole/phase = 3 and conductors/slot = 12.

Calculate (*i*) Distribution factor, (*ii*) Useful flux per pole.　　　　**[RTU 2002]**

Solution: Given: $f = 50$ Hz, $N_S = 500$ rpm, $E_{line} = 4000$ V, $m = 3$,

$$N_S = \frac{120f}{P}, \text{ i.e. } 500 = \frac{120 \times 50}{P}$$

$\therefore$ 　　　$P = 12$

$$n = \text{slots/pole} = m \times 3 = 3 \times 3 = 9$$

$$\text{slots} = n \times P = 9 \times 12 = 108$$

$$\beta = \frac{180°}{\eta} = \frac{180°}{9} = 20°$$

$$K_d = \frac{\sin\left(\dfrac{m\beta}{2}\right)}{m\sin\left(\dfrac{\beta}{2}\right)} = \frac{\sin\left(\dfrac{3 \times 20}{2}\right)}{3\sin\left(\dfrac{20}{2}\right)} = 0.9598$$

$$Z = \text{Total conductors} = \text{slots} \times \text{conductors/slot}$$

$$= 108 \times 12 = 1296$$

$$Z_{ph} = \frac{Z}{3} = \frac{1296}{3} = 432$$

$$T_{ph} = \frac{Z_{ph}}{2} = \frac{432}{2} = 216$$

Assume full pitch coils hence $K_c = 1$

$$E_{ph} = \frac{E_{line}}{\sqrt{3}} = \frac{4000}{\sqrt{3}} = 2309.40107 \text{ V}$$

$$E_{ph} = 4.44\, f\phi T_{ph} \times K_c \times K_d$$

$$2309.40107 = 4.44 \times 1 \times 0.9598 \times \phi \times 50 \times 216$$

$$\boxed{\phi = 0.05017 \text{ Wb}}$$

Example 6.3: A 3-phase, 4 pole, 50 Hz star connected alternator has 60 slots with 2 conductor per slot and having armature winding of the double layer type coils are short pitched such that if one coil side lies in slot number 1 then other side lies in slot number 13. Find the useful flux per pole required to induced a line voltage of 6.6 kV.

　　　　[RTU 2003]

Solution: Given: $f = 50$ Hz, $P = 4$, 2 conductors/slot, 60 slots, $E_{line} = 6.6$ kV

$$Z = \text{Total conductors} = 60 \times 2 = 120$$

$$Z_{ph} = \frac{Z}{3} = \frac{120}{3} = 40 \text{ hence } T_{ph} = \frac{Z_{ph}}{2} = 20$$

Coil span = $13 - 1 = 12$ slots

$$n = \text{slots/pole} = \frac{60}{4} = 15$$

$$\beta = \frac{180°}{n} = \frac{180°}{15} = 12°$$

$$\therefore \quad \text{Coil span} = 12 \times \beta = 12 \times 12° = 144°$$

$$\alpha = 180 - \text{coil span} = 180° - 144° = 36°$$

$$m = \text{slot/phase} = \frac{n}{3} = \frac{15}{3} = 5$$

$$K_d = \frac{\sin\left(\dfrac{m\beta}{2}\right)}{m\sin\left(\dfrac{\beta}{2}\right)} = 0.9567 \text{ and } K_c = \cos\frac{\alpha}{2} = \cos 18°$$

$$K_c = 0.95105$$

$$E_{ph} = 4.44 K_c K_d \phi f T_{ph}$$

$$3810.5117 = 4.44 \times 0.9567 \times 0.95105 \times \phi \times 50 \times 20$$

$$\boxed{\phi = 0.9432 \text{ Wb flux per pole}}$$

Example 6.4: A 3-phase, 12 pole, star connected aternator has 180 slots with 10 conductors per slot and conductors of each phase are connected in series. The coil span is 144° electrical. Determine the phase and line values of emf. if the machine runs at 600 rpm and the flux per pole is 0.06 Wb, distributed sinusoidally over the pole. **[RTU 2005]**

Solution: Given: $P = 12$, $N_S = 600$ rpm, 180 slots, 10 conductors/slot $\phi = 0.06$ Wb.

$$N_S = \frac{120f}{P}, \text{ i.e. } 600 = \frac{120f}{12} \quad \therefore f = 60 \text{ Hz}$$

$$Z = \text{slots} \times \text{conductors/slots} = 180 \times 10 = 1800$$

$$\therefore \quad Z_{ph} = \text{slots/phase} = \frac{Z}{3} = \frac{1800}{3} = 600$$

$$\therefore \quad T_{ph} = \frac{Z_{ph}}{2} = \frac{600}{2} = 300 \text{ (two conductors-one turns)}$$

$$n = \frac{\text{slots}}{\text{pole}} = \frac{180°}{12} = 15$$

$$\beta = \frac{180°}{n} = \frac{180°}{15} = 12°$$

$$m = \text{slot/phase} = \frac{\eta}{3} = \frac{15}{3} = 5$$

$$\alpha = 180° - \text{coil span} = 180° - 144° = 36°$$

$$K_d = \frac{\sin\left(\dfrac{m\beta}{2}\right)}{m\sin\left(\dfrac{\beta}{2}\right)} = 0.95667 \text{ and } K_c = \cos\frac{\alpha}{2} = 0.95105$$

$$E_{ph} = 4.44 K_c K_d \phi T_{ph}$$
$$= 4.44 \times 0.95667 \times 0.95105 \times 0.06 \times 60 \times 300$$
$$E_{ph} = 4362.8994 \text{ V phase emf}$$
$$E_{line} = \sqrt{3} E_{ph} = \sqrt{3} \times 4362.8994 = 7556.763$$

$$\boxed{E_{line} = 7556.763 \text{ V}}$$

Example 6.5: A 3-φ star connected alternator supplies a load of 1000 kW at a power factor 0.8 lags with a terminal voltage of 11 kV. Its armature resistance is 0.4Ω per phase while synchronous reactance is 3Ω per phase. Calculate the line value of emf generated and the regulation at this load.

Solution: Given: $P = 1000$ kW, $\cos\phi = 0.8$, $V_t = 11$ kV, $R_a = 0.4$ Ω, $X_S = 3$ Ω, $E_L = ?$.

$$E_{ph} = \sqrt{(V_{pt}\cos\phi + I_a R_a)^2 + (V_{pt}\sin\phi + I_a X_s)^2}$$

$$E_{ph} = \sqrt{(6350.8 \times 0.8 + 65.6 \times 0.4)^2 + (6350 \times 0.6 + 65.6 \times 3)^2}$$

$$E_{ph} = 6491.47 \text{ volt.}$$

$$V_{ph} = \frac{V_{pt}}{\sqrt{3}} = \frac{11 \times 10^3}{\sqrt{3}} = 6350.85 \text{ V}$$

$$P = \sqrt{3} V_L I_L \cos\phi = 1000 \times 10^3$$

$$I_L = \frac{1000000}{\sqrt{3} \times 11000 \times 0.8} = 65.6 \text{ A}$$

In Y connected, $\quad I_L = I_{ph} = I_a = 65.6$A

$$E_{ph} = \frac{E_L}{\sqrt{3}}$$

∴

$$E_L = \sqrt{3} \times E_{ph}$$

$$E_L = \sqrt{3} \times 6491.47 \text{ V}$$

$$\boxed{E_L = 11.24 \text{ kV}}$$

$$VR = \frac{E_p t - V_p t}{E_p t} = \frac{6491 - 6350}{6491} \times 100 = 2.214\%$$

$$\boxed{VR = 2.214\%} \text{ Ans.}$$

Example 6.6: A 1200 kVA, 6600 V, 3-φ star connected alternator has its armature resistance as 0.25 Ω per phase and its synchronours reactance as 5 Ω per phase. Calculate its regulation if it delivers a full load at (*i*) 0.8 lags (*ii*) 0.8 leading pf.

Solution: Given: $S = 1200$ kVA, $V = 6600$ V, $R_a = 0.25$, $X_S = 5$ Ω.

$$I_L = \frac{1200 \times 10^3}{\sqrt{3} \times 6600}$$

$$= 104.97 \text{ A}$$

$$I_L = I_{ph} = I_a = 104.97 \text{A}$$

$$V_{ph} = \frac{V_L}{\sqrt{3}} = \frac{6600}{\sqrt{3}} = 3810.512 \text{ V}$$

(i) For lagging pf

$$E_{ph} = \sqrt{(V_{ph} \cos\phi + I_a R_a)^2 + (V_{ph} \sin\phi + I_a \times s)^2}$$

$$= \sqrt{(3810.51 \times 0.8 + 104.97 \times 0.25)^2 + (3810.5 \times 0.6 + 104 \times 5)^2}$$

$$E_{ph} = 4166.06 \text{ V}$$

$$\% \text{ VR} = \frac{E_{ph} - V_{ph}}{E_{ph}} = \frac{4166 - 3810.512}{4166} \times 100$$

(ii) For leading, 0.8 pf

$$E_{ph} = \sqrt{(3810.51 \times 0.8 + 104.9 \times 0.25)^2 + (3810.5 \times 0.8 - 104.97 \times 5)}$$

$$= 3543.47 \text{ volt.}$$

$$\% \text{ VR} = \frac{E_{ph} - V_{ph}}{V_{ph}} = \frac{3543.47 - 3810}{3810.5} \times 100$$

$$= -7\% \text{ Ans.}$$

Example 6.7: In a 3-φ star connected alternator there an 2 coil sides per slots and 16 turns per coils. Armature has 288 slots on its periphery. When driven at 250 rpm, it produces 6600 V, between the lines at 50 Hz. The pitch of the coil is 2 slots less than the full pitch. Calculate the flux per pole.

Solution: Given: $N = 250$.

$$\frac{120f}{P} = 250$$

$$P = \frac{120 \times 50}{250} = 24$$

$$\boxed{P = 24}$$

$$n = \frac{288}{24} = 12$$

$$m = \frac{12}{3} = 4$$

$$\beta = \frac{180}{12} = 15$$

$$K_d = \frac{\sin\left(\frac{4 \times 15}{2}\right)}{4\sin(7.5°)}$$

$$= 0.9576$$

$$\alpha = 2 \times 15 = 30°$$

$$K_C = \cos\frac{\alpha}{2}$$

$$= \cos 15° = 0.9659$$

$$Z = 288 \times 16 \times 2$$

$$T_{ph} = \frac{288 \times 16 \times 2}{2 \times 3} = 1536$$

$$E_{ph} = \frac{E_{line}}{\sqrt{3}} = \frac{6600}{\sqrt{3}} = 3810.51$$

$$3810.51 = 4.44 \times \phi \times 50 \times 1536 \times 0.9576 \times 0.095$$

$$\therefore \quad \boxed{\phi = 0.012\text{wb} = 12\text{mWb}} \quad \textbf{Ans.}$$

Example 6.8: A single phase 1500 rpm, 4 pole alternator has 8 conductor per slots with total of 24 slots. The winding is short pitched by 1/6th of full pitch. Assume distributed winding with $\phi = 0.05$ Wb. Calculate induced emf.

Solution:

$$N_S = 1500 = \frac{120 \times 50}{P} \quad (f = 50)$$

$$\boxed{P = 4}$$

$$n = \frac{24}{4} = 6, \, m = 6$$

$$\beta = \frac{180°}{6} = 30°, \, K_d = \frac{\sin\left(\frac{6 \times 30}{2}\right)}{6\sin\left(\frac{30}{2}\right)}$$

$$K_d = 0.6439$$

$$= \frac{1}{6} \times 6 = 1 \text{ slots}$$

Full pitch $= n = 6$ slots.

$$\alpha = 30°$$

$$K_c = \cos 15° = 0.9659$$

$$T = \frac{8 \times 24}{2} = \frac{192}{2} = 96$$

$$E_{ph} = 4.44 \times 0.9659 \times 0.643 \times 50 \times 96 \times 0.05$$

$$\boxed{E_{ph} = 662.70 \text{ volts}} \text{ Ans.}$$

Example 6.9: A 3-phase star connected alternator has a slots per pole, carrying full pitch winding. The alternator has 48 poles and driven at 125 rpm. The winding is double layer having 4 terms in a coil. If flux per pole is 51.75 mWb. Determine the value of induced emf between the line.

Solution: Given: $P = 48$, $N = 125$ rpm.

$$125 = \frac{120 \times f}{48} \quad \therefore \ f = 50 \text{ Hz}$$

$$n = 9, \ m = 3$$

$$\beta = \frac{180°}{9} = 20°$$

$$K_d = \frac{\sin\left(\dfrac{3 \times 20}{2}\right)}{3 \sin 10°} = 0.9597$$

$$Z = 9 \times 48 \times 8$$

$$T = \frac{Z}{3 \times 6} = \frac{4 \times 48 \times 8}{6} = 576$$

$$E_{ph} = 4.44 \phi K_c K_d T_{ph} f$$
$$= 4.44 \times 51.75 \times 10^{-3} \times 1 \times 576 \times 50 \times 0.9597$$
$$= 6350.69 \text{ V}$$

$$E_L = E_{ph} \times \sqrt{3}$$
$$= 6350.69 \times \sqrt{3} = 11 \text{ kV Ans.}$$

Example 6.10: The open circuit and short circuit test is conducted on a 3-ϕ star connect, 866 V, 100 kVA alternator.

The OC test result are:

I_f	1	2	3	4	5	6	7
V_{OC} line	173	310	485	605	728	790	840

The field current of 1A, produce a short circuit current of 25A. $R_a = 0.15$ W. Calculate regulation at 0.8 lagging pf.

Solution: $V_L = 866$ V, kVA $= 100$.

$$I_L = I_{ph} = \frac{100 \times 10^3}{\sqrt{3} \times 866} = 66.67 \text{ A}$$

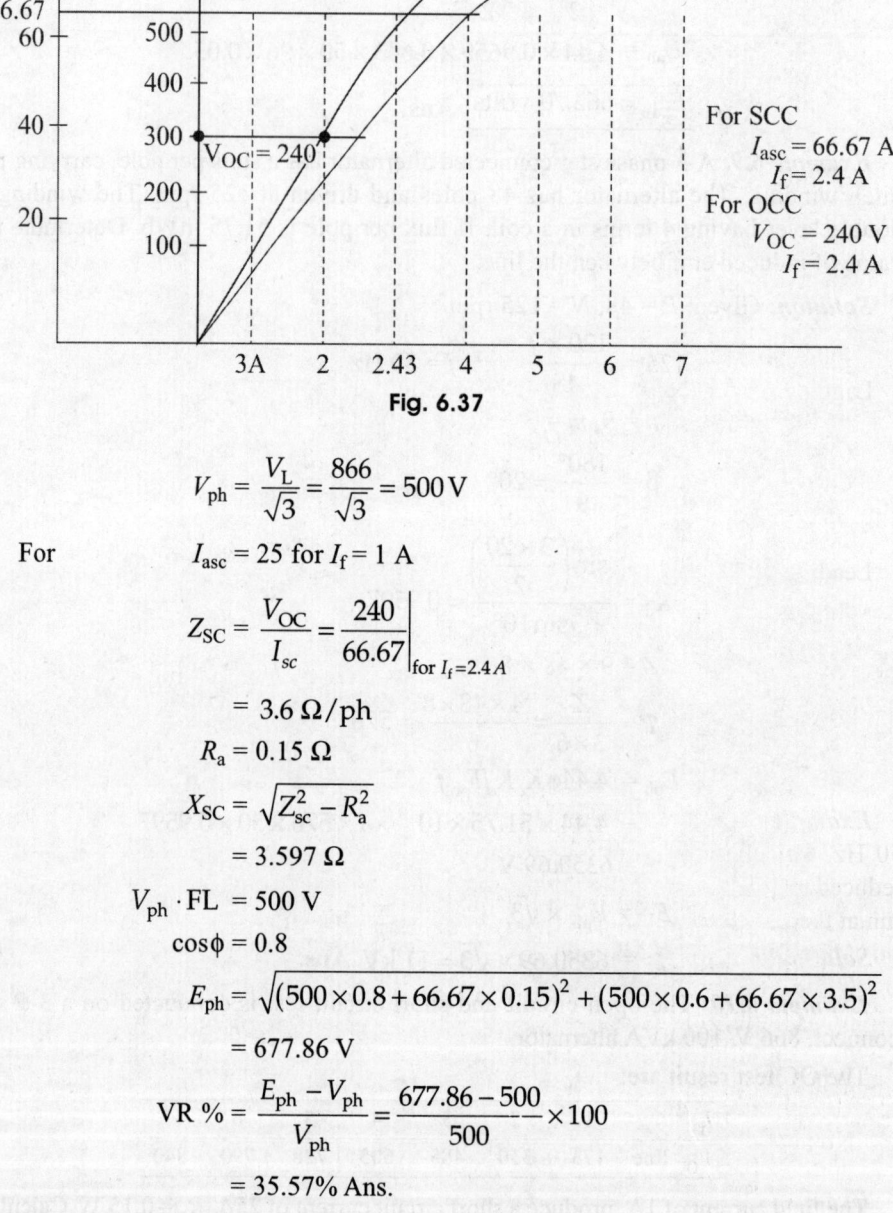

For SCC
$I_{asc} = 66.67$ A
$I_f = 2.4$ A
For OCC
$V_{OC} = 240$ V
$I_f = 2.4$ A

Fig. 6.37

$$V_{ph} = \frac{V_L}{\sqrt{3}} = \frac{866}{\sqrt{3}} = 500 \, \text{V}$$

For

$$I_{asc} = 25 \text{ for } I_f = 1 \text{ A}$$

$$Z_{SC} = \frac{V_{OC}}{I_{sc}} = \frac{240}{66.67}\Big|_{\text{for } I_f = 2.4 A}$$

$$= 3.6 \, \Omega/\text{ph}$$

$$R_a = 0.15 \, \Omega$$

$$X_{SC} = \sqrt{Z_{sc}^2 - R_a^2}$$

$$= 3.597 \, \Omega$$

$$V_{ph} \cdot \text{FL} = 500 \text{ V}$$

$$\cos\phi = 0.8$$

$$E_{ph} = \sqrt{(500 \times 0.8 + 66.67 \times 0.15)^2 + (500 \times 0.6 + 66.67 \times 3.5)^2}$$

$$= 677.86 \text{ V}$$

$$\text{VR \%} = \frac{E_{ph} - V_{ph}}{V_{ph}} = \frac{677.86 - 500}{500} \times 100$$

$$= 35.57\% \text{ Ans.}$$

Example 6.11: A 230 V, 3-ϕ star connected alternator gives on open circuit, emf of 230 V. For a field current of 0.38A. The same field current on short circuit causes an armature current of 12.5 A. The armature resistance measured between two lines is 1.8 Ω. Find the regulation for the current of 10 A at 0.8 pf lagging and 0.8 leading pf.

Solution: Given: $V_L = 230$ V, $R_{Ry} = 1.8$ Ω, $V_{OC} = 230$ V, $I_{ac} = 12.5$ A, for same $I_4 = 0.38$A.

$$Z_S = \frac{V_{oc}ph}{I_{ac}ph} = \frac{230/\sqrt{3}}{12.5} = 10.623 \ \Omega/ph$$

$$2R_a = 1.8$$

$$R_a = \frac{1.8}{2} = 0.9 \ \Omega/ph$$

$$X_S = \sqrt{Z_s^2 - R_a^2} = \sqrt{(10.623)^2 - (0.9)^2}$$

$$X_S = 10.585 \ \Omega$$

$$\boxed{I_a = 10 \ A}$$

Lagging pf:
$$E_{ph} = \sqrt{(132.79 \times 0.8 + 10 \times 0.9)^2 + (132.7 \times 0.6 + 10 \times 10.58)^2}$$

$$= 218.39 \ V$$

$$\% \ R_{eg} = \frac{218.39 - 132.79}{132.79} \times 100 = 64.46\%$$

Leading pf:
$$E_{ph} = \sqrt{(132.79 \times 0.8 + 101 \times 0.9)^2 + (132.79 \times 0.6 - 10 \times 10.585)^2}$$

$$\boxed{E_{ph} = 118.168 \ V}$$

$$\% \ R_{eg} = \frac{118.168 - 132.79}{132.79} \times 100$$

$$\boxed{VR = -11.01\%}$$ **Ans.**

Example 6.12: A 4-pole AC generator is running and producing the frequency of 50 Hz. Calculate the revolulations per minute of the generator. If the frequency is reduced to 15 Hz, how many number of poles will be required if the generator is to be run at the same speed.

Solution: Given: Frequency generated, $f = 50$ Hz, Number of poles, $P = 4$.

$$\text{Speed}, \ N = \frac{120f}{P} = \frac{120 \times 50}{4} = 1500 \ \text{rpm} \ \textbf{Ans.}$$

When frequency, $\qquad f' = 15$ Hz

Number of poles required, $P' = \dfrac{120f'}{N} = \dfrac{120 \times 15}{1500} = 1.2 \cong 2$ **Ans.**

Example 6.13: A 3-phase, 50 Hz, 12-pole alternator has a star-connnected with 140 slots and 10 conductors per slot. The flux per pole is 0.08 Wb, sinusoidally distributed. Determine the phase and line voltages.

Solution: Let us consider full pitch coil. Thus,

$$\alpha = 0°, \ k_c = \cos \alpha/2 = \cos 0° = 1$$

$$m = \frac{\text{slots}}{\text{poles} \times \text{phase}} = \frac{140}{12 \times 3} = 3.889 \cong 4$$

$$\beta = \frac{180 \times \text{poles}}{\text{slots}} = \frac{180 \times 12}{140} = 15.428$$

$$k_d = \frac{\sin \dfrac{m\beta}{2}}{m \sin \dfrac{\beta}{2}} = \frac{\sin \dfrac{4 \times 15.28}{2}}{4 \sin \dfrac{15.428}{2}} = 0.954$$

Total number of conductors = conductors per solts × number of slots
$$= 10 \times 140 = 1400$$

Conductors per phase, $E_p = 2.22 \times k_c \times k_d \times f \times \phi \times Z_p$
$$= 2.22 \times 1 \times 0.954 \times 50 \times 0.08 \times 460$$
$$= 3896.89 \text{ V}$$

Generated line voltage, $E_L = \sqrt{3} \times E_p$
$$= \sqrt{3} \times 3896.89$$
$$= 6749.61 \text{ V}$$

Example 6.14: A a field excitation of 10 A in a certain alternator gives a current of 200 A on short-circuit and a terminal voltage of 1 kV (phase value) on open-circuit, find the internal voltage drop with a load current of 90 A.

Solution: Generally alternators are star connected.

Synchronous impedance, $Z_S = \dfrac{\text{o.c. volts/phase}}{\text{s.c. current/phase}}$

$$= \frac{1000}{200} = 5 \, \Omega$$

When the load current, $I_a = 90$ A, then

Internal voltage drop $= I_a Z_s = (90 \times 5) = 450$ V

Example 6.15: A 2000 kVA, 6800 V, 3-phase star connected alternator with a resistance of 0.6 Ω and reactance of 8 Ω per phase delivers full load current at power factor 0.75 lagging and normal rated votlage. Estimate the terminal voltage for the same excitation and load current at 0.08 p.f. leading.

Solution: Full load current, $I_a = \dfrac{2000 \times 10^3}{\sqrt{3} \times 6800}$

$$= 169.80 \cong 170 \, \text{A}$$

Terminal voltage/phase, $V = \dfrac{6800}{\sqrt{3}} 392.98 \cong 3926 \, \text{volts}$

$$E_0 = \sqrt{(V \cos\phi + I_a R_a)^2 (V \sin\phi + I_a X_s)^2}$$

$I_a R_a$ drop $= 170 \times 0.6 = 102$ volts

I_aX_s drop $= 170 \times 8 = 1360$ volts

$$E_0 = \sqrt{(3926 \times 0.75 + 102)^2 + (3926 \times 0.66 + 1360)^2}$$

$$= \sqrt{(3046.5)^2 + (3951.1)^2}$$

$$= \sqrt{24892353.46}$$

$$= 4989.22 \cong 4989 \text{ volts/phase}$$

Since excitation remains constant, no-load emf in the second case will also be 4989 volt/phase. Let the new terminal voltage/phase be V'.

$$E_0 = \sqrt{(V' \cos\phi + I_a R_a)^2 + (V' \sin\phi - I_a X_s)^2}$$

$$4989 = \sqrt{(V' \times 0.8 + 102)^2 + (V' \times 0.6 - 1360)^2}$$

$$(4989)^2 = (V' \times 0.8 + 102)^2 + (V' \times 0.6 - 1360)^2$$

$$(4989)^2 = (0.8\ V')^2 + 2\ V' \times 0.8 \times 102 + (102)^2 + (0.6V')^2$$
$$- 2V' \times 0.6 \times 1360 + (1360)^2$$

$$(4989)^2 = (V')^2 + 1860004$$

$$V' = \sqrt{23030117} = 4798.97 \cong 4800 \text{ volts/phase}$$

Line to Line terminal voltage $= \sqrt{3} \times 4800$

$$= 8313.84 \cong 8314 \text{ V}$$

EXERCISE

1. Draw neat diagram of synchronous alternator and derive the emf equation.
   [2003–2004]
2. Explain the following: (a) synchronizing of alternators (b) armature reaction in alternator. **[2003–2004]**
3. Define form factor, chording factor and breadth factor.
4. Draw phasor diagram of a nonsalient pole machine supplying full load at: (a) lagging power factor and (b) leading power factor. **[2004–2005]**
5. Discuss the synchronous impedance method of calculating the voltage regulation of an alternator at different load power factors. Give phasor diagrams.
   [2005–2006]
7. Explain the terms coil span factor and distribution factor in connection with alternator armature windings and deduce the emf equation of an alternator incorporating the effects of these factors. **[2005–2006]**
7. Describe experimental method of determination of voltage reulgation of a 3 phase synchronous generator following synchronous impedance method.
   [2006–2007]
8. Explain the constructional details of a synchronous machine giving reasons for making two different types of rotors. **[2007–2008]**

Synchronous Motors

7.1 SYNCHRONOUS MOTORS

A synchronous motor is electrically identical with an alternator or an AC generator. However, when used as synchronous motor, the three phase AC supply is connected to the stator and the external excitation is connected to the rotor field. The synchronous machine will then operate as a motor to deliver rotational mechanical energy to the load. The speed of rotation is called as *synchronous speed* N_s which is directly proportional to supply frequency because

$$N_s = \frac{120f}{P}$$

7.1.1 Important Characteristics of Synchronous Motors

A synchronous motor is electrically identical with an alternator or an AC generator. Some characteristic features of a synchronous motor are as follows:
- It runs either at synchronous speed or not at all, i.e. while running it maintains a constant speed.
- The only way to change its speed is to vary the supply frequency.
  (Because synchronous speed $N_s = 120f/P$)
- It is not inherently self starting. It has to run upto synchronous (or near synchronous) speed by some means before it can be synchronized to the supply With electronic controller used, this problem can be overcome.
- It is capable of being operated under a wide range of power factor both lagging and leading. Hence it can be used for the power factor correction purpose in addition to supplying torque to drive loads.

7.2 CONSTRUCTION OF THREE PHASE SYNCHRONOUS MOTOR

Since the alternators can be used a synchronous motor, the construction of synchronous motor is same as that of an alternator. It has two windings namely stator or armature and rotor or field winding. The armature (stator) winding is placed in the slots which are distributed over the entire yoke as shown in Fig. 7.1.

The stator winding is connected to the 3 phase AC supply. The field winding is placed in the rotor slots and it is connected to the DC supply via two slip rings. The

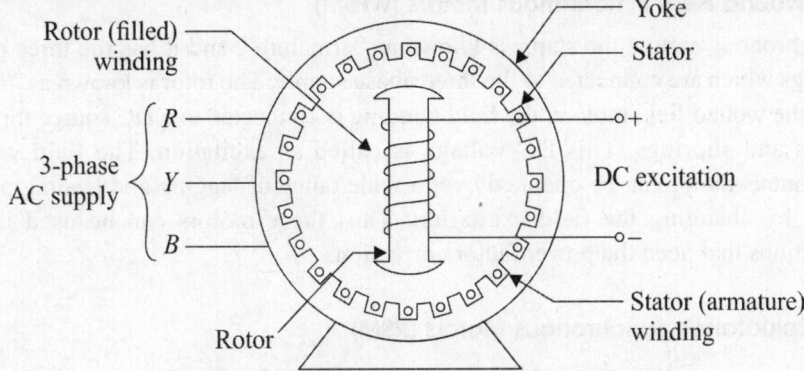

Fig. 7.1: Construction of synchronous motor

DC supply is also called as excitation and it is obtained from a DC shunt generator mounted on the same shaft. The rotor can be a salient-projected type or cylindrical type. Normally the salient pole rotor is preferred.

7.3 ROTATING MAGNETIC FIELD (RMF)

When the stator winding (star or delta connected) is connected to a 3-phase AC supply, a rotating magnetic field (rmf) is produced in the air gap between the stator and rotor. The RMF rotates at a speed called synchronous speed N_s which is given by

$$N_s = \frac{120f}{P}$$

where f = Supply frequency.

P = Number of poles for which the stator is wound.

7.4 TYPES OF SYNCHRONOUS MOTORS

There are various types of synchronous motors. The most widely used are:
- Permanent magnet synchronous motors.
- Wound field synchronous motor.
- Reluctance type synchronous motor.

7.4.1 Permanent Magnet Synchronous Motors (PMSM)

These motors do not have the field (rotor) winding. Instead they have a permanent magnet rotor. No external DC excitation is needed. This will eliminate the need for the slip rings and brushes. This reduces the winding losses and necessity of maintenance. PMSM cannot produce very high torque and hence preferred in low power applications only. Such motors are often referred to as brushless DC motor or electrically commutated motors. Because of the elimination of "field winding" the ability to control the power factor is lost.

7.4.2 Wound Field Synchronous Motors (WFSM)

In synchronous motors the stator is known as "armature", and it has the three phase windings which are connected to the three phase supply. The rotor is known as "field" and in the wound field motors, the field winding is connected to a DC source through brushes and sliprings. This DC voltage is called as excitation. The field wound synchronous motor can be operated over a wide range of lagging and leading power factors by changing the field excitation. Thus these motors can be used in the applications that need the power factor correction.

7.4.3 Reluctance Synchronous Motors (RSM)

These motors also do not have the field (rotor) winding, but they have salient poles on the rotor. The flux is the airgap between the stator and the rotor is produced only due to the current drawn by the stator (armature) winding of the motor. The torque producing capacity is therefore low and the motor has low power factor. It is therefore used for applications requiring low torque and power.

7.5 OPERATING PRINCIPLE OF SYNCHRONOUS MOTORS

The fact the synchronous motor has no self starting torque can be easily explained.

1. Consider a 3-phase synchronous motor having two rotor poles $N_R S_R$. Then the stator will also be wound for two poles $N_S S_S$. The motor has direct voltage applied to the rotor winding and 3 ph voltage applied to the stator winding. The stator winding produces a rotating field which revolves around the stator at synchronous speed N_S (= 120 f/P). The direct (or zero frequency) current sets up a two pole field which is situation so long as the rotor is not turning. Thus we have a situation in which there exists a pair of revolving armature poles (i.e. N_S–S_S) and a pair of stationary rotor poles (i.e. N_R–S_R).

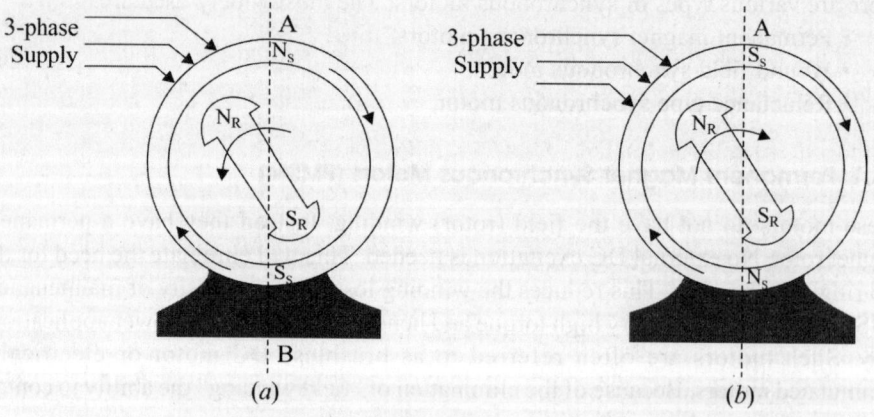

(a)   (b)

Fig. 7.2

2. Suppose at any instant, the stator poles are at positions A and B as shown in Fig. 7.2(a). It is clear that poles N_S and N_R repel each other and so do the poles S_S and S_R. Therefore the rotor tends to move in the anticlockwise direction. After a period of half-cycle $\left(\text{or } \dfrac{1}{2}f = \dfrac{1}{100}\sec\right)$, the polarities of the stator poles are reversed but the polarites of the rotor poles remain the same as shown in Fig. 7.2(b). Now S_S and N_R attract each other and So do N_S and S_R. Therefore, the rotor tends to move in the anticlockwise direction. Since the stator poles change their polarities rapidly, they tend to pull the rotor fast in one direction and then after a period of half cycle in the other. Due to high inertia of the rotor, the motor fails to start.

7.5.1 How to Get Contineous Unidirectional Torque?

If the rotor poles are rotated by some external means at such a speed that they interchange their position along with the stator poles, then the rotor will experience continuous undirectional torque.

This can be understand from the following discussion.

(i) Suppose the stator field is rotating in the clockwise direction and the rotor is also rotated clockwise by some external means at such a speed that the rotor poles interchange their position along with the stator poles.

(ii) Suppose at any instant the stator and rotor poles are in the position shown in Fig.7.3(a). It is clear that torque on the rotor will be clockwise. After a period of half cycle, the stator poles reverse their polarities and at the same time rotor poles also interchange their position in shown in Fig 7.3(b). The result is that again the torque on the rotor is clockwise direction. Under this condition, poles on the rotor always face poles of opposite polarity on the stator and a strong magnetic attraction is set up between them. This mutual attraction locks the rotor and stator together and the rotor is virtually pulled into step with the speed of revolving flux (i.e. synchronous speed).

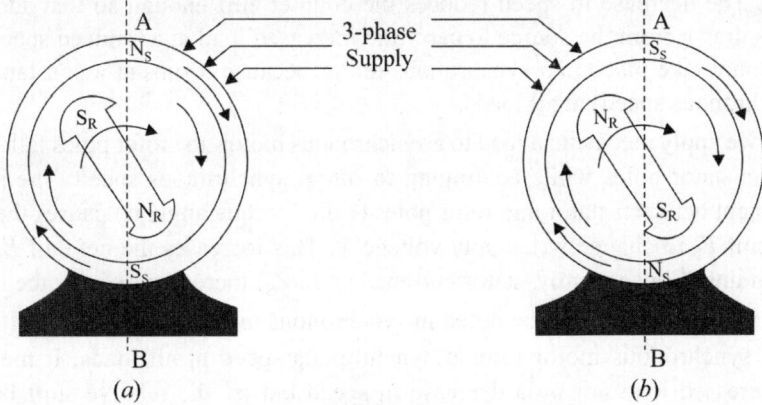

Fig. 7.3

(*iii*) If now the external prime mover driving the rotor is removed, the rotor will continue to rotate at synchronous speed in the clockwise direction because the rotor poles are magnetically locked up with the stator poles.

7.5.2 Making Synchronous Motor Self Starting

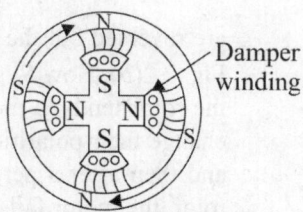

A synchronous motor cannot start by itself. In order to make the motor self starting, a squirrel cage winding (also called damper winding) is provided on the rotor. The damper winding consists of copper bars embedded in the pole faces of the salient poles of the rotor as shown in Fig. 7.4. The bars are short-circuited at the ends to the form in effect a partial squirrel cage winding. The damper winding serves to start the rotor.

Fig. 7.4

(*i*) To start with, 3-ph supply is given to the stator winding while the rotor field winding is left unenergised. The rotating stator field induces currents in the damper or squirrel cage winding and the motor starts as an induction motor.

(*ii*) As the motor approaches the synchronous speed, the rotor is excited with direct current. Now the resulting poles on the rotor face poles of opposite polarity on the stator and a strong magnetic attraction is set up between them. The rotor poles locks in with the poles of rotating flux. Consequently, the rotor revolves at the same speed as the stator field, i.e. at synchronous speed.

(*iii*) Because the bars of squirrel cage portion of the rotor now rotate at the same speed as the rotating stator field, those bars do not cut any flux and therefore, have no induced currents in them. Hence squirrel cage portion of the rotor is in effect, removed from the operation of the motor.

7.6 MOTOR ON LOAD

In DC motors and induction motors, an addition of load causes the motor speed to decrease. The decrease in speed reduces the counter emf enough so that additional current is drawn from the source to carry the increased load at a reduced speed. This action cannot take place in a synchronous motor because it runs at a constant speed (i.e. synchronous speed) at all loads.

When we apply mechanical load to a synchronous motor, the rotor poles fall slightly behind the stator poles while continuing to run at synchronous speed. The angular displacement between stator and rotor poles (called torque angle α) causes the phase of back emf E_b to change wrt supply voltage V. This increases the net emf E_r in the stator winding. Consequently, stator current $I_a (= E_r/Z_S)$ increases to carry the load.

The following points may be noted in synchronous motor operation:

(*i*) A synchronous motor runs at synchronous speed at all loads. It meets the increased load not by a decrease in speed but by the relative shift between stator and rotor poles, i.e. by the adjustment of torque angle α.

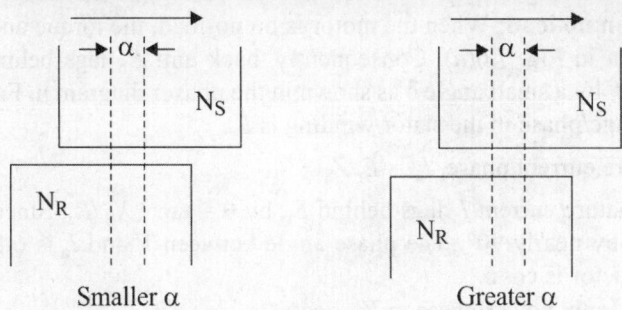

Smaller α   Greater α

Fig. 7.5

(*ii*) If the load on the motor increases, the torque angle a also increases (i.e. rotor poles lag behind the stator poles by a greater angle) but the motor continues to run at synchronous speed. The increase in torque angle α causes a greater phase shift of back emf E_b wrt supply voltage V. This increases the net voltage E, in the stator winding. Consequently, armature current $I_a(= E_r/Z_S)$ increases to meet the load demand.

(*iii*) If the load on the motor decreases, the torque angle α also decreases. This causes a smaller phase shift of E_b wrt V. Consequently, the net voltage E, in the stator winding decreases and so does the armature current $I_a(= E_r/Z_S)$.

[α is in mechanical degree while δ is in electrical degree]

7.7 PULL–OUT TORQUE

There is a limit to the mechanical load that can be applied to a synchronous motor. As the load increases, the torque angle δ also increases so that a stage is reached when the rotor is pulled out of synchronism and the motor comes to a standstill. This load torque at which the motor pulls out of synchronism is called pull-out or breakdown torque. Its value varies from 1.5 to 3.5 times the full-load torque.

When a synchronous motor pulls out of synchronism, there is a major disturbance on the line and the circuit breakers immediately trip. This protects the motor because both squirrel cage and stator winding beat up rapidly when the machine ceases to run at synchronous speed.

7.8 PHASOR DIAGRAM OF MOTOR

Consider an underexcited star-connected synchronous motor ($E_b < V$) supplied with fixed excitation, i.e. back emf E_b is constant.

Let V = supply voltage/phase

E_b = back emf phase

Z_S = synchronous impedance/phase

(i) **Motor on no load:** When the motor is on no-load, the torque angle α is small as shown in Fig. 7.6(*a*). Consequently, back emf E_b lags behind the supply voltage V by a small angle δ as shown in the phasor diagram in Fig. 7.6(*c*). The net voltage/phase in the stator winding is E_r.

Armature current/phase, $I_a = E_r/Z_S$

The armature current I_a lags behind E_r, by $\theta = \tan^{-1} X_S/R_a$. Since $X_S \gg R_a$, I_a lags E_r, by nearly 90°. The phase angle between V and I_a is φ so that motor power factor is cos φ.

Input power/phase $= VI_a \cos\phi$

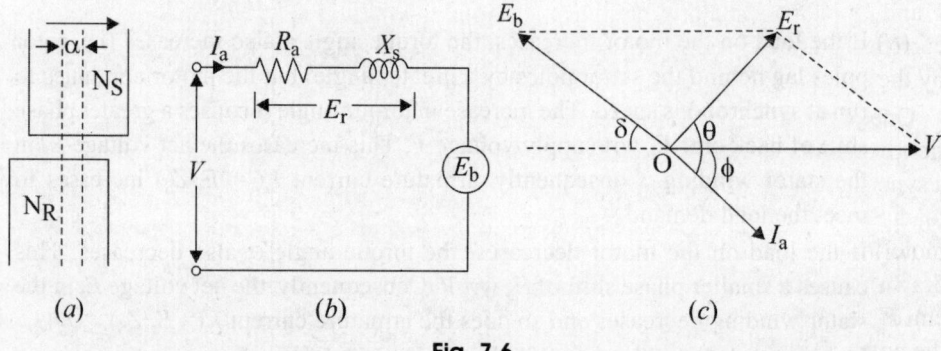

Fig. 7.6

Thus at no load, the motor takes a small power $VI_a \cos \phi$/phase from the supply to meet the no-load losses while it continues to run at synchronous speed.

(ii) **Motor on load:** When load is applied to the motor, the torque angle α increases as shown in Fig. 7.7(*a*). This causes E_b (its magnitude is constant as excitation is fixed) to lag behind V by a greater angle as shown in the phasor diagram in Fig.7.7(*b*). The net voltage/phase E_r in the stator winding increases. Consequently, the motor draws more armature current I_a ($= E_r/Z_S$) to meet the applied load.

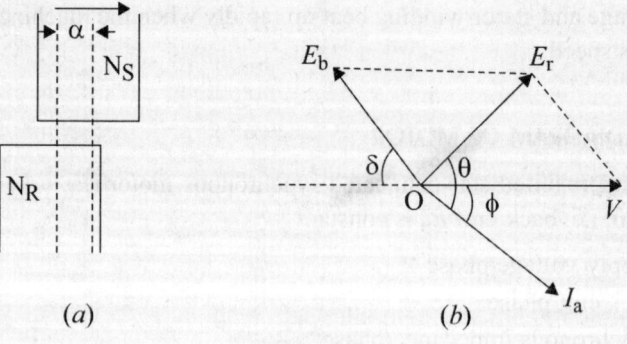

Fig. 7.7

Again I_a lags E, by about 90° since X_S R_a. The power factor of the motor is cosϕ.

Input power/phase, $P_i = VI_a\cos\phi$

Mechanical power developed by motor/phase is

$$P_m = E_b \times I_a \times \text{cosine of angle between } E_b \text{ and } I_a$$
$$= E_bI_a\cos(\delta - \phi)$$

7.9. EFFECT OF CHANGING FIELD EXCITATION AT CONSTANT LOAD

In a DC motor, the armature current I_a is determined by dividing the difference between V and E_b by the armature resistance R_a. Similarly, in a synchronous motor, the stator current (I_a) is determined by dividing voltage-phasor resultant (E_r) between V and E_b by the synchronous impedance Z_S.

One of the most important features of a synchronous motor is that by changing the field excitation, it can be made to operate from lagging to leading power factor. Consider a synchronous motor having a fixed supply voltage and driving a constant mechanical load. Since the mechanical load as well as the speed is constant, the power input to the motor (= 3 VI_a cos ϕ) is also constant. This means that the in-phase component I_a cosϕ drawn from the supply will remain constant. If the field excitation is changed, back emf E_b also changes. This results in the change of phase position of I_a wrt V and hence the power factor cos ϕ of the motor changes. Fig. 7.8 shows the phasor diagram of the synchronous motor for different values of field excitation. Note that extremities of current phasor I_a lie on the straight line AD.

(*i*) **Underexcitation:** The motor is said to be underexcited if the field excitation is such that $E_b < V$. Under such conditions, the current I_a lags behind V so that motor power factor is lagging as shown in Fig. 7.8(*a*). This can be easily explained. Since $E_b < V$, the net voltage E_r is decreased and turns clockwise. As angle θ (= 90°) between E_r and I_a is constant, therefore, phasor I_a also turns clockwise, i.e. current I_a lags behind the supply voltage. Consequently, the motor has a lagging power factor.

(*ii*) **Normal excitation:** The motor is said to be normally excited if the field excitation is such that $E_b = V$. This is shown in Fig. 7.8(*b*). Note that the effect of increasing excitation (i.e. increasing E_b) is to turn the phasor E_r, and hence I_a in the anticlockwise direction, i.e. I_a phasor has come closer to phasor V. Therefore, pf increases though still lagging. Since input power (= 3 VI_a cosϕ) is unchanged, the stator current I_a must decrease with increase in pf.

Suppose the field excitation is increased until the current I_a is in phase with the applied voltage Y, making the pf of the synchronous motor unity. For a given load, at unity pf the resultant E_r, and, therefore, I_a are miniumum.

(*iii*) **Overexcitation:** The motor is said to be overexcited if the field excitation is such that $E_b > V$. Under such conditions, current I_a leads V and the motor power

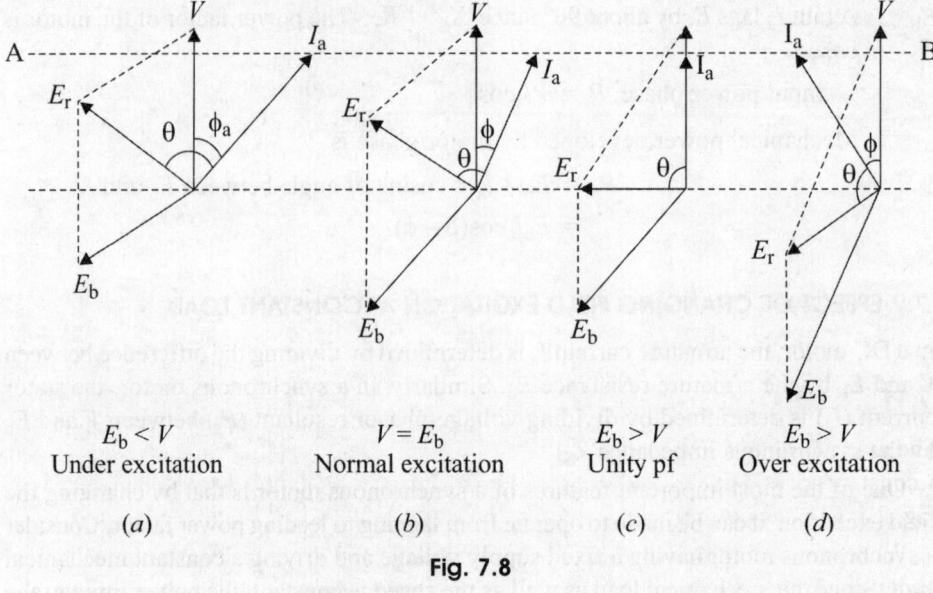

$$E_b < V$$
Under excitation

$$V = E_b$$
Normal excitation

$$E_b > V$$
Unity pf

$$E_b > V$$
Over excitation

(a)   (b)   (c)   (d)

Fig. 7.8

factor is leading as shown in Fig. 7.8(c). Note that E, and hence I_a further turn anticlockwise from the normal excitation position. Consequently, I_a leads V.

From the above discussion, it is concluded that if the synchronous motor is underexcited, it has a lagging power factor. As the excitation is increased, the power factor improves till it becomes unity at normal excitation. Under such conditions, the current drawn from the supply is minimum. If the excitation is further increased (i.e. overexcitation), the motor power factor becomes leading. **Note:** The armature current (I_a) is minimum at unity pf and increases as the power factor becomes poor, either leading or lagging.

7.10 PHASOR DIAGRAMS WITH DIFFERENT EXCITATIONS

Figure 7.9 shows the phasor diagrams for different field excitations at constant load. Figure 7.9(a) shows the phasor diagram for normal excitation ($E_b = V$), whereas Fig. 7.9(b) shows the phasor diagram for underexcitation. In both cases, the motor has lagging power factor.

Figure 7.9(c) shows the phasor diagram when field excitation is adjusted for unity pf operation. Under this condition, the resultant voltage E, and, therefore, the stator current I_a are minimum. When the motor is overexcited, it has leading power factor as shown in Fig. 7.9(d). The following points may be remembered:

(i) For a given load, the power factor is governed by the field excitation; a weak field produces the lagging armature current and a strong field produces a leading armature current.

(ii) The armature current (I_a) is minimum at unity pf and increases as the pf becomes less, either leading or lagging.

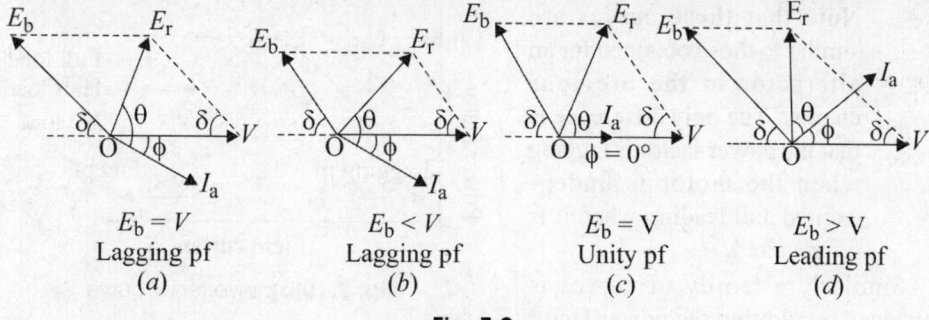

$E_b = V$
Lagging pf
(a)

$E_b < V$
Lagging pf
(b)

$E_b = V$
Unity pf
(c)

$E_b > V$
Leading pf
(d)

Fig. 7.9

7.11 CURVES FOR SYNCHRONOUS MOTOR

The graph between armature current and field current (I_a) of a synchronous motor for a constant load is called its V curve. The curve is so called because it is V-shaped. Fig. 7.10(a) shows V curves for three different loads.

When the level of excitation of a synchronous motor is changed gradually from under-excitation to overexcitation for a constant load, the following effects are observed:

(i) When the motor is underexcited (i.e. $E_b < V$), the power factor is lagging. In this case, the motor behaves like an inductive load.

(ii) When the motor is normally excited (i.e. $E_b = V$), the power factor is unity. In this case, the armature current is minimum and is in phase with the terminal voltage.

(iii) When the motor is overexcited (i.e. $E_b > V$), the power factor is leading. In this case, the motor behaves like a capacitive load. Therefore, an overexcited synchronous motor not only delivers load torque but also improves the power factor of the 3-phase power supply.

When the magnitude of armature current (I_a) is plotted as a function of field current (I_f) for a constant load, we get the V curve as shown in Fig. 7.10(a).

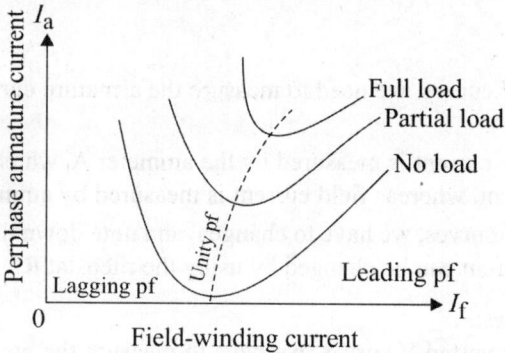

Fig. 7.10(a): V curves for synchronous motor

Note that these curves are similar to those obtained for an alternator in the previous chapter. The only difference is that the power factor is lagging when the motor is under–excited and leading when it is overexcited.

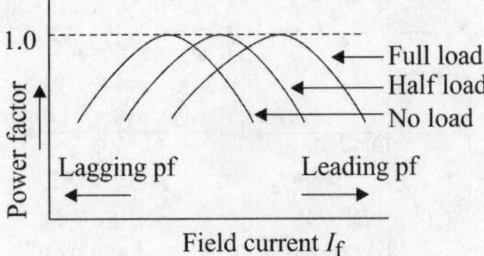

Fig. 7.10(b): Inverted V curve

Similarly a family of curves is obtained by plotting the power factor versus field current. These are inverted V curves in Fig. 7.10(b).

A comparison between synchronous motor and 3φ induction motor is given in Table 7.1.

Table 7.1: Comparison of synchronous motor and 3-φ induction motor

S. no.	Particular	Synchronous motor	3-φ Induction motor
1.	Speed	Remains constant from no-load to full load	Decreases with load
2.	Power factor	Can be made to operate from lagging to leading power factor	Operates at lagging pf
3.	Excitation	Requires DC excitation	No excitation for the rotor
4.	Economy	Economical for speeds below 300 rpm	Economical for speeds above 600 rpm
5.	Self-starting	No self starting torque auxilary means have to be provided for starting	Self starting
6.	Construction	Complicated	Simple
7.	Starting Torque	More	Less

7.11.1 Setup to Plot the V Curves and Inverted V Curves

The experimental set up for plotting the V curves and the inverted V curves is showing in Fig. 7.11.

V curves:
- To plot the V curves, we need to measure the armature current I_a and the field current I_f.
- The armature current is measured by the ammeter A_1 which actually measures the line current, whereas field current is measured by ammeter A_2.
- To plot the V curves, we have to change I_f and note down the corresponding I_a. The field current can be changed by using the rheostat R.

Inverted V curves:
- To plot the inverted V curves, we have to measure the power factor cos φ for different values of I_f.

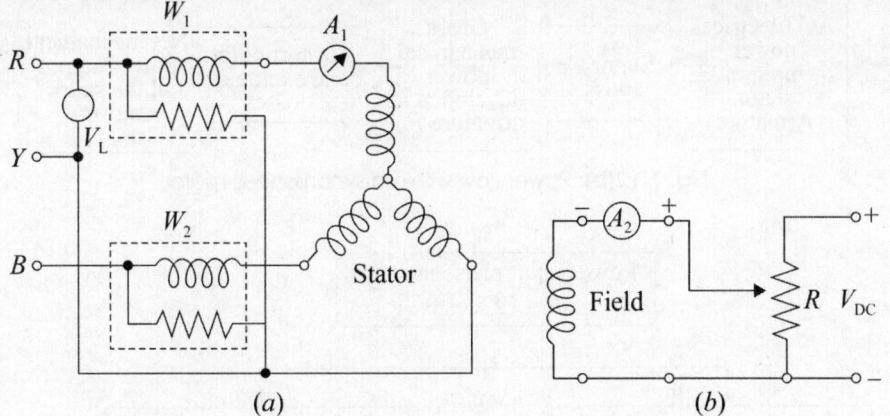

(a)   *(b)*

Fig. 7.11: Set up to plot the V curves and inverted V curve

- Here $\cos \phi$ is measured by using the two wattmeter method.
- Measure W_1 and W_2, i.e. take the readings of the two wattmeters and calculate $\cos \phi$ as follows:

$$\cos \phi = \cos\left\{ \tan^{-1}\left[\frac{\sqrt{3}(W_1 - W_2)}{(W_1 + W_2)} \right] \right\}$$

- Thus we can measure $\cos \phi$ for different values of field current I_f and plot the inverted V characteristics.

7.12 POWER FLOW WITHIN A SYNCHRONOUS MOTOR

- The power flow within a synchronous motor is as shown in Figs 7.12(*a*) and (*b*).
- The input power per phase in the stator is the sum of armature copper loss and mechanical power in the armature.
- The mechanical power in the armature (P_m) is equal to the sum of iron, excitation and friction losses and the output power P_{out}.
- This is as shown in Fig. 7.12(*c*).

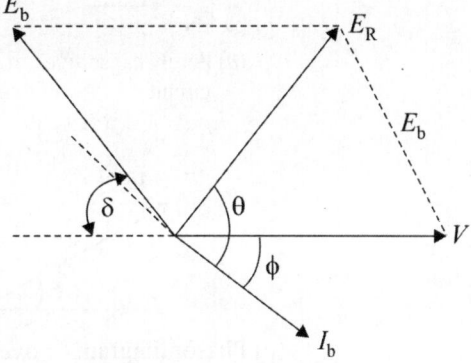

Fig. 7.12(*a*): Phasor diagram

7.13 SYNCHRONOUS CONDENSER

An overexcited synchronous motor running on no-load is known as *synchronous condenser*. The phasor diagram for a purely capacitive circuit is shown in Fig. 7.13(*b*). It shows that the current I leads the supply voltage V by 90°.

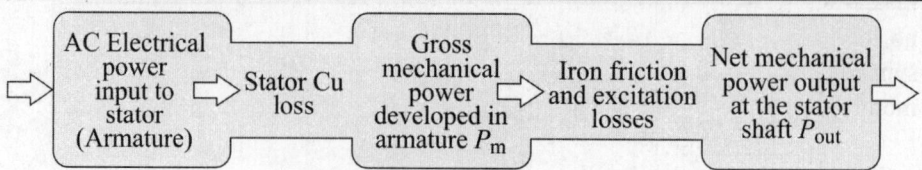

Fig. 7.12(b): Power flow within a synchronous motor

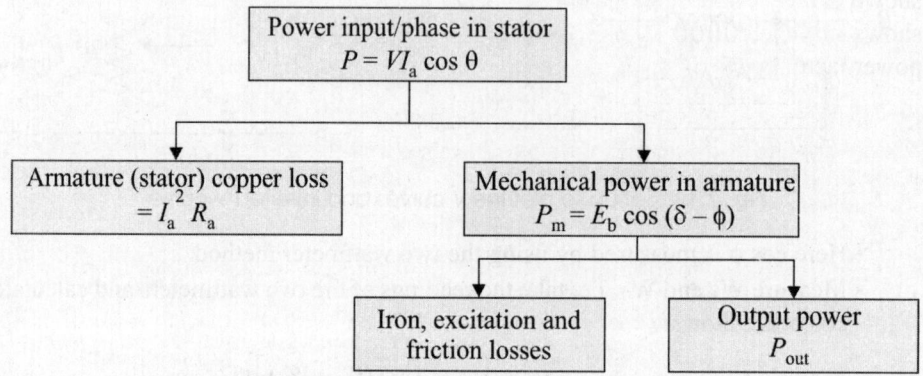

Fig. 7.12(c): Power flow within the synchronous motor

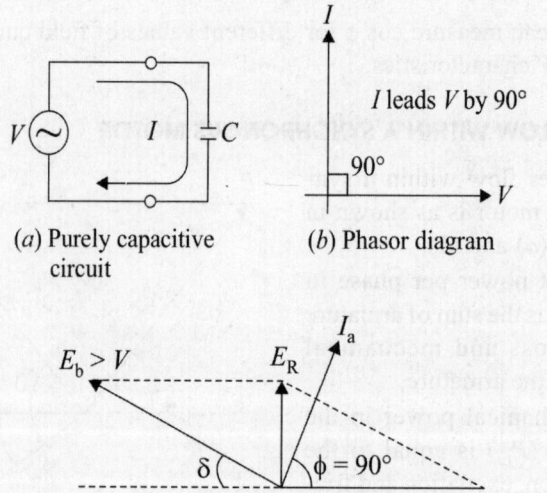

(a) Purely capacitive circuit

(b) Phasor diagram

(c) Phasor diagram of over-excited synchronous motor

Fig. 7.13

Now let us see the phasor diagram of a synchronous motor when overexcited. If the value of δ is small (i.e. no load) and if the motor is overexcited then the phasor diagram is shown in Fig. 7.13. Note that I_a leads V by an angle ϕ which is close to 90°. This is similar to the phasor diagram of a capacitor. Hence a synchronous motor operating at no load and overexcited condition is equivalent to a capacitor and hence called as synchronous condenser. Due to this property of drawing leading current the synchronous motors can be used for the purpose of power factor correction.

7.13.1 Power Factor Correction using Synchronous Condenser

Consider a lagging power factor load connected across a single phase AC supply as shown in Fig. 7.14(a). The phasor diagram for this load is shown in Fig. 7.14(b) which shows that the current I_1 lags the supply voltage V by a large angle ϕ_1. Hence the power factor cos is ϕ_1 small, i.e. poor.

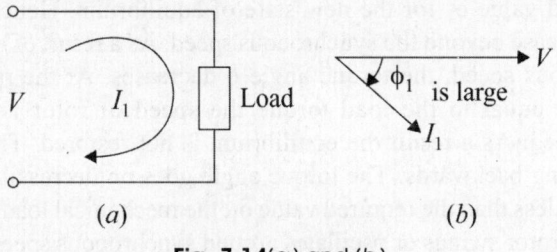

(a)   (b)

Fig. 7.14(a) and (b)

To improve the power factor we have to reduce the angle ϕ towards zero. This can be achieved by connecting a synchronous condenser across the load as shown in Fig. 7.14(c). The synchronous motor is on no load and overexcited. So it is taking a leading current I_2. This current leads the supply voltage by an angle ϕ_2. The net current I drawn from the supply is equal to the vector sum of I_1 and I_2 as shown in Fig. 7.14(d).

$$\bar{I} = \bar{I_1} + \bar{I_2}$$

From Fig. 7.13(d) it is evident that the angle ϕ between V and I is much less than ϕ_1. Hence the value of cos ϕ will be higher than that of cos ϕ which means that the power factor improvement has taken place.

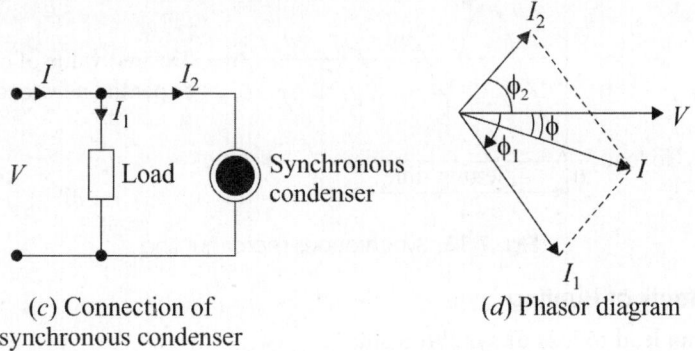

(c) Connection of synchronous condenser   (d) Phasor diagram

Fig. 7.14(c) and (d)

7.14 HUNTING IN SYNCHRONOUS MOTOR

The electromagnetic torque is equal and opposite to the load torque in a condition of equilibrium for a synchronous motor when it operates at steady state. The rotor runs at synchronous speed at steady state and maintains a constant value of the torque angle δ. If there is a sudden change in the load torque, the equilibrium is disturbed, and there is a resulting torque which changes the speed of the motor. When there is a sudden increase in the load torque, the motor slows down temporarily and the torque angle δ is adequately increased to restore the torque equilibrium and the synchronous speed. As δ is increased, the electromagnetic torque increases. Therefore the motor is accelerated. When the rotor reaches synchronous speed, the torque angle δ is larger than the required value δ_1 for the new state of equilibrium. Hence, the rotor speed continues to increase beyond the synchronous speed. As a result of motor acceleration above synchronous speed, the torque angle δ decreases. At the point where motor torque becomes equal to the load torque the speed of rotor is greater than the synchronous speed. As a result the equilibrium is not restored. Therefore, the rotor continues to swing backwards. The torque angle goes on decreasing. When the load angle δ becomes less than the required value δ_1, the mechanical load angle is increased again. Thus, the rotor swings or oscillates around synchronous speed and the required value δ_1 of the torque angle before reaching the new steady state (Fig. 7.15). Likewise, the motor responds to a decreasing load torque by a temporary increase in speed, and thereby, a reduction of the torque angle δ. The rotor swings or oscillates around synchronous speed and the new required value δ_2 of the torque angle before reaching the new equilibrium position (steady state). This phenomenon of oscillation of the rotor about its final equilibrium position is called **hunting**. Since during rotor oscillations, the phase of the phasor E_b changes relative to phasor V, hunting is also known as **phase swinging.**

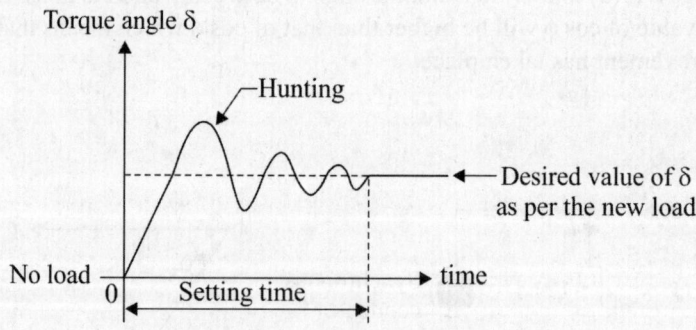

Fig. 7.15: Synchronous motor hunting

7.14.1 Effects of Hunting

1. It can lead to loss of synchronism.
2. It can cause variation of the supply voltage producing undesirable lamp flicker.
3. It increases the possibility of resonance.

4. Large mechanical stresses may develop in the rotor shaft.

5. The machine losses increase and the temperature of the machine rise.

How to reduce of hunting? The following are some of the techniques used to hunting:

1. Damper windings (Fig. 7.16).

2. Use of flywheels.

3. By designing synchronous machines with suitable synchronizing power coefficient.

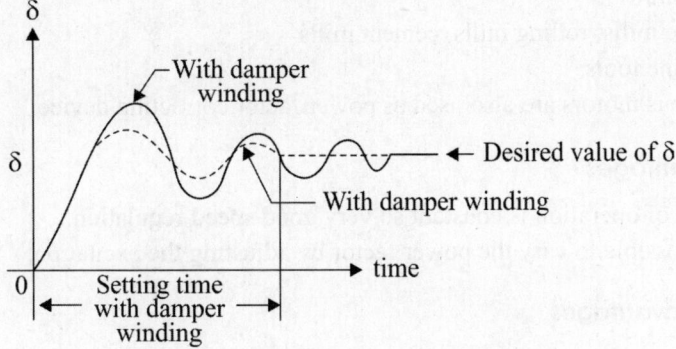

Fig. 7.16: Effects of damper winding

7.14.2 Phasor Diagram of Synchronous Motor While Hunting

• The phasor diagram under the hunting situation is shown in Fig. 7.17.

• Let δ the desired value of delta and let δ' and δ'' be the two maximum values above and below δ.

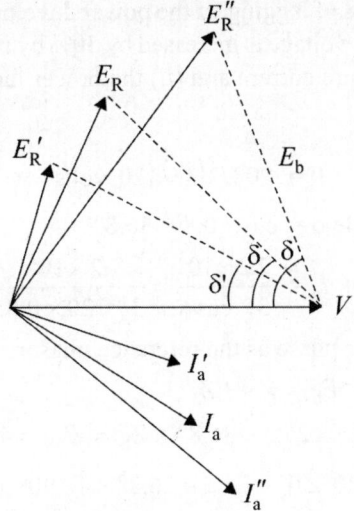

Fig. 7.17: Phasor diagram of synchronous motors under hunting situation

7.15 APPLICATIONS, ADVANTAGES AND DISADVANTAGES OF SYNCHRONOUS MOTOR

7.15.1 Application

The constant speed of operation makes the synchronous motor suitable for the following applications:

1. Fans, blowers
2. Centrifugal pumps
3. Grinders
4. Textile mills, rolling mills, cement mills
5. Machine tools.

Synchronous motors are also used as power factor correcting device.

7.15.2 Advantages

1. Speed of operation is constant so very good speed regulation.
2. It is possible to vary the power factor by adjusting the excitation.

7.15.3 Disadvantages

- It is not self starting.
- It needs frequent maintenance.
- External DC source is necessary for providing excitation.
- Additional damper winding is necessary.
- Hunting takes place if load is changed suddenly.

Example 7.1: A 208 V star-connected, 3-phase synchronous motor has a synchronous reactance of 4 Ω/phase and negligible armature winding resistance. At a certain load, the motor takes 7.2 kW at 0.8 pf lagging. If the power developed by the motor remains the same while the excitation voltage is increased by 50% by raising the field excitation. determine (*i*) the new armature current and (*ii*) the power factor.

Solution:

Supply voltage/phase, $\quad V = 208/\sqrt{3} = 120$ volts

$$\text{pf angle } \phi = \cos^{-1} 0.8 = 36.8°$$

$\therefore \qquad$ Armature current, $I_a = \dfrac{7.2 \times 10^3}{3V \cos\phi} = \dfrac{7.2 \times 10^3}{3 \times 120 \times 0.8} = 25\,\text{A}$

Taking supply voltage per phase as the reference phasor, we have,

generated voltage, $\qquad E_b = V - I_a Z_S$

Here $V = 120\angle 0°$ volts; $I_a = 25\angle -36.8°\,A; Z_S = jX_s = 4\angle 90°\,\Omega$

$$E_b = 120°\angle 0° - 25\angle -36.8° \times 4\angle 90°$$

$$= 120°\angle 0° - 100\angle -53.2° \times 100\angle 53.13° \text{ volts}$$

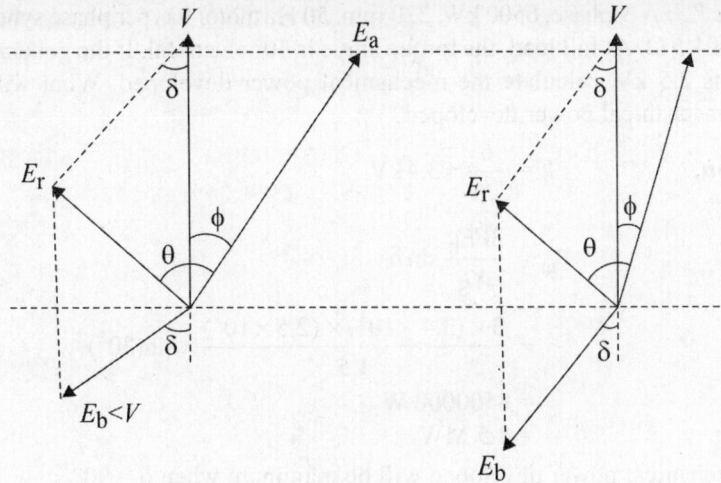

Fig. 7.18

(*i*) New generated voltage,

$$E_b = 1.5 \times E_b = 1.5 \times 100 = 150 \text{ volts}$$

The new torque angle δ' can be found as under:

$$P_m = \frac{3VE_b'}{X_s}\sin\delta \quad \text{or} \quad 7.2 \times 10^3 = \frac{3 \times 120 \times 150}{4}\sin\delta'$$

$$\therefore \quad \sin\delta' = \frac{7.2 \times 10^3 \times 4}{3 \times 120 \times 150} = 0.533 \text{ or } \delta' = 32.23°$$

New armature current, $\quad I_a' = \dfrac{V - E_b'}{X_s}$

$$= \frac{120\angle 0° - 150\angle -32.23°}{j4} = 20\angle 4.92° \text{A}$$

(*ii*) New power factor $= \cos 4.92° = 0.996$ leading

Example 7.2: A 3-phase synchronous motor has 8 poles and operates from 400 V, 50 Hz supply mains. Calculate its speed. If it takes a line current of 150 A at 0.6 power factor leading, what torque the motor will be developing? Neglect all losses.

Solution:

Motor speed $\quad N_S = \dfrac{120f}{P} = \dfrac{120 \times 50}{8} = 750 \text{ rpm}$

Input power $\quad P_{in} = \sqrt{3} \times 400 \times 150 \times 0.6 = 62353.82 \cong 62354 \text{ W}$

as losses are neglected, we have,

Output power $\quad P_{out} = P_i = 62354 \text{ W}$

Motor torque $\quad T = 9.55 \times \dfrac{P_{out}}{N_s}$

$$= \frac{9.55 \times 62354}{750} = 793.97 \cong 794 \text{ N m}$$

Example 7.3: A 3-phase, 6500 kW, 220 rpm, 50 Hz motor has per phase synchronous reactance of 1.5 Ω. At full load, the torque angle is 30° electrical. If the generated back emf/phase is 2.5 kV, calculate the mechanical power developed. What will be the maximum mechanical power developed?

Solution:
$$V = \frac{6}{\sqrt{3}} = 3.4 \, kV$$

$$P_m = \frac{3VE_b}{X_S} \sin\delta$$

$$= \frac{3 \times (3.4 \times 10^3) \times (2.5 \times 10^3)}{1.5} \times \sin(30°)$$

$$= 8500000 \, W$$
$$= 8.5 \, MW$$

The mechanical power developed will be maximum when $\delta = 90°$.

$$P_{max} = \frac{3VE_b}{X_S}$$

$$= \frac{3 \times (3.4 \times 10^3)(2.5 \times 10^3)}{1.5} = 17000000 \, W = 17MW$$

Example 7.4: A 3-phase, 6500 kW, 8 kV, 250 rpm, 50 Hz synchronous motor has a synchronous reactance of 1.6 Ω per phase. At full load, the rotor poles are displaced by a mechanical angle of 2.3° from their no-load position. If the line to neutral excitation voltage $E_b = 3.1$ kV, calculate the mechanical power developed .

Solution: Induced voltage/phase, $E_b = 3.1$ kV

Terminal voltage/phase, $V = 8/\sqrt{3} = 4.61 kV$

Number of poles, $P = \dfrac{120f}{N_S} = \dfrac{120 \times 50}{250} = 24$

Torque angle, $\delta = \dfrac{P \cdot \alpha}{2} = \dfrac{24 \times 2.3}{2} = 27.6° \cong 28° \, (\text{electrical})$

Power developed/phase, $P_d = \dfrac{VE_b}{X_S} \sin\delta$

$$= \frac{3.1 \times 10^3 \times 4.61 \times 10^3}{1.6} \times \sin(28°)$$

$$= 4184165.303 \, W = 4.1 \, MW$$

Total power developed = $(3 \times 4.1) = 12.3$ MW

Example 7.5: A 15 HP 240 V, 50 Hz, 3-phase, star-connected sychronous motor delivers full load at a pf of 0.8 leading. The synchronous reactance of the motor is $j6\Omega$/phase. The rotational loss is 200 W and the field winding loss is 50 W. Calculate (i) efficiency (ii) generated voltage of the motor. Neglect armature winding resistance.

Solution: (*i*) Power output, $P_{out} = 15 \times 746 = 11190$ W

Rotational loss $= 200$ watt

Power developed, $P_d = P_{out} + $ Rotational losses

$$= 11190 + 200 = 11390 \text{ W}$$

Since, there is no copper loss in the armature winding power supplied by the AC source is 11390 W.

Armature current, $I_a = \dfrac{11390}{\sqrt{3} \times 240 \times 0.8} = 34.25 \text{ A}$

Total input power, $P_{in} = 11390 + 50 = 11440$ W

Motor efficiency, $\eta = \dfrac{P_{out}}{P_{in}} \times 100\% = \dfrac{11190}{11440} \times 100\% = 97.81\%$

(*ii*) Taking supply voltage/phase ($V = 240/\sqrt{3}$) $= 138.56$ V as the reference phasor, we have

Generated voltage, $E_b = V - I_a \times X_S$ ($\because R_a = 0$)

Here, $V = 138.56 \angle 37°$, $X_S = j6\Omega$ and $I_a = 34.25 \angle 36.86° = 34.25$ & $37°$A

[pf is 0.8 leading. Thus, $\cos \phi = 0.8$ so, $\phi = \cos^{-1}(0.8) = 36.86° \cong 37°$. The current I_a leads the voltage V by $37°$.]

Thus, $E_b = 138.56 \angle 0° - 34.25 \angle 37° \times 6 \angle 90°$

$$= 138.56 - 205.5 \angle 127°$$

$$= 138.56 - (-123.67 + j164.11)$$

$$= 262.23 - j164.11$$

$$= 309.34 - \angle -32.03° \text{ V}$$

EXERCISES

1. Explain the operating principle of synchronous motor.

2. What is the effect of variation of load?

3. Define the synchronous impedance.

4. Explain the phasor diagram for an ideal synchronous motor on no load.

5. Explain the operation of synchronous motor on load.

6. Draw the torque-angle characteristics.

7. State various types of torques in synchronous motor.

8. Draw the power flow diagram for a synchronous motor.

9. Compare synchronous motor and induction motor.

10. What are the advantages and disadvantages of synchronous motor?

11. Explain: V-curves of synchronous motor. **(2003–2004)**

12. Explain the operation of synchronous motors. Also describe their industrial applications. **(2004–2005)**

13. Draw and explain the phasor diagram of a synchronous motor operating at lagging power factor, leading power factor and unity power factor. What is a synchronous condenser and where is it used? **(2005–2006)**

14. Explain V-curves, their origin and significance with regard to synchronous motor. **(2005–2006)**

15. Explain construction, starting and operation of synchronous motor. Also show the effect of excitation current on the motor armature current. **(2006–2007)**

16. Explain the effect of change of excitation of a synchronous motor on its:
    1. Armature current
    2. Power factor
    Also draw the phasor diagram. **(2007–2008)**

8

General Introduction of Special Machines

8.1 INTRODUCTION

In this chapter, we will deal with some special purpose electric machines. Although the basic principle of operation of all electrical machines is same, However, some features of these special machines that distinguish them from conventional machines. For example: a stepper motor rotates by a special number of degrees (e.g. $-2°$, $2.5°$, $5°$ or $7.5°$) in response to an input electrical signal and is widely used in digital control system. Similarly, brushless DC motors eliminate mechanical commutator and brushes. Instead, these motors use electric commutators. Brushless DC motors offer excellent characteristics for applications that require constant speed drives. It is not possible to deal with all types of special purpose machines in one chapter. However, we shall introduce the basic operating principles of some special purpose machines that are being used exclusively in homes, recreational and industrial applications.

8.2 STEPPER MOTOR

Stepper motors are also known as stepping motors or step motors. A stepper motor is an electromagnetic motor (i.e. it converts electrical energy to mechanical energy) that rotates by a specific number of degrees in response to an input electric signal. Typical step size is $2°$, $2.5°$, $7.5°$ and $15°$ for each electric pulse. The rotor does not rotate continuously like a conventional electric motor since there is no continuous energy conversion from electrical energy to mechanical energy. In a stepper motor, proportionate mechanical movement is obtained from given electric pulses. In a stepper motor, a series of definite individual steps complete the revolution of each stepper motor. A '**step**' is defined as the angular rotation (in degrees) of the motor each time it receives the electric pulse. Such a step control is required in many applications. Figure 8.1 illustrates a simple application for a stepper motor. Each time the controller receives an input electric signal, the paper is driven to a certain incremental distance. Stepper motors are relatively cheap and simple in construction and can be made to rotate in steps in either direction. These motors are excellent machines for such applications as electric type-writers, control of floppy disc drives, numerical control of machine tools, etc. The two most popular types of stepper motors are:

   (i) Permanent-magnet (PM) stepper motor

   (ii) Variable-reluctance (VR) stepper motors

Stator windings are housed in the stator of a stepper motor (PM or VR) and they are energized from a DC source to create two or more stator poles. The stator poles are also called stator teeth. The rotor of a stepper motor may be a permanent magnet (in case of PM motor) or a soft-iron material (in case of VR motor). There can be two or more rotor poles in the rotor and they are also called rotor teeth.

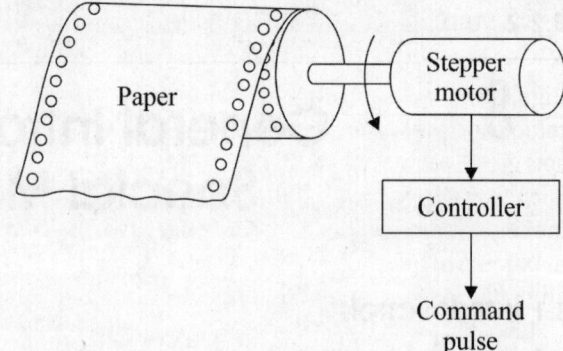

Fig 8.1 Simple application for a stepper motor

The coils in the stator are energized in groups referred to as phases. The stator windings can be arranged as two-phase, three-phase or four-phase windings. For providing DC excitation, these phase windings are brought out to the terminals. Some introductory remarks about these motors shall be provided now.

8.2.1 Permanent Magnet (PM) Stepper Motor

Figure 8.2 shows a two-pole, single phase permanent magnet (PM) stepper motor. The excitation torque acts on the rotor (PM) as the stator is being energized and the rotor moves to a position of zero excitation torque, i.e. there will be parallel alignment of the rotor and the stator field.

Figure 8.3 shows the variation of excitation torque with the position of rotor in case of a PM rotor type. Note that when there is a displacement of 90° or 270° between the rotor and the stator, the maximum torque is developed. However, zero torque is experienced when there is parallel alignment of the rotor and the stator field.

Fig 8.2: Two pole, single phase permanent magnet stepper motor

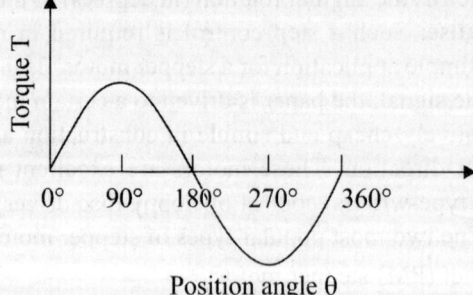

Fig 8.3: Excitation torque curve varies with rotor position

8.2.2 Variable Reluctance (VR) Stepper Motor

Figure 8.4 shows a two-pole, single phase variable reluctance (VR) stepper motor. Reluctance torque acts on the rotor (soft-iron material) when the stator is energized. The rotor will move to a position of maximum air gap flux, i.e. where the reluctance is maximum which means that the rotor teeth will align with the energized stator poles.

Fig 8.4: Two pole, variable reluctance permanent magnet stepper motor

Figure 8.5 shows how reluctance torque varies with the rotor position for a VR soft-iron rotor. There is no torque developed with the rotor being at zero degree or 90°. Maximum torque is developed at 45° and 135° which is the position where the rotor is forced to move to a position of minimum reluctance by the reluctance torque.

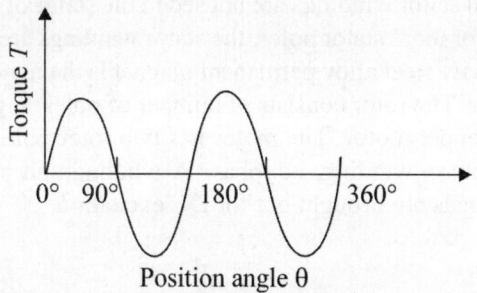

Figure 8.5: Reluctance torque varies with rotor position

8.3 DEFINITIONS ASSOCIATED WITH STEPPER MOTOR

Step angle: The angle through which the motor shaft rotates for each command pulse is called step angle. It can be shown that for any PM or VR stepper motor, the step angle can be found from the following two relations:

(i) In terms of stator poles (N_s) and rotor poles (N_r), the step angle (α) is given by

$$\text{Step angle, } \alpha = \frac{(N_s - N_r)}{N_s \times N_r} \times 360°$$

α = step angle in degrees

N_s = number of stator poles (or teeth)

N_r = number of rotor poles (or teeth)

(ii) In terms of stator phases (m) and rotor poles (N_r), the step angle is given by

$$\text{Step angle } \alpha = \frac{360°}{mN_r}$$

Stepping rate: An important specification of a stepper motor is the stepping rate. The number of steps per second is known as stepping rate or stepping frequency (f). The actual speed of stepper motor depends on the step angle (α) and stepping frequency (f) and is given by

$$\text{Speed of stepper motor, } N = \frac{\alpha f}{6}$$

where N = Motor speed in rpm

f = Stepping frequency i.e. steps/second

8.4 PERMANENT MAGNET (PM) STEPPER MOTOR

One of the popular types of stepper motors in the permanent-magnet (PM) stepper. It operates on the principle of interaction between a permanent magnet rotor and an electromagnetic field.

8.4.1 Construction

Steel laminations and stator windings are housed in the stator of the PM stepper motor and for creating two or more stator poles, the stator windings are energized from a DC source. High-retentivity steel alloy permanent magnet is the material used to make the rotor of such a motor. The rotor consists of number of poles. Figure 8.6 shows a two-phase, 2-pole PM stepper motor. The motor has two rotor poles. The stator coils are grouped to form 2-phase winding, i.e. phase-A winding and phase-B winding. The phase winding terminals are brought out for DC excitation.

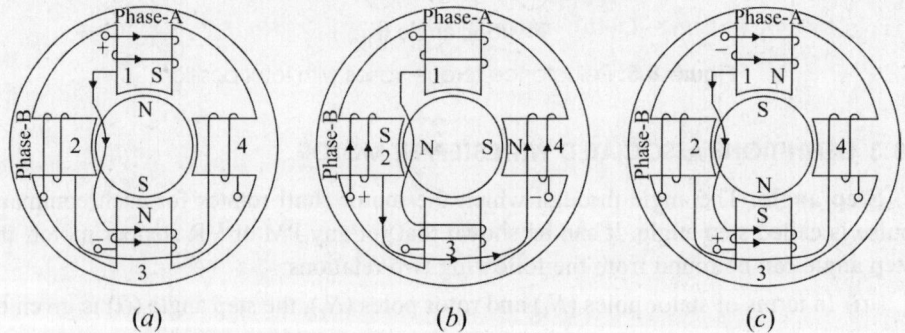

Fig 8.6: Operation of permanent magnet stepper motor

8.4.2 Operation

For this PM stepper motor, the number of rotor poles, $N_r = 2$ and number of phases, $m = 2$

$$\text{Step angle, } \alpha = \frac{360°}{mN_r} = \frac{360°}{2 \times 2} = 90°/\text{step}$$

(i) When a constant current is used to excite the phase-A winding only as shown in Fig. 8.6 (a) South Pole is attained at stator tooth 1 which ensures that the

north pole of the PM rotor aligns parallel with the South Pole (stator tooth) of the stator. This position of the rotor will remain intact until phrase A winding remains energized. The first row of truth table in Fig. 8.7 shows that only phase-A winding is excited while phase B winding remains unexcited. Under this condition, step angle θ = 0°. The applied voltage waveforms in Fig. 8.8 also tally with facts shown in the truth table.

(ii) When we de-energize the phase A winding and energizing phase B winding as shown in Fig 8.6(*b*), stator tooth 2 becomes south pole and as a consequence of which the north pole of the PM and the south pole (stator tooth 2) of the stator get aligned parallel. Thus, there is a displacement of 90° in the anti-clockwise direction of the rotor. The reader may see the truth table and applied voltage waveform for this switching sequence.

(iii) If the phase-B winding is de-energized and excitation of phase A with a reverse current is done [i.e. current in it is opposite to the case in above], further rotation in the anti-clockwise direction by 90° is obtained as shown in Figure 8.6(*c*). Now the north pole of PM motor aligns with the stator tooth 3.

Cycle	Phase		Position δ°
	A	B	
+	1	0	0
	0	1	90
−	−1	0	180
	0	−1	270
+	1	0	360

The natations 1, -1, 0 correspond to positive, negative and zero current in a phase winding respectively

Fig. 8.7: Truth table

(iv) So far, the rotor has completed one-half revolution. However, if we continue to appropriate switching [see truth table in Fig.8.7, the rotor will complete one revolution in 90° steps.

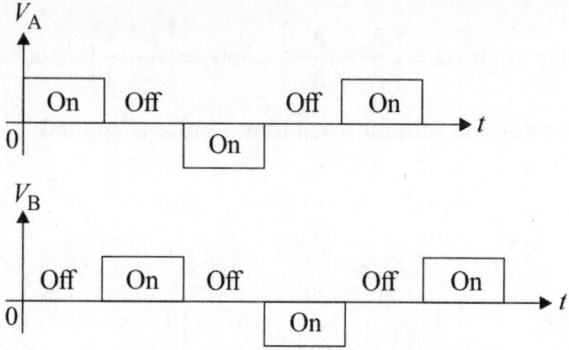

Fig 8.8: Applied voltage waveform

We can change the step angle (α) of a PM stepper motor by changing the number of rotor poles (N_r) and the number of phases (*m*). Thus, for a 3-phase, 24-pole PM stepper motor, the step angle

$$θ = 360°/m \, N_r = 360°/3 × 24 = 5°/\text{step}$$

8.4.3 Limitations

Some of the drawbacks of the PM stepper motor are as followed

(i) Since making a small permanent magnet rotor with a large number of poles is difficult, thus PM motors have a large step angle in the range of 30° to 90°.

(ii) Because of permanent magnet rotor they have high inertia and hence their acceleration is slow with a maximum step rate (stepping frequency) of 300 steps/second.

(iii) On account of high stepping angle, these motors have high speed and hence for a given power output the torque is low.

8.5 VARIABLE-RELUCTANCE (VR) STEPPER MOTOR

The principle of operation of the variable reluctance (VR) stepper motor is same as that of the reluctance motor. Thus, if we place a piece of ferro-magnetic material which is free to rotate, it is brought to the position of minimum reluctance in the path of the magnetic flux by virtue of the torque acting on the material.

8.5.1 Construction

The construction of the stator is the same for both the VR stepper motor and PM stepper motor. Phase windings of the stator are wound on each stator tooth and soft steel is used to make the rotor with tooth and slots. The basic Variable–Reluctance stepper motor Fig. 8.9 shows. In this circuit, the number of tooth of the rotor is fewer than that of the stator which ensures that at a particular instant, a single set of stator rotor tooth will align. In Fig. 8.9, the stator has six teeth and rotor has four teeth. The stator has 3 phases A, B and C with teeth 1 and 4, 3 and 6 and 2 and 5 respectively. For this VR stepper motor

$$\text{Step angle, } \alpha = \left[\frac{N_s - N_r}{N_s \times N_r}\right] \times 360° = \left[\frac{6-4}{6 \times 4}\right] \times 360° = 30°/\text{step}$$

Therefore, the rotor will turn 30° each time a pulse is applied.

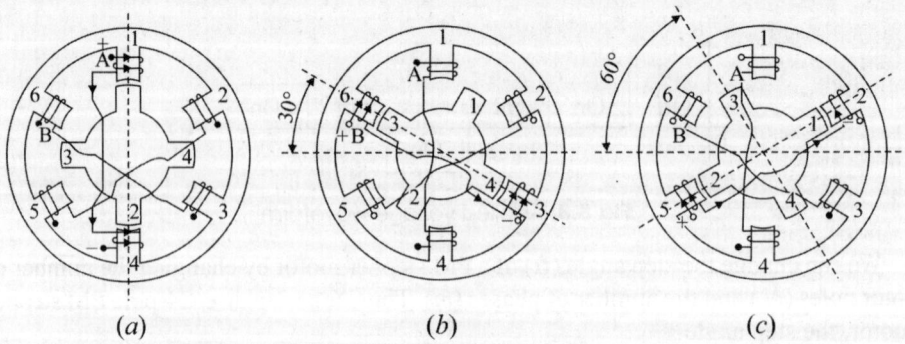

(a) (b) (c)

Fig. 8.9: Operation of variable reluctance stepper motor

8.5.2 Operation

On energizing the phase winding, the energized stator poles get aligned with the stator teeth.

(i) When constant current is used to energize phase A, the position of the rotor is depicted in Fig. 8.9(*a*). The rotor will stay stationary until the phase A is kept energized. Note that in this condition, the rotor teeth 1 and 2 and the energized stator teeth 1 and 4 get aligned together and the step angle θ = 0°. The truth table (Fig. 8.10) and the applied voltage waveform (Fig. 8.11) can also be referred.

Cycle	Phase			Position
	A	B	C	
1	ON	OFF	OFF	0°
	OFF	ON	OFF	30°
	OFF	OFF	ON	60°
2	ON	OFF	OFF	90°
	OFF	ON	OFF	120°
	OFF	OFF	ON	150°
3	ON	OFF	OFF	180°
	OFF	ON	OFF	210°
	OFF	OFF	ON	240°
4	ON	OFF	OFF	270°
	OFF	ON	OFF	300°
	OFF	OFF	ON	330°
5	ON	OFF	OFF	360°

Fig. 8.10: Truth table

(ii) On switching off phase A and energizing phase B, 30° clockwise rotation of the rotor takes place and the rotor teeth 3 and 4 and the energized stator teeth 6 and 3 get aligned together.

(iii) On de-energizing phase B and with an energized phase C the effect taking place is shown in Fig. 8.9(*c*). In this circuit, further 30° clockwise rotation of

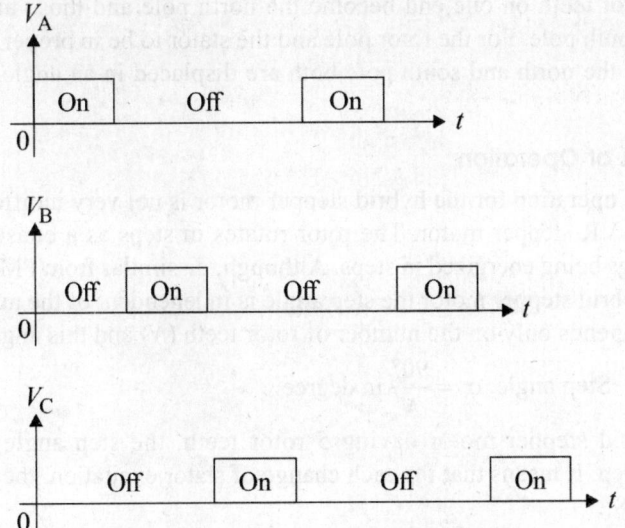

Fig 8.11: Applied voltage waveforms

the rotor takes place so that rotor teeth 1 and 2 and the energized stator teeth 2 and 5 get aligned together.

(iv) The completion of one cycle of the step sequence takes place when the rotor has displaced by 60° in the clockwise direction from the starting point. A complete 360° rotation of the motor consisting of six and for rotor and stator poles respectively has a switching sequence which has been depicted in the truth table of Fig. 8.10.

The direction of rotation will be reversed if the switching sequence is in the order of A, C and B. For this particular motor, applied voltage must have at least five cycles for one revolution.

8.6 HYBRID STEPPER MOTOR

The etymology of the word 'hybrid' implies that a 'hybrid' stepper motor is the one which delivers the combined features of a PM and a VR stepper motor simultaneously. The developed torque of such a motor surpasses that of the PM or VR types.

8.6.1 Construction

Figure 8.12 depicts the rudimentary design aspects of a hybrid stepper motor with a stator construction similar to that of a VR or a PM motor. However, the constructional design of the rotor is a combination of that of the VR and PM motors both. Twin identical stacks of soft iron and an axially magnetised round permanent magnet are the design elements of the rotor of a hybrid stepper motor. The north and south poles of the permanent magnet have soft iron stacks attached to them as shown in Fig. 8.12.

Along the soft iron stacks, the rotor tooth is machined on, as a consequence of which the rotor teeth on one end become the north pole and those at the other end become the south pole. For the rotor pole and the stator to be in proper alignment, the rotor tooth of the north and south pole both are displaced in an angle as depicted in Fig. 8.12.

8.6.2 Modes of Operation

The modes of operation for the hybrid stepper motor is not very indifferent from that of the PM or VR stepper motor. The rotor rotates in steps as a consequence of the phase windings being energized in steps. Although, dissimilar from PM or VR stepper motor, in a hybrid stepper motor the step angle is independent of the number of stator phases and depends only on the number of rotor teeth (N) and this angle is given by

$$\text{Step angle, } \alpha = \frac{90°}{N} \text{ in degree}$$

For a hybrid stepper motor having 5 rotor teeth, the step angle $\alpha = 90°/N_r = 90°/5 = 18°$/step. It means that for each change of stator excitation, the rotor will turn by a step of 18°.

It should be noted that the combined principles of the permanent magnet and variable reluctance stepper motors is of the hybrid stepper motor's operating principle. Thus,

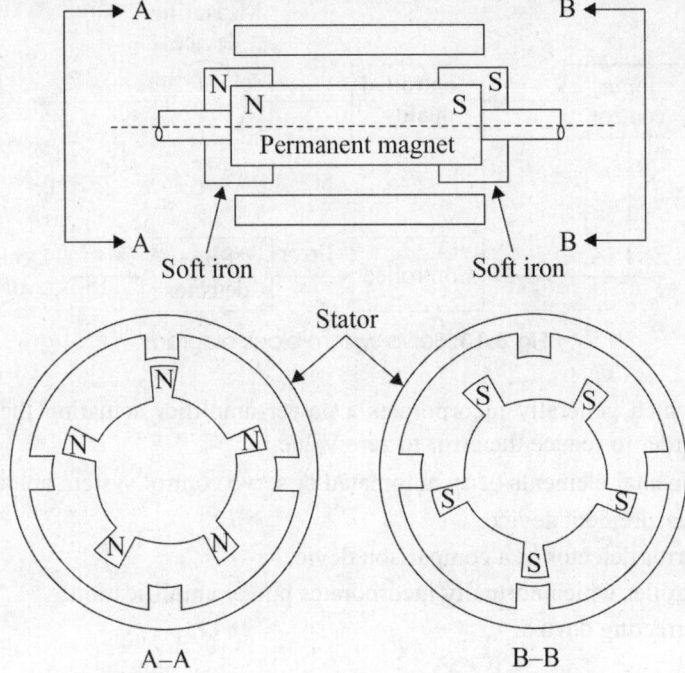

Fig 8.12: Various views of hybrid stepper motor

the developed torque of a hybrid stepper motor consists of excitation torque and reluctance torque both, and hence as a consequence of which the resultant torque development by such a stepper motor surpasses that of the PM or VR stepper motor.

8.7 SERVO MECHANISM

Servo mechanism is the automated control of a particular physical quantity of a machine such as angular displacement or speed of motor. The disadvantages of human operator such as fatigue, slow reaction time (about 0.3s) and limited power is the cause which has led to the development of an automated control system or servo system. The servo system involves information feedback about the controlled physical quantity in such a way that if that desired value of the quantity differs, an error is observed and is transmitted to the control system which then operates to reduce the error to zero. Such kind of a control system is termed as a closed loop control system mechanism.

The block diagram representing an automated control system is shown Fig 8.13 with the controllable quantity being angular position or speed of shaft or any other physical quantity. Electrical measurement of the actual value of the physical quantity compared with a desired value in the error detector which is a comparison device. The quantity sense any type of difference in between the desired and actual values of the output which is considered to be an error signal and this error signal is then fed to a

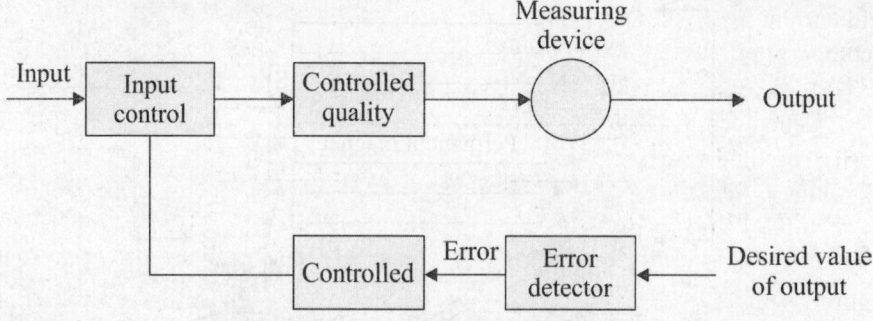

Fig 8.13: Servo system block diagram

controller which generally incorporates a power amplifier actuating the correcting devices in order to reduce the error to zero value.

The rudimental elements of an automated or servo control system are as follows:

(*i*) A measurement device.

(*ii*) An error detector or a comparison device.

(*iii*) Controller which normally incorporates power amplification.

(*iv*) A correcting device.

8.8 DC SERVO MOTORS

A DC motor that is used in servo mechanism is called a DC servo motor. The three main types of DC servo motors are:

(*i*) Field-controlled DC servo motor

(*ii*) Armature-controlled DC servo motor

(*iii*) Permanent magnet armature-controlled DC servo motor

8.8.1 Field-Controlled DC Servo motor

Figure 8.14 demonstrates the field-controlled DC servomotor. Here the field winding is constrained by an electronic amplifier. The armature is provided by a consistent current source. The error voltage represents the contrast between the measured and the desired signals. The torque delivered by engine is zero when no field excitation is provided by the DC error amplifier. Since armature current is constantly steady, the torque is straightforwardly relative to the field motion and furthermore it is legitimately corresponding to handle current up to immersion ($T \propto I_a$). In the event that the

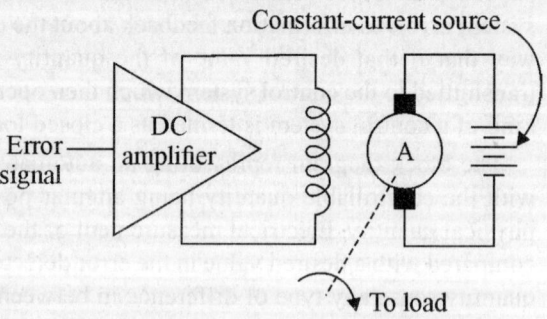

Fig 8.14: Field-controlled DC servomotor

extremity of the field is turned around, the engine heading switches. The control of field current by this strategy is utilized distinctly in little servo motors in light of the accompanying reason:

(i) It is not recommended to supply a huge and fixed armature current as would be demanded by large DC servomotors.

(ii) The time constant of the exceedingly inductive field circuit is exceptionally long. Accordingly, the dynamic reaction of the engine will be moderate.

8.8.2 Armature-Controlled DC Servomotor

Figure 8.15 shows the armature-controlled DC servomotor. Here the armature circuit is controlled by an electronic amplifier. The field winding is supplied by a constant current source. The error voltage represents the difference between the measured signal and the desired signal. A sudden large or small change in armature voltage pro-duced by an error signal will cause

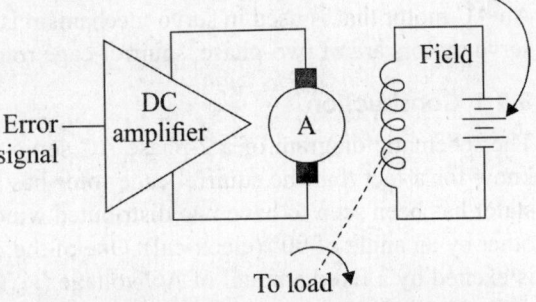

Fig 8.15: Armature-controlled DC servomotor

an almost immediate response in torque. It is because the armature circuit is essentially resistive compared to the highly inductive field circuit of field-controlled DC servomotor. If the error signal and the polarity of the armature voltage are reversed, the motor reverses its direction.

8.8.3 Permanent-Magnet Armature-Controlled DC Servomotor

Figure 8.16 shows the permanent magnet (PM) armature-controlled DC servo motor. It uses permanent magnets for constant field excitation instead of a constant current source. These types of motors use either alnico or ceramic magnets to produce the magnetic field. Permanent magnets have several advantages over wound-field servomotors including increased efficiency, reduced frame size and high accelerating torque. The speed of a PM DC servo motor is varied by changing the voltage applied

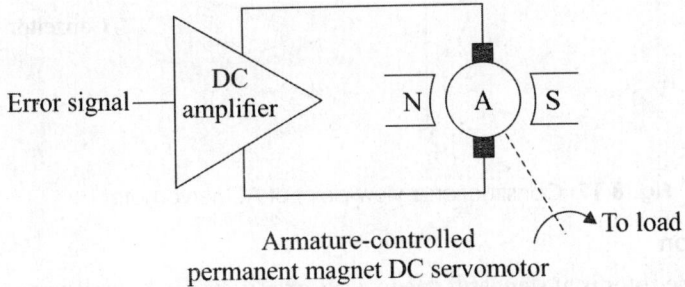

Armature-controlled
permanent magnet DC servomotor

Fig. 8.16: Permanent-magnet armature controlled DC servomotor

to the armature. Conventional wound-field motors are often equipped with interpoles in order to improve commutation and minimize armature reaction. The high coercive force of permanent magnet materials used in PM motors eliminates the need for inter-poles due to their excellent commutation characteristics, PM servo motors are used in a wide variety of applications including electric vehicles and cordless power tools. The popularity of these motors is due to their high efficiency, compact design and good commutation.

8.9 AC SERVOMOTOR

An AC motor that is used in servo mechanism is called an AC servo motor. Most AC servo motors are of two-phase, squirrel-cage rotor type induction motors.

8.9.1 Construction

The schematic diagram of a 2-phase AC servo motor is shown in Fig. 8.17. As we know for a fact that the squirrel cage rotor has high resistance and low inertia. The stator has been seen to have two distributed windings which are displaced from each other by an angle of 90° (electrical). One of the stator winding called **main winding** is excited by a fixed amount of AC voltage (V_m). The other stator winding known as the **control winding** is provided by the output voltage (V_a) of an amplifier. The voltages namely V_m and V_a must be synchronised which means that they both must be derived from the same AC source. These both voltages are made to have a phase difference of 90° by connecting a suitable capacitor in series with the main winding as shown in Fig. 8.17.

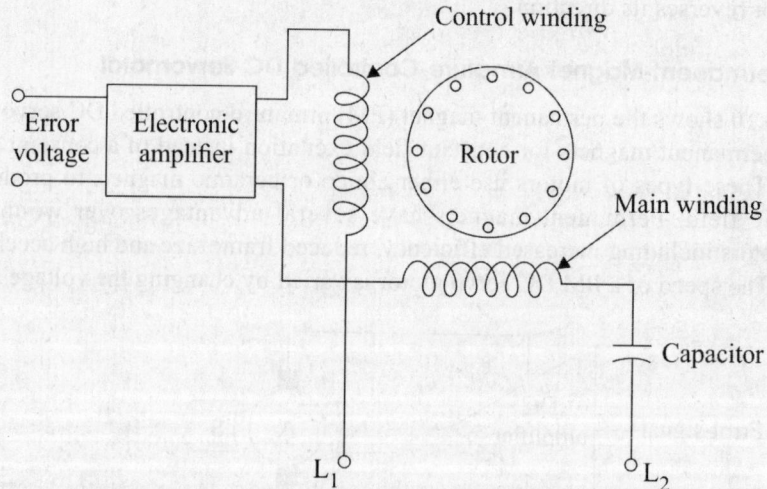

Fig. 8.17: Constructional view-point of AC servomotor

8.9.2 Operation

The squirrel-cage rotor is at standstill due to zero error voltage. A small error signal is amplified by the AC amplifier and is led to the control winding. The servomotor rotates with a speed that is proportional to this voltage and in a direction that will help to reduce the error signal. The motor ceases to rotate when zero error signals is produced at the control winding which seems quiet logical. Servo motors are designed to generate large values of torque at low speeds. Since these motors are mainly used in positional control devices, it is a prime concern that the torque is reduced at high speed to prevent the motor from overshooting from its desired position. To reduce overshooting, the squirrel cage rotor is made longer and thinner than in the ordinary induction motor. The rotor bars have a high value of resistance that helps in preventing the inductive reactance of the motor affecting the torque. As a result, the torque of the motor decrease as the speed increases which is the desired result.

8.10 SWITCHED RELUCTANCE MOTOR (SRM)

8.10.1 Construction

A switched reluctance motor (SRM) works on the same basic principle as a variable reluctance motor. The schematic diagram of a switched reluctance motor is shown in Fig. 8.18. Both the stator and rotor of SRM have salient poles which is not the case in the conventional synchronous motor. The stator carries coils on each pole. The diametrically opposite coils are connected in series to form a phase. The four stator coils in Fig. 8.18 are grouped together to form two phase viz. phase A and phase B. The rotor is made of soft magnetic material and carries no windings and both stator and rotors are laminated.

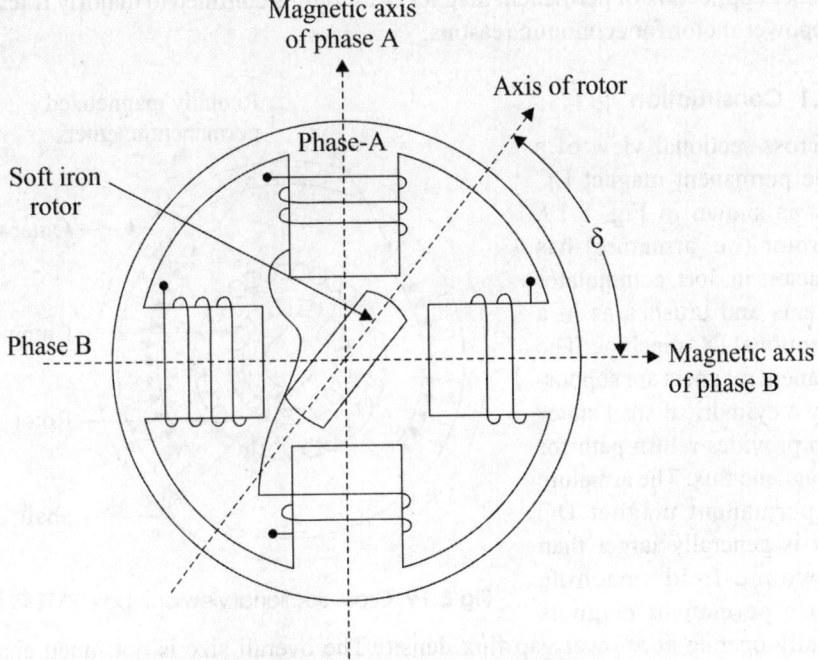

Fig 8.18: Schematic of switched reluctance motor

8.10.2 Working

The stator coils are energised with a single pulse of current at high speed sequentially.The reluctance torque is developed due to the attraction between the rotor and stator poles. Consequently, the rotation of the rotor takes place. In order to develop torque in the motor, the rotor position should be correctly identified by sensors so that excitation timing of the phase windings is apparent. Very high speeds can be obtained by using high-frequency pulses to energise the poles sequentially.

8.10.3 Advantages

(i) The rotor consists of no windings or slip rings. This makes the motor go very robust.

(ii) It has very high efficiency (70% to 80%).

(iii) The motor works well in harsh environment where it may be exposed to high temperatures and severe kind of vibrations. It can be run at high speeds (10000 to 30000 rpm).

8.11 PERMANENT MAGNET DC MOTOR

A permanent magnet (PM) DC motor is similar to an ordinary DC shunt motor except for the fact that its field is provided by permanent magnets instead of wound-field circuit. Although DC motors up to75 HP have been designed with permanent magnets, the major application of permanent magnet DC motor is confined to majorly fractional-horsepower motor for economic reasons.

8.11.1 Construction

The cross-sectional view of a 2-pole permanent magnet DC motor is shown in Fig. 8.19. The rotor (i.e. armature) has conductors in slots, commutator segments and brushes as in a conventional DC machine. The permanent magnets are suppor-ted by a cylindrical steel stator which provides return path for the magnetic flux. The armature of a permanent magnet DC motor is generally larger than the wound-field machine because permanent magnets

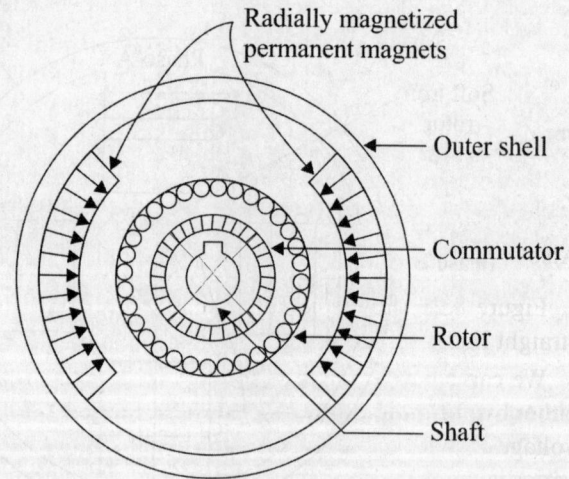

Fig 8.19: Cross-sectional view of 2-pole PM DC motor

generally operate at a lower gap flux density.The overall size is not much changed because of permanent magnets occupy less space.

8.11.2 Working

Most of permanent magnet DC motors run on 12 V (or less) DC supply obtained from mainly batteries or rectifiers. The interaction that takes place between the flux of permanent magnets and current-carrying conductors is the phenomenon that produces the driving torque. Consequently, the rotation of the rotor (i.e. armature) takes place.The starting torque is about 6 times the full-load torque. Hence we can infer from it that the motor cannot be run conti-nuously anywhere near its maxi-mum torque level. However, it is possible to start the motor directly from the supply without a conventional motor starter.

8.11.3 Motor Analysis

The analysis of operation of Perma-nent Magnet DC motor is always comparable with ordinary DC shunt machine. Figure 8.20 shows the schematic diagram of a permanent magnet DC motor. Note that there are no field connections.The flux is provided by the permanent mag-nets.

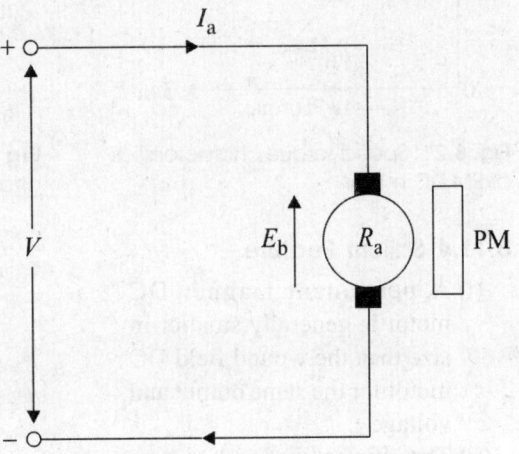

Fig 8.20: Schematic diagram of PM DC motor

$$\text{Motor speed, } N \propto \frac{E_b}{\phi}$$

or $$N \propto \frac{V - I_a R_a}{\phi} \quad \text{(Neglecting armature reaction effect)}$$

$\therefore$ $$N \propto V - I_a R_{Ea} \quad (\ \phi \text{ is constant})$$

Also $$T \propto I_a \quad (\ T \propto \phi I_a \text{ and } \phi \text{ is constant})$$

Figure 8.21 shows the speed-torque and speed-current characteristics which is a straight line.

We can control the speed-torque characteristic of a permanent magnet DC motor either by changing (i) the effective resistance of the armature circuit or (ii) the supply voltage V .With the change in the supply voltage the no-load speed (N_o) of the motor varies without affecting the slope of the characteristic curve. As a result of it, Fig. 8.22 shows the speed-torque characteristics Corresponding to different set of supply voltages (V_1, V_2, V_3, etc) .With the variation in the effective armature resistance the slope of the speed-torque characteristics changes without armature affecting the no load speed (N_o) as shown in Fig. 8.23.

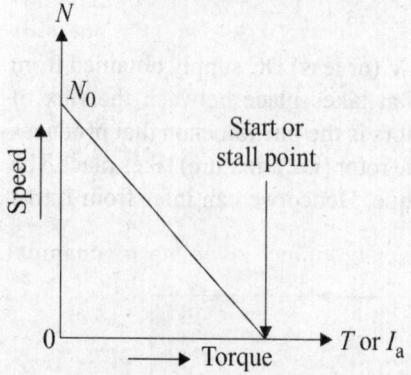

Fig. 8.21: Speed-torque characteristics of PM DC motor

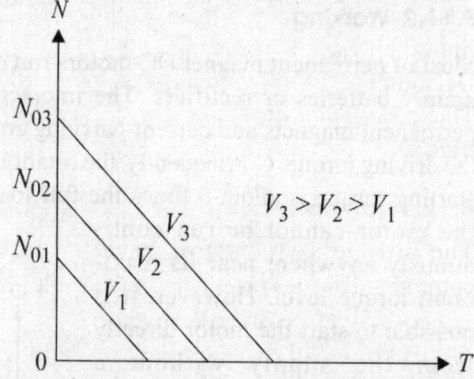

Fig 8.22: Parallel set of speed torque characteristic of PM DC motor

8.11.4 Salient Feature

(i) A permanent magnet DC motor is generally smaller in size than the wound field DC motor for the same output and voltage.

(ii) The efficiency of such motors is generally higher than that of the wound-field motors because of the fact that field excitation is not required in permanent magnet DC motor.

(iii) The life of very small motors is in between 1000–2000 hours, which is determined by commutators and brushes.

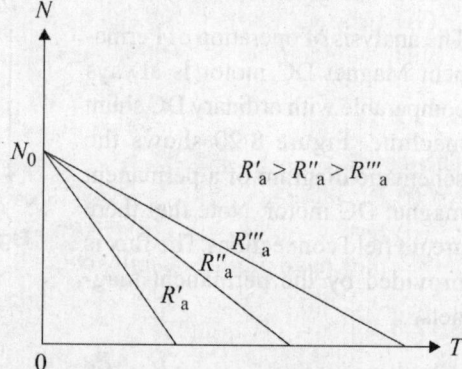

Fig. 8.23: Change in the effective armature resistance of PM DC motor

(iv) We use these motors mainly for low voltages (< 12 V) and small power ratings (0.1 to 4 W). We make a compromise between initial cost, life span and working efficiencies while designing these small motors. For battery-driven motors up to 4 W, the efficiency can be as high as 80%.

8.11.5 Applications

A huge number of little permanent magnet DC motors working on 12V (or less), 0.1–4 W are created each year. These engines are utilised for toys, shavers, cameras, versatile vacuum cleaner,wind-shieldwipers,automobile-heaters,fans,radio-antennas-wheel-chairs, cordless power instruments and so on.

8.12 BRUSHLESS DC MOTORS

The ordinary DC motors depend on mechanical commutation. These motors have two vital issues. To begin with, they require occasional support to fix and supplant brushes.

Also, electric bends delivered by the mechanical commutator brush course of action limit the working rate and voltage of the motor. A motor that holds the quantities of a DC motor and wipes out the commutator and the brushes is called brushless DC motor. These motors utilize electronic commutators. An electronic commutator comprises transistors and thyristors and plays out the capacity of its mechanical partner.

8.12.1 Advantages

The brushless DC motors have the following advantages over the mechanical commutator-brush type DC motors

(i) Very low or no maintenance.

(ii) Very long operating life.

(iii) Little to no arcing in the motors so the possibility of an explosion is reduced.

(iv) They are more efficient than the conventional DC motors.

(v) Can be used in situations where combustible fluids gases are present.

(vi) They are very reliable and their efficiencies may exceed 75%.

(vii) These are capable of rotating at speeds more than 40000 rpm.

8.12.2 Disadvantages

The disadvantages of brushless DC motors are:

(i) DC motors are much cheaper than these motors for a similar output power.

(ii) Due to the requirement of additional equipment for electronic commutation, they have greater overall size.

8.12.3 Operating principle

A multiphase winding is wound on a nonsalient stator and a permanent magnet (P M) rotor in case of a brushless DC motor as represented in Fig. 8.24. The individual phase windings are supplied with direct voltage or alternating voltage by utilizing a sequential switching for achieving the required commutation for achieving the rotation of the motor. Power transistors or thyristors are used to achieve electronic switching.

The operation of brushless DC motor shall now be elaborated. On energizing winding 1, the rotor of the PM gets aligned with the magnetic field produced by itself. On switching of winding 1 and energizing winding 2, the rotor is made to rotate to align itself with the produced magnetic field of winding 2.And in such a way, by application of voltage through a switching circuit to individual phase windings, the rotation of the motor can be achieved. The speed torque characteristics of a brushless DC motor depend on and can be achieved by the magnitude and rate of switching of the phase currents.

8.12.4 Applications

On account of high reliability and low maintenance requirements, the brushless DC motor has achieved popularity in the fields of biomedical and aerospace industries. Such motors are used as motors for artificial heart pumps, used in satellites, gyroscopes

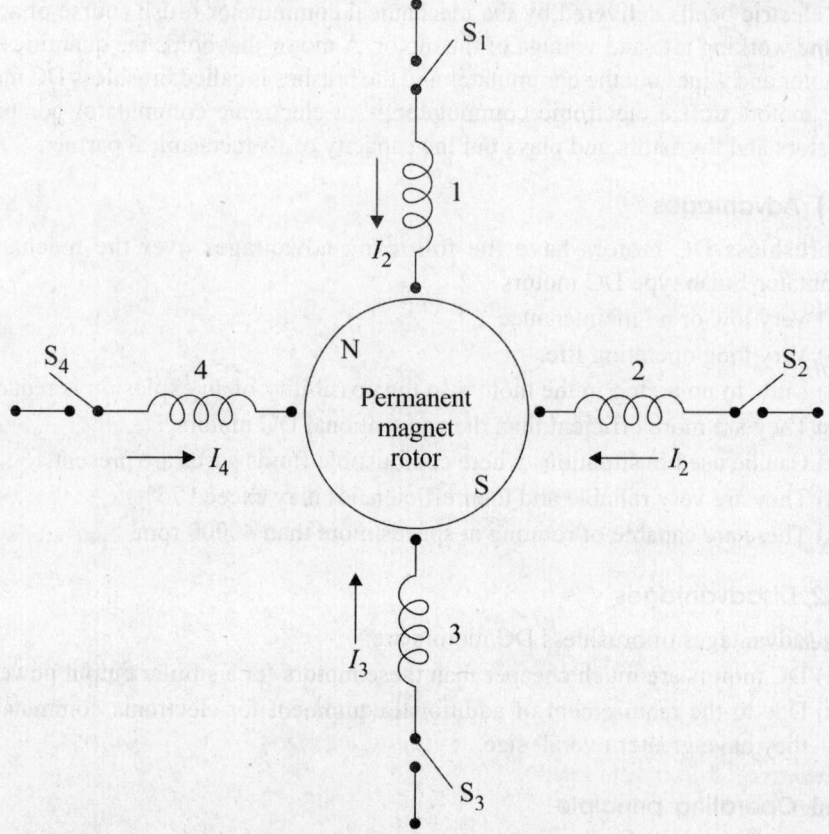

Fig. 8.24: Operation of brushless DC motor

and high efficiency robotic systems with several other applications including disc drive motors, video recorders and tape transport systems.

SOLVED EXAMPLES

Example 8.1: Determine the step angle of a variable-reluctance stepper motor with 16 teeth in the stator and 10 rotor teeth.

Solution:

Number of stator teeth, $N_s = 16$

Number of rotor teeth, $N_r = 10$

$$\text{Step angle,} \quad \frac{N_s - N_r}{N_s N_r} \times 360° = \frac{(16-10)}{16 \times 10} \times 360° = 13.5°/\text{step}$$

Example 8.2: Calculate the stepping angle for (i) a 3-phase, 16-tooth rotor VR motor (ii) 3-phase, 20 poles PM motor.

Solution: Step angle, $\theta = \dfrac{360°}{mN_r}$

(i) Here $m = 3$, $N_r = 10$ $\therefore$ $\theta = \dfrac{360°}{3 \times 10} = 12°/\text{step}$

(ii) Here $m = 3$, $N_r = 24$ $\therefore$ $\theta = \dfrac{360°}{3 \times 20} = 6°/\text{step}$

Example 8.3: A stepper motor with a step-angle of 20° has a stepping frequency of 200 steps/ second what is the motor speed?

Solution: Motor speed, $N = \dfrac{\alpha f}{6}$

Here $\alpha = 20°$, $f = 200$ steps/second

$\therefore$ $\alpha = \dfrac{20° \times 200}{6} = 666.67 \text{ rpm}$

Example 8.4: A stepper motor has a step angle of 20° and is required to rotate at 300 r.p.m. Determine the pulse rate (steps per second) for this motor.

Solution: Motor speed, $N = \dfrac{\alpha f}{6}$

Pulse rate, $f = \dfrac{6 \times N}{\alpha} = \dfrac{6 \times 300}{20} = 90 \text{ step/second}$

Example 8.5: A hybrid stepper motor has 60 variable-reluctance rotor teeth. Calculate the step angle in degrees.

Solution: Step angle, $\theta = \dfrac{90°}{N_r} = \dfrac{90°}{60} = 1.5°/\text{step}$

Example 8.6 A hybrid stepper motor has 40 variable reluctance rotor teeth and 4 stator phases. Calculate the step angle.

Solution: Step angle, $\theta = \dfrac{90°}{N_r} = \dfrac{90°}{60} = 2.25°/\text{step}$

EXERCISES

Descriptive Questions

1. With the help of neat labelled diagram discuss the application of stepper motor.
2. Define step-angle and stepping-rate of a stepper motor.
3. Explain with the help of neat labelled sketch the operation as well as working of a stepper motor.
4. What are the limitations faced by a permanent magnet stepper motor?
5. With the help of neat labelled diagram discuss the operational view point of variable reluctance stepper motor.

6. Discuss the constructional and operational part of hybrid stepper motor.
7. With the help of servo system block diagram explain servomechanism.
8. Discuss the working of different forms of DC servomotor.
9. Explain with neat labelled sketch the construction and working principle of AC servomotor.
10. Discuss the construction as well as operation of switch reluctance motor.
11. What are the utilities of switch reluctance motor?
12. With the help of neat labelled diagram explain the construction as well as working features of permanent magnet DC motor.
13. Discuss the motor analysis of permanent magnet DC motor by sketching suitable characteristics curve.
14. State the salient features as well as application of permanent magnet DC motor.
15. State the advantages and disadvantages of a brushless DC motor.
16. State the working features of brushless DC motor. Also discuss its main application field.

Numerical Type Questions

1. Determine the step angle of a variable–reluctance stepper motor with 20 teeth in the stator slots and 10 teeth in the rotor slots.
2. Calculate the stepping angle and stepping rate for (i) a 1-phase, 40 tooth rotor VR motor (ii) 1-phase, 30 poles PM motor.
3. A stepper motor with a step-angle of 25.3° has a stepping frequency of 300 steps/second. Calculate the revolution/second of the motor.
4. A hybrid stepper motor has 90 variable reluctance rotor teeth. Calculate the step angle for the motor in radian.

Index